GENESIS AND EVOLUTION OF

HORTICULTURAL CROPS

Prof. (Dr) K V Peter is a horticulturist, a plant breeder and a University Professor. He is an acknowledged and decorated scientist and science manager. A post graduate from G.B. Pant University of Agriculture and Technology, he did post doctoral research at BARC Beltsville Maryland USA and worked at Laboratories at AVRDC Taiwan and Guadeloupe (French West Indies). He was associated with development and release of improved and high yielding varieties/hybrids in tomato, brinjal, chilli, bittergourd, melons, amaranths and cowpea. Sources of resistance to bacterial wilt in tomato, chilli and brinjal; aphids in cowpea and viral leaf curl in chilli were located by him and are now used in breeding programmes. A vegetable seed production complex established by him first at Pantnagar and later at Kerala Agricultural University in 1980 continue to supply quality seeds to farmers even today. Prof. Peter provided managerial support to the ICAR Indian Institute of Spices Research Calicut to possess the World's largest collection of black pepper and cardamom germplasm. The technology package for protected cultivation of bush pepper would make available green pepper throughout the year. A prolific writer and academic editor he authored/edited 72 books published both in India and abroad, 16 technical bulletins, 110 research papers, 22 short research notes, 15 scientific reviews, 111 papers in symposia/seminars, 85 chapters in books, 307 popular articles and 17 radio/TV programmes. He received 7 scholarships and fellowships at various stages of education. He was member in the Board of Examiners in 12 Universities, Chairman/member in 122 committees of Government/research institutes, and membership in 18 scientific societies. His students occupy important positions in civil services, research institutes and other sectors. His publishers include Taylor and Francis USA, Elsevier USA, National Book Trust, I.C.A.R., New India Publishing Agency, Astral International Pvt. Ltd; Studium Press(USA) and Universities Press Hyderabad. Now Commissioning Editor New India Publishing Agency New Delhi, he is Director, World Noni Research Foundation, Chennai since 2008. He was Director, ICAR IISR Calicut; Director of Research KAU and the Vice-Chancellor KAU. The awards received include Rafi Ahmed Kidwai Award 1996-1998 for outstanding research in Horticulture-ICAR New Delhi; Recognition Award 2000–National Academy of Agricultural Sciences, New Delhi; Dr. M.H. Marigowda National Award for the Best Horticulturist-2000; Silver Jubilee Medal for outstanding contributions in Vegetable Research, Indian Society of Vegetable Science, Varanasi 1998; Dr. Harbhajan Singh Award 1993 instituted by the Indian Society of Vegetable Science, Varanasi for the Best Paper in the Journal Vegetable Science 1993; Silver Jubilee Memonto awarded in appreciation of services rendered to the Indian Society of Vegetable Science 1998; Biotech Product and Process Development Award 2003 awarded by Department of Biotechnology, Ministry of Science and Technology, Government of India; ICAR, New Delhi awarded a set of books for being one of the Best UG Students during 1966-1969; Awarded National Scholarship (1966-69), Junior Research Fellow, ICAR (1969-1971)-only one Jr. Research Fellowship in Horticulture in India during 1969, Senior Research Fellow, ICAR (1971-1975), Scholarship for Study Abroad(1981-1982); Shiva Sakthi-HSI National Award for Life Time Achievement in Horticulture-2008; NAAS Dr. K. Ramiah Memorial Award for outstanding contribution to Crop Improvement-2009; Best Institution Award to IISR, Calicut for 1994-1999 instituted by ICAR, New Delhi; Sardar Patil Award 2003 for the best ICAR institution to Kerala Agricultural University conferred on 19/10/2004; KRLCC Award for contributions to Education and Science-2015 and Suganda Bharati award by Indian Society of Spices, Calicut. He is an elected Fellow of National Academy of Sciences, Allahabad; National Academy of Agricultural Sciences New Delhi and National Academy of Biological Sciences, Chennai. The Parish of Sacred Heart Church Presented a shield of recognition at the Altar on a Sunday-a divine event.

Family: Vimala is wife, Anvar and Ajay sons, Anu and Cynara Daughers-in-law and Kuruppacharil Antony Ajay Peter, the grand son and the grand daughter Anna Vimala Anvar. Parents are Late Kuruppacharil Devassey Varkey and Late Rosa Varkey.

GENESIS AND EVOLUTION OF
HORTICULTURAL CROPS

Editor
Professor K V Peter

Kruger Brentt
Publishers
2017

Kruger Brentt Publishers UK. LTD.
Company Number 9728962

Regd. Office: 68 St Margarets Road, Edgware, Middlesex HA8 9UU

Disclaimer:

Every possible effort has been made to ensure that the information contained in this book is accurate at the time of going to press, and the publisher and author cannot accept responsibility for any errors or omissions, however caused. No responsibility for loss or damage occasioned to any person acting, or refraining from action, as a result of the material in this publication can be accepted by the editor, the publisher or the author. The Publisher is not associated with any product or vendor mentioned in the book. The contents of this work are intended to further general scientific research, understanding and discussion only. Readers should consult with a specialist where appropriate.

Every effort has been made to trace the owners of copyright material used in this book, if any. The author and the publisher will be grateful for any omission brought to their notice for acknowledgement in the future editions of the book.

Library of Congress Cataloging-in-Publication Data

Genesis and evolution of horticultural crops / editor, Professor KV Peter.
pages cm
Contributed articles.
Includes bibliographical references.
ISBN 978-1-78715-001-0 (Hardbound)

1. Horticultural crops--Genetics. 2. Horticultural crops--Evolution.
I. Peter, K. V., editor. II. Title: Horticultural crops.

SB318.G46 2017 DDC 635 23

For information on all our publications visit our website at http://krugerbrentt.com/

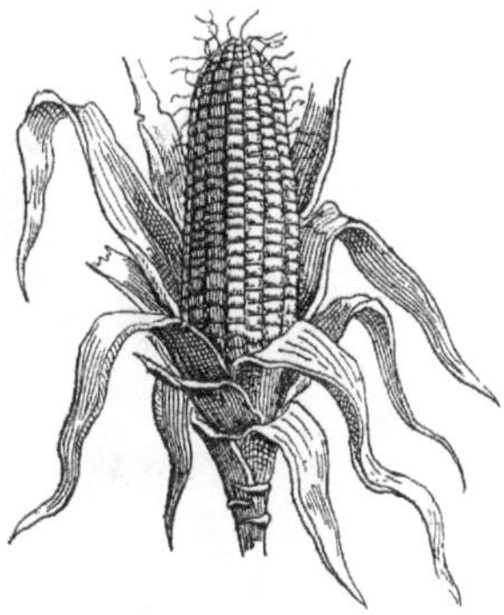

Dedication

The Genesis and Evolution of Horticultural Crops Vol. I is dedicated to my Mastoral thesis guide Dr R D Singh and Doctoral thesis guide Dr Bupendra Rai at G B Pant University of Agriculture and Technology, Pantnagar during 1969-1975. Dr Raymond E Web at Vegetable Laboratory, Beltsville Agricultural Research Centre USDA Maryland USA supervised my post doctoral research during 1981-1982. The above three mentors were great academicians and researchers who left golden imprints in horticultural research and education. Memories of them motivate me to do better.

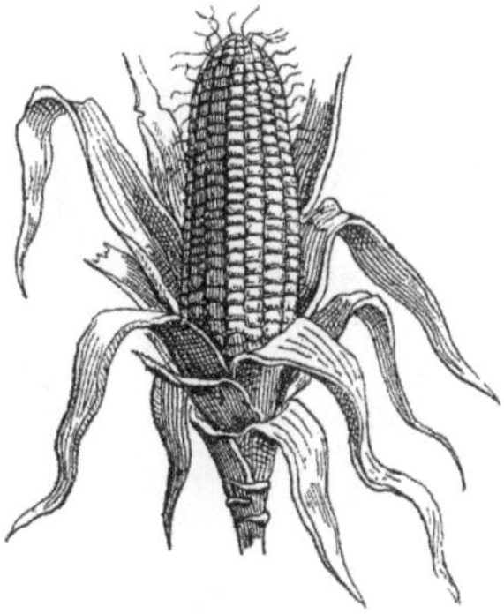

Acknowledgement

Forty two active scientists from 18 Research Institutes and Universities toiled hard to contribute 21 chapters to the first volume of **The Genesis and Evolution of Horticultural Crops**. Prof M S Swaminathan Father of Green Revolution and one among the three legendary Indians- others being M K Gandhi and Rabindra Nath Tagore-wrote the Foreword. Preface to the book is by Dr Panjab Singh, Chancellor, Rani Lakshmi Bai Central Agricultural University, Jhansi, Uttar Pradesh. I am grateful to all of my mentors.My wife Vimala, sons Anvar and Ajay, daughters-in-law Anu and Cynara, grand son Antony and grand daughter Anna were sources of inspiration. The Kruger Brentt Publishers, UK. LTD., has done an excellent work in getting a near error free text in pleasing way to readers. I also acknowledge Prof P I Peter, Chairman NoniBiotech, Chennai who provided all the facilities and Dr Kirti Singh, Chairperson World Noni Research Foundation, Chennai for the academic guidance.

Prof. M. S. Swaminathan
Founder Chairman, M S Swaminathan Research Foundation
Third Cross Street, Taramani Institutional Area
Chennai – 600 113 (India)
Email: swami@mssrf.res.in

Foreword

The statement of mine "there is a horticultural remedy for every nutritional malady" is widely quoted for its immediate relevance to the Food Security Act-2013. As member of Upper house of Indian Parliament(Rajya Sabha) I was fortunate to join the debate and vote for the bill. My apprehension was "there is no wellness and health security with out nutritional security. In a country like India where 60% of women and children are anaemic and underweight, role of enriched food with adequate consumption of fruits and vegetables (400 g/day/capita) becomes relevant.Horticulture-vegetables, fruits, ornamentals, spices, plantation crops, medicinal and aromatic plants–is core to Indias heritage and culture. Two of Vavilovian megabiodiversity centres–Indo–Malayan centre comprising Indo–China and Malay Archipelago and Hindustan comprising Assam and Burma (present Myanmar) are centers of diversity and wild species.The Botanical Survey of India made extensive documentation of plants. The series Wealth of India published by CSIR New Delhi documented and published volumes on plant wealth of India.

During my stint as Director General of ICAR, the National Bureau of Plant Genetic Resources with regional stations all over India was established. The Division of Plant Introduction at IARI New Delhi with Late Dr Harbhajan Singh was the prelude to NBPGR. Later crop based Research Institutes took the responsibility to survey, collect, maintain and document the germplasm. The NBPGR. New Delhi became the national depository and the nodal agency for exchange. The National Biodiversity Authority was established in 2004 to implemtnt the Biological Diversity Act.The Protection of Plant Varieties and Farmers Right Authority was established in 2002 to ensure right of plant varieties to farmers.

The future of Horticulture is challenging in India and a series on **Genesis and Evolution of Horticultural Crops** edited by Dr K V Peter will be a reckonor to students, teachers and scientists.The Volume I covers 21 crops authored by 42 academicians from 18 Institutes and Universities. I also compliment the Kruger Brentt Publishers, UK. LTD. for a worthy serial.

(M S Swaminathan)

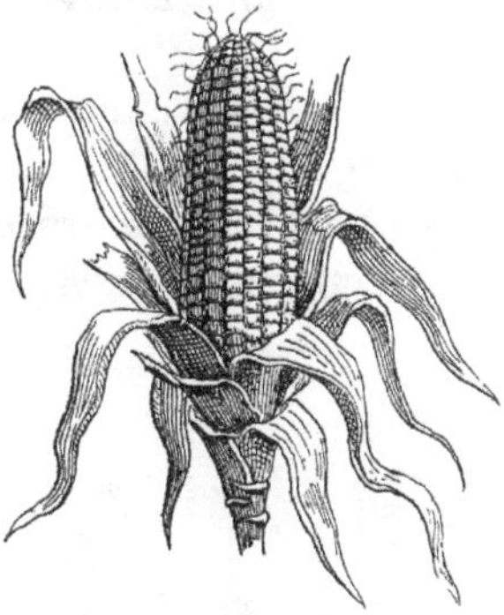

Preface

Evolution is "a process of change in a certain direction" as per Websters New Collegiate Dictionary. It was Charles Robert Darwin (12 Feb. 1809 to 19 April 1882) who published the theory of evolution in 1859 in his classical book "On the origin of Species", debated even now for the logic and science behind it." Change is a must and we should change and accept changes" as per books of wisdom. In plants, changes are brought about by human interference, climate change, natural disasters and survival of the fittest. Human food needs and new life styles are putting pressure on plants to evolve to meet his changing requirements for food, shelter and clothing, the basic needs and novel plants to meet his livelihood. Urbanization of horticulture-vegetables, fruits, ornamentals, spices, medicinal and aromatic plants and plantation crops- is getting plants evolved to meet newer requirements. Climate change leading to higher temperature,higher atmospheric carbon content and urban pollution are putting pressure on flora to change and evolve. Dwarf plant stature is becoming common due to diminishing space, water and energy. The boundaries among temperate, sub-tropical and tropical horticultural crops are getting vanished. Cabbages, cauliflower, capsicum carrots and radishes are now grown in tropical climate. Mangoes originally a sub-tropical fruit is now a major fruit of the tropics. Bush black pepper and interspecific grafts in pepper are becoming common.

The series **Genesis and Evolution of Horticultural Crops** will cover crops of health, nutrition and wellness, aesthetic values, industrial uses , medicinal values and of gardens. The volume I covers 21 crops authored by 42 working scientists. The Editor is Prof. K V Peter Former Vice-Chancellor Kerala Agricultural University and presently Director World Noni Research Foundation, Chennai. Publisher is Kruger Brentt Publishers, UK. LTD. I congratulate all of them for the academic contribution.

(Panjab Singh)

Contents

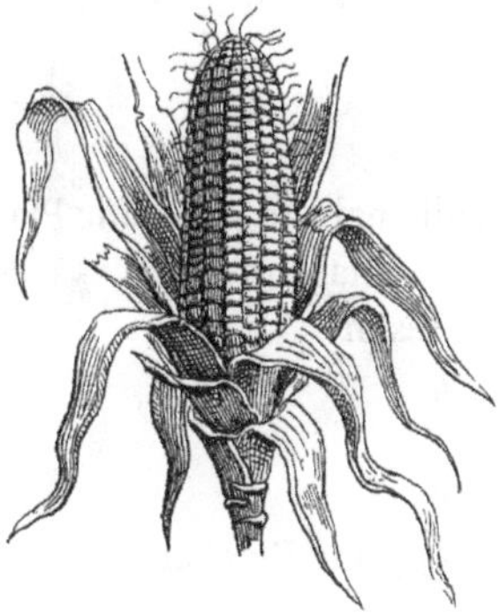

Contributors

1. **Chapters I and VII**

   K R M Swamy
   Former Principal Scientist and Head, Division of Vegetable Crops
   ICAR-Indian Institute of Horticultural Research
   P O Hessarghatta Lake
   Bengaluru-560089, Karnataka
   E-mail: krmswamy@gmail.com

2. **Chapter II**

   V P Neema & P M Ajith
   Pepper Research Station, Kerala Agricultural University
   P O Panniyur-670142, Kannur, Kerala
   E-mail:prspanniyur@kau.in

3. **Chapter III**

   J P Shahi & Varsha Gayatonde
   Department of Genetics and Plant Breeding, Institute of Agricultural Sciences
   Banaras Hindu University, Varanasi-221005, Uttar Pradesh
   Email: jpshahi1@gmail.com

4. **Chapter IV**

   Verughese George, Palpu Pushpangadan & T P Ijinu
   Amity Institute of Phytochemistry and Phytomedicine 3, Ravi Nagar
   P O Peroorkada, Thiruvanandapuram-695005, Kerala
   E-mail:georgedrv@yahoo.co.in;palpuprakulam@yahoo.co.in

5. **Chapter V**

Sanchita Ashok Sharma
Biotechnology Division
CSIR-Central Institute of Medicinal and Aromatic Plants
P O CIMAP, Lucknow-226015, Uttar Pradesh
E-mail: ashoksharma@cimap.res.in

6. **Chapter VI**

P K Ghosh & P Bhattacharjee
Department of Food Technology and Biochemical Engineering Jadavpur University
Kolkatta-73002, India
Email: pb@ftbe.jdvu.ac.in

R S Singhal
Institute of Chemical Technology
Nathalal Parekh Marg, Matunga, Mumbai-400019, Maharashtra

7. **Chapter VIII**

Avijit Kr Dutta
School of Agriculture and Rural Development
Ramakrishna Mission Vivekananda University
F/C IRTDM, Ramakrishna Mission
Ashrama, Morabadi, Ranchi-834008, Jharkhand
E-mail: avijitkumardutta@gmail.com

8. **Chapters IX and XIV**

R K Singh
Assistant Director(Horticulture)

R P Gupta
Director
National Horticulture Research and Development Foundation
Chitegaon Phata
P O Darna Sangavi, Nasik-422303, Maharashtra
E-mail: nasik@nhrdf.com

V Mahajan
ICAR-Directorate of Onion and Garlic Research
Rajgurunagar, Pune-410505, Maharashtra
E-mail: vijbmaha@yahoo.com

9. **Chapter X**

Alok Nandi
Professor(AICRP on Vegetables)
Orissa University of Agriculture and Technology
Bhubeneswar-751003, Odisha
E:mail: alok_nandi@hotmail.com

10. **Chapter XI**

P M Hadankar, Y R Parulekar & S M Sawratkar
Department of Horticulture
Dr Balasaheb Sawant Konkan Krishi Vidyapeeth
Dapoli-415 712, Maharashtra
E-mail: parag5663@rediffmail.com

11. **Chapters XII and XIII**

Amish Kumar Sureja & Anilabh D. Munshi
Division of Vegetable Science
ICAR-Indian Agricultural Research Institute
Pusa,New Delhi-110012
Email: anilabhm@yahoo.co.in

12. **Chapter XV**

Vijay Mahajan
ICAR-Directorate of Onion and Garlic
Rajgurunagar, Pune-410505, Maharashtra
E-mail: vijbmaha@yahoo.com

13. **Chapter XVI**

P P Joy & R Anjana
Pineapple Research Station
Kerala Agricultural University
P O.Vazhakulam, Ernakulam-686675, District, Kerala
E-mail: prsvkm@kau.in

14. **Chapter XVII**

K Dhinesh Babu, N V Singh, H B Shilpa, A Maity, N Gaikwad & R K Pal
ICAR-NRC on Pomegranate
Solapur-413255, Maharashtra
Email: dr_babu@yahoo.co.in

M Sankaran
ICAR-Indian Institute of Horticultural Research
P O Hessarghatta, Bengaluru-560089

15. **Chapter XVIII**

Shirin Akhtar
Department of Horticulture(Vegetable and Floriculture)
Bihar Agricultural University
Sabour, Bhagalpur-813210, Bihar, India
Email: shirin.0410@gmail.com

Abhishek Naik
Area Manager(Produce Development)
Kolkata Zone
UPL-Advanta Ltd.

16. Chapter XIX

Ali Izanloo & Mohammad Ali Behdani
Saffron Research Group
University of Birjand
Birjand Iran
E-mail:mabehdani@birjand.ac.ir

17. Chapter XX

Pradip Baruah & N Muraleedharan
Tocklai Tea Research Institute
Tea Research Association
Jorhat-785008, Assam
E-mail: director@tocklai.net

18. Chapter XXI

B R Choudhary
ICAR-Central Institute for Arid Horticulture
Bikaner-334006, Rajastan
E-mail: choudharybr71@gmail.com

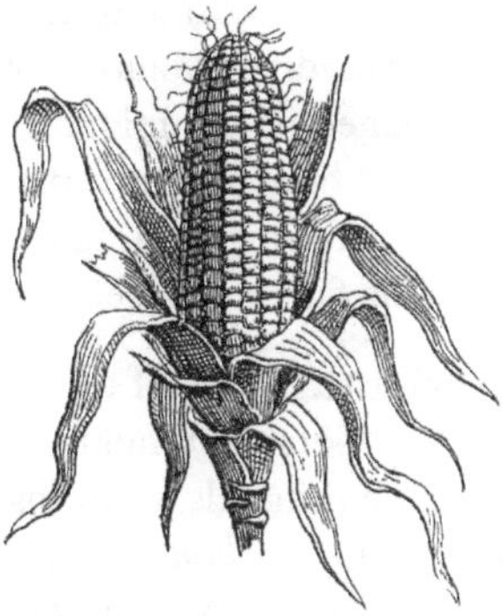

Introduction

Webster's New Collegiate Dictionary defines Genesis as "Origin" and Evolution as "a process of continuous change from a lower, simpler, or worse to a higher, more complex, or better state". Charles Darwin(1809-1882) wrote the classic "The Origin of species" in 1859 and "The variation of plants and animals under domestication" in 1883. He demonstrated that selection could have major effects and that these effects were heritable. Human selection to meet varying requirements under both biotic and abiotic stresses led to refining of plants. Alphonse de Candole (1806-1893) is considered as the father of the study of crop evolution. He proposed types of evidences in evolution as botanical,archaeological,historical and linguistic. He published the much read Origin of Cultivated Plants in 1883. Nicolai I. Vavilov(1887-1943) established a very active program of study of crop diversity in what is now the Vavilov Institute of Plant Industry in St.Petersburg,Russia. In 1926, he published "Studies on the origin of Cultivated Plants". The region of maximum variation, usually including a number of endemic forms and characteristics as well can usually be considered as the centre of type formation" to quote Vavilov." The center of origin of a cultivated plant is often correlated with the center of associate pathogens". Seven primary centres and several secondary centers of diversity were proposed by him. Jack R Harlan(1917-1998) further persued research on evolution and in 1973 published the much read "Evidence for origin and Dispersal of Cultivated Plants" along with colleague de Wet JMJ. They listed experimental taxonomy,geographic distribution,ecological behavior,genetic systems, variation patterns, morphology and genetic reconstruction as evidence for origin and dispersal of living plants and archaeology, palynology and palyobotany for dead plants. J. Harlan agreed broadly with Vavilov's idea that just a few geographical locations are crucial for generating much of the diversity on which plant breeders depend. Harlan preferred the term Center of diversity to Vavilov's term Center of Origin because while the centers of crop diversity are known and mapped , the origin of crops cannot be definitely pinned down. His publications Crops and Man (1976) and The living Fields: Our Agricultural Heritage (1995) are classics in understanding evolution of crops and plants. Evolutionary processes like mutation, selection, genetic drift, migration, recombination and modes of reproduction-self, often and cross- and human interventions like preferences and choices, aestheticity, novelty, distinctiveness, uniformity and stability have played roles in evolution. Domestication is always an influencing factor in accelerating evolutionary processes.

Codice Florentino (16 century) listed foods of the Aztecs - amaranth, beans, dragon fruit (prickly pear), bell peppers, sweet potato, pumpkins and squashes, tomato and peanut *etc.* Amaranth is a leaf vegetable rich in minerals, vitamins and fibre. The edible species are *A.tricolour, A.viridis, A. dubious* and *A.hypochondriacus.* There are diploid and tetraploid types with varying leaf and stem colours and spiny/nonspiny. Grown through out tropics and sub-tropics amaranth has transformed from wild stage to edible stages over a period of time. Amaranth grain being rich in essential amino acids is a basic raw material for baby foods and food for expectant mother and the aged. There are also ornamental types grown in edges, pots and in hanging gardens.Human selection to meet his designs and plans has made amaranth a short duration crop still in evolution.Presence of anti-nutrient factors like free oxalates and nitrates in amaranth leaves is compelling selection for anti-nutrient free lines. The Tamil Nadu Agricultural University, Coimbatore has released several high yielding lines of amaranth named Co1 to Co5.

Black pepper (*Piper nigrum* L.)-king of spices- originated in the Western Ghats of India. Once grown in forests and berries collected by the forest inhabitants, its value as a spice and herbal medicine was understood by Chinese travellors and later the Portugal,Egyptions, French and The British. The Italian Explorer Christopher Columbus was voyaging in search of black pepper but landed in New World –present North American continent- on 5 December 1492. The Portuguese Navigator Vasco de Gama landed in Kappad Beach in Kozhikode primarily in search of black gold-pepper on 20 May 1498. The rest is history of trade, colonization and rule by French and British. The oriental practices of health care like Ayurveda, Sidha, Unani and Tribal medicine took into uses of spices especially black pepper for natural medicine.The species of Piper *i.e., P. nigrum, P. longum, P. attenuatum, P. chaba, P. colubrinum etc.* gained importance.Being propagated by cuttings, the intra clonal variability is limited in black pepper. The ICAR-Indian Institute of Spices Research, Calicut (Kozhikode) Kerala has the world's largest collection of pepper germplasm.Protocol for rapid multiplication of black pepper is available. Panicle bearing laterals when rooted give bush pepper flowering through out the year once irrigated. Blackpepper has many diseases-Phytophthora foot rot and viral leaf curl and little leaf- and pests-scales, mealy bugs, nematodes. The cultivated pepper was evolved by undergoing stresses both biotic and abiotic and meetings requirements of oils and oleoresin industry like high oleoresin lines and high piperine lines. Pepper Research Station, Panniyur, Kannur is one of the oldest research stations under Kerala Agricultural University which released Panniyur 1, the first hybrid in black pepper.

Corn (maize) is a cereal of the Andes and called "maiz" in Spanish and Sara in Quechus-language of the Incas.Based on usages, there are dent corns, sweet corns, pop corns,flour corn,flint corn and baby corns-meeting the requirement of a vegetable. Corn based ethanol is a bio-fuel and called green fuel. As a cattle feed, corn has its global value.

Other uses are fodder, chemicals, ornamentals and bio-fibre. Being a highly cross pollinated crop due to separation of male and female flowers with in a comb,the capacity of the crop to withstand and resilient vagaries of weather is spectacular. Varieties, hybrids-single cross, double cross, composites- and multilineal hybrids are now available. Historians believe corn was domesticated in the Tehuacan Valley of Mexico. Mexico based CYMMIT (Centro Internacional de Malormiento de Maize Y Trigo-Centre for Improvement of Maize and Wheat) is spearheading research on corn. In India ,the Indian Institute of Maize Research and All India Co-ordinated Research Project on Maize are undertaking research on this economic crop. Baby corn is listed as a vegetable for its nutritive value, rich mineral and fibre content and taste.

Among medicinal herbs, basil(thai basil,sweet basil) belonging to family Lamiaceae (mints) occupies important place in culinary and in cuisine.Botanically *Ocimum basilicum*, the half-hardy annual plant has nativity to India. It is also known as Saint Joseph's Wort in Italy, UK and USA. Basil plays a major role in South East Asian cuisines of Indonesia, Thailand, Malaysia, Vietnam ,Cambodia, Laos and Taiwan.The Central Institute of Medicinal and Aromatic Plants (CIMAP), Lucknow under CSIR New Delhi maintains germplasm of basil. The Directorate of Medicinal and Aromatic Plants Research (DOMAPR), Anand (Gujarat) under ICAR New Delhi and All India Co-ordinated Research Project on Medicinal and Aromatic Plants under ICAR have research programmes on basil.

Bay laurel (*Laurus nobilis*) (sweet bay, bay tree, true laurel, Gracian laurel, laurel tree, laurel) is an aromatic evergreen tree with green, glossy leaves, native to the Mediterranean region. It figures prominently in Greek, Roman and Biblical culture. The genus *Laurus* includes three species whose diagnostic key characters often overlap. The laurel is dioceous with separate male and female plants. Considerable genetic diversity was found in *L.nobilis*. The dioceous nature reveals evolution from the primitive hermaphrodite floral nature. Cactus pears is an arid desert crop popular in South East Asian countries Indonesia, Malaysia, Vietnam, Cambodia, Laos, Singapore and China. The most common culinary species is the Indian Fig Opuntia (*Opentia ficus-indica*) Prickly pears grow with flat rounded cladodes (platyclades) armed with two kinds of spines-large, smooth-. It is used in earmarking boarders and protecting crops from wild animals.

Cashew (*Anacardium occidentale*) is a tropical evergreen plantation crop, whose nuts are processed to cashew kernels, rich in minerals, fats and aminoacids. Introduced from Brazil to India for soil conservation, the tree got adopted to tropical and sub-tropical conditions. The cashew apple is processed to several value added products-ready to serve drinks, jam and jelly, dehydrated apple, cashew fenny; cashew honey and cashew nut rind oil. India has the largest processing facility for cashew nut. Domestic production is not adequate to meet the processing capacity and cashew nuts are imported from Costa Rica, Kenya, Brazil, Nigeria, Vietnam, Argentina and South Africa. Cashew nut kernel is further processed to bits, powders and scented and salted packed items. Cashew nut soup is a delicacy. The Directorate of Cashew Research at Puttur (Karnataka) maintains germplasm of cashew. The All India Co-ordinated Research Project on Cashew with its centres in cashew growing states also maintain germplasm. In addition the All India Co-ordinated Research Project on Cashew under ICAR, has research centres all over India. There is an exclusive Cashew Export Promotion Council under Ministry of Commerce,

Government of India. There are many Cashew Research Stations in India-Madakkathara (Kerala), Pilicode (Kerala), Arabhavi (UHS, Bagalkot), Baptla (Dr YSRHU), Bhubaneswar (OUAT), Hogalagere, Daisal (BAU, Ranchi), Goa (Central Institute for Coastal Agriculture), Jagadalpur (IGKV, Raipur), Jhargram (BCKVV, Kalyanai), Paria (GAU, Navsari), Vengurla (Dr BSKKV, Dapoli), Vridhachalam (TNAU) and Tura (ICAR Complex Barapani).

French bean (green bean, string bean, snap bean,haricot verts, navy beans, bush beans, pole beans-all based on specific characters) is a leguminous vegetable alone or in combination much liked through out the globe. Being a main source of protein and fiber to dietary, French beans are cherished and demanded. Fresh beans, canned beans, frozenbeans, canned bean bits and stringless beans in cans are a few products much in demand. An essential vegetable in nutrition garden/kitchen garden/homestead garden, French beans are easy to be grown throughout the year.It comes up well under protected cultivation. There is much improvement in French bean varieties in terms of height, pod colour, stringless/ stringed and resistance to pests and diseases. Hyacynth bean (lab-lab bean, bonavist bean/ pea, dolichos bean, seim bean, egyption kidney bean, Indian bean, bataw and Australian pea) is only species (*Lablab purpureus*) in the monotypic genus Lablab. There are both pole and bush types of lab lab bean. Being a short day plant, it flowers and sets fruits under a day length of less than 12 hours. Both seeds and immature pods are edible. Luffa gourd (ridge, smooth and satputia) is monoecious and yields fruit vegetables. Satputia carries hermaphrodite flowers and bears fruits in clusters. There are two species-*Luffa aegyptiaca* and *L. acutangula* grown for fruit vegetable and to make sponges for body cleansing. *Momordica* genus consists of *M.charantia* (bitter gourd), *M.dioica* (Round gourd) and *M.muricata* (spiny gourd). Hypoglycemic property of bitter gourd has been established. Being monoecious, cross pollination is the rule.There are variations in fruit size, colour and level of bitterness.

Onion (*Allium cepa* var. *cepa* L.) is a bulb vegetable cultivated and used around the world. A monocot belonging to family Alliaceae(earlier Amaryllidaceae), it is propagated through seeds.Being bee pollinated, cross pollination is the rule. Genetic and cytoplasmic male sterility is reported and used to develop hybrids. There are red, yellow, purple and white onions . There are short day, long day and day neutral onions. There are small, medium and large sized onion varieties differing in pungency-allicin-. The Pune (Maharashtra) based ICAR- Directorate of Onion and Garlic Research; National Horticultural Research Foundation, Nashik and All India Co-ordinated Research Project on Vegetables, Varanasi maintain germplasm of onion. Potato onion (Multiplier onion) (*Allium cepa* cultivar Aggregatum) is a distinct group propagated through bulblets and very popular in southern states of India-Kerala, Tamil Nadu, Karnataka, Andhra Pradesh and Telengana. Garlic (*Allium sativum*)is another spice with a history of use of over 7,000 years. Garlic is native to central Asia and is a staple in the Mediterranian region and a seasoning in Asia,Africa and Europe. Garlic is known to ancient Egyptions and has uses both in culinary and medicine.

The pineapple (*Ananas comosus* Family:Bromeliaceae) is a much cherished tropical fruit grown in Hawai (USA), Kew (London), Philippines, Java (Indonesia) and through out Indian sub-continent. The edible multiple fruit is sliced for table purpose and for canning. The pineapple leaves yield the much needed fibre for textiles. The All India Co-ordinated Research Project on Tropical Fruits based at ICAR-Indian Institute of Horticultural Research,

Bengaluru collects germplasm of pineapple and conducts trials on varieties, spacing and package of practices. Pineapple Research Station at Vazhakulam(Ernakulam) under Kerala Agricultural University conducts experiments to improve production, productivity and quality of pine apple. Pomegranate (*Punica granatum*; Family: Lythraceae) is an arid tropical fruit originated in the modern day Iran and cultivated throughout Mediterranean region and India. The wild types bear small fruits and have several medicinal uses. Mention about this fruit appears in Old Testament and the forbidden fruit tree in "Eden"is said Pomegranate tree. The ICAR-Central Institute of Arid Horticulture, Bikaner maintains germplasm. The evolution of pome granate from wild to cultivated is mostly by selection. Airlayering is practiced to propagate pome granate in addition to seedlings.

Radishes (*Raphanus sativus*; Family: Brassicaceae) are root vegetables with variations in root colour, shape and weight;oriental and occidental;temperate,sub-tropical and tropical and leafy and semi-leafy.Leaves are also edible and cooked like spinach or beet leaf.Immature seed pods are edible and cooked like any legume. Indian "mooli"is slightly pungent and used to prepare local dishes in Jaunpur(UP)-"Jaunpur mooli".

Saffron (*Crocus sativus*) is the costliest among spices derived from the flower stigmas native to Greece or Southwest Asia. It was cultivated first in Greece and slowly progressed to parts of North Africa, North America and Oceania. The saffron unknown in the wild descends from *Crocus cartwrightianus*, originated in Crete, *Crocus thomasii* and *C.pallasi* being the possible precursors. Saffron is grown in Kashmir Valley (India) and the ICAR-Central Institute of Temperate Horticulture conducts research and development on this crop. Spain is another European country growing saffron and Spanish saffron is unique in its chemical profile. Tea(*Camellia sinensis*) is a beverage used world over. Egyptian tea, Chinese Tea, Darjeeling Tea, Kenyan Tea, Sri Lankan Tea and Nilgiri Tea are ecotypes varying in taste and flavor. The Chinese Tea is the most popular, China being the centre of origin. In India tea is a plantation crop and covers large area in Assam (Assam Tea), Nilgiri (Nilgiri tea) and Darjeeling (Darjeeling tea). Many plantation companies like Harrison Malayalam Ltd, Tata Tea Plantations *etc* have large areas under tea. Leaf tea, tea powder, speciality tea, green tea *etc.* are traded under different brands. Introduced in 59 BC, the tea shrub has acclimatized into modern plant type ideal for plucking. The United Planters Association of South India conducts research and development in tea through Tea Research Institute Valparai Tamil Nadu. The Tea Research Institute –Toklai-in Assam maintains germplasm in tea.

Watermelon (*Citrullus lanatus* var. *lanatus*, Family: Cucurbitaceae) is a fruit which quenches thirst during summer. *Citrullus colocynthis* is the ancestor species originated in Southern Africa.Being a monoecious plant with separate male and female flowers, the plant is cross pollinated and variability in morphological traits-leaf, stem, vines,flowers, fruits-are distinct and explicit. The ICAR-Indian Institute of Vegetable Research Varanasi(UP) and the All India Co-ordinated Vegetable Improvement Project conduct research on water melon. Many private seed companies (Indam Seeds,Bengaluru; Namdhari Seeds, Bengaluru etc.) have research projects which led to evolution of seedless water melons and hybrid melons.

The Series **Genesis and Evolution of Horticultural Crops** aim at study of evolution of horticultural plants from a stage of wilderness to the present cultivated and future plants determined by climate change, life style,urbanization and processing requirements. In a situation of limited space, water,energy and labour and changes in soil micro-flora and microbes the evolution of horticultural crops will be a continuous process. The FOREWORD to the series is by Professor Monkombu Sadasivan Swaminathan the Father of Green Revolution in India.

(K V Peter)

1

Amaranths

K R M Swamy

Amaranth is the common name for more than 60-70 different species of *Amaranthus*, which are usually very tall plants with broad green leaves and impressively bright purple, red, or gold flowers. The name for amaranth comes from the Greek *amarantos*, "one that does not wither," or "the never-fading." True to form, amaranth's bushy flowers retain their vibrancy even after harvesting and drying, and some varieties of ornamental amaranth forego the production of fancy flowers in favor of flashy foliage, sprouting leaves that can range from deep blood-red to light green shot with purple veining. Although several species can be viewed as little more than annoying weeds, people around the world value amaranths as leaf vegetables, cereals, and ornamental plants (*Anon.*, 2015b).

In all fairness to whole grains everywhere, we need to "out" amaranth as a bit of an imposter. It is not a true cereal grain in the sense that oats, wheat, sorghum, and most other grains are. "True cereals" all come from the Poaceae family of plants, while amaranth (among others) is often referred to as a "pseudo-cereal", meaning it belongs to a different plant species. So why are these invaders almost always included in the whole grain roundup? Because their overall nutrient profile is similar to that of cereals, and more importantly, pseudocereals like amaranth have been utilized in traditional diets spanning thousands of years in much the same way as the "true cereals" have been (*Anon.*, 2015b).

The genus *Amaranthus* is rather unique in having species which are used for grain, vegetable and ornamental purposes. Recently the potential of microcrystalline (1-3 μm) starch granules for possible replacement of talc in the cosmetic industry and edible dyes has emerged. The most common use in the major regions of its cultivation (Peru, Bolivia, Ecuador, Mexico, India, Nepal and Bhutan) is in the form of cakes or balls (laddoos) prepared by binding the popped seed in jaggery or sugar. The vegetable amaranths are used as pot herbs in most tropical countries of the world. While the related foliage ornamental types add colour to the otherwise drab garden surroundings in summer months (Pal, 1999). Vegetable amaranth is widely grown in the tropics and is one of the most important leafy vegetables in the lowlands of Africa and Asia. Amaranth is an annual, fast growing herb that can be cultivated easily in home gardens and at commercial scale (Ebert *et al.*, 2011). **(Figs. 1 to 5)**

Fig. 1: Amaranth Growing in a Field

Fig. 2: Amaranth Ready for Harvest

Fig. 3a: Amaranth Grains

Fig. 3b: Amaranth Grains

Fig. 4: Popped/ puffed Amaranth

Fig. 5: Amaranth flour

Amaranth today is enjoyed many ways. In Mexico and India the seeds are popped and mixed with sugar to make a confection. In Mexico they are roasted for the traditional drink "atole." Peruvians use the grain to make a beer. Elsewhere it is used to treat toothache and fevers or to color maize and quinoa. Women performing native dances often wear the red amaranth flower as rouge. In many countries the leaves are used boiled or fried. In Nepal the seeds are made into a gruel. Although amaranth seed and flour can be found in health food stores (Green Deane, 2012).

Nutritional Value of Amaranth Grains

Amaranth plant is valued for the positive chemical composition of seed that does not contain gluten. Amaranth is a very interesting crop from the point of its high production potential. It grows intensively, photosynthesizes fast and effectively, does not suffer from major diseases and is tolerant to various extreme conditions (*Anon.*, 2015a). According to Howard (2013) amaranth is a broad-leaved, bushy plant that grows about 1.8 m tall. It produces a brightly colored flower that can contain up to 60,000 seeds. The seeds are nutritious and can be made into a flour. Not a true grain, amaranth is often called a pseudocereal, like its relative quinoa (*Chenopodium quinoa).* Both plants belong to a large family that also includes beets, chard, spinach, and lots of weeds. There are around 60-70 different species of amaranth, and a few of them are native to Mesoamerica. "Obesity is a devastating problem in Mexico," "Amaranth may be part of the solution. It is a whole, healthy food that can be produced locally, and it may create the possibility of change".

Amaranth is gluten free and its seeds contain about 30 percent more protein than rice, sorghum, and rye, according to a USDA Forest Service report. It is also relatively high in calcium, iron, potassium, magnesium, and fiber. "Amaranth's amino acid profile is as close to perfect as you can get for a protein source. The plant contains eight essential amino acids and is particularly high in the amino acid lysine, which is largely lacking in corn and wheat. Amaranth has been recommended by the World Health Organization as a well-balanced food and recommended by National Aeronautics and Space Administration (NASA) for consumption in space missions. The variety of amaranth consumed in Mexico has 16 to 18 percent protein, compared with 14 percent protein in wheat and 9 to 10 percent protein in corn. Some studies have shown that amaranth also contains beneficial omega-3s and may help reduce blood pressure (Howard, 2013). Amaranth contains more than three times the average amount of calcium and is also high in iron, magnesium, phosphorus, and potassium. It is also the only grain documented to contain Vitamin C.

Amaranth seeds have high content of proteins, essential amino acids and minerals. All compounds are usually in following ranges: 14-19 % of protein, 5-8 % of lipids, 62-69 % of starch, 2-3 % of total carbohydrates and 4-5 % of fibre. The seed composition is comparable with seed composition of oat. In Amaranth, bioavailability of protein reaches 78 %. Seed does not contain gluten causing celiac disease to sensitive individuals. In contrast to cereals amaranth has higher content of amino acids mainly lysine, methionine, treonin and cysteine. Starch is the major part of carbohydrates and starch granules are small (1-3 Ìm) easily degradetable by alpha-amylases. Amaranth starch is highly stabile during freezing and highly resistant to mechanical stress. Lipid content in amaranth seed ranges from 5 to 8 %. Most of it is placed in embryo as linoleic, oleic and palmitic acid. According to the seed composition amaranth oil is similar to the ones made from cotton or maize but has lower digestibility. Amaranth oil contains about 8 % of squalen, a sterol precursor, used in medicine and cosmetic industry. Content of minerals depends on species, and growing conditions. Amounts of calcium and magnesium are higher than amounts in other cereals. Seeds are a good source of vitamins mainly ascorbic acid and B-complex, and antioxidants alpha-tocopherol and beta- and gamma tocotrienols. Amaranth is used as a food ingredient primarily in bread, pasta, baby's food, instant drinks, *etc.* For such purposes, seeds have to undergo various processing technologies - boiling, swelling, flaking, extrusion, puffing, roasting, grinding, sprouting, etc. The most common product is flour, whole amaranth seeds are used in breads, müsli bars, breakfast food and porridges, pastas and biscuits and cookies. Leaves and stems are interesting vegetables suitable for soups, salads or other meals. As a fodder Amaranth is a valuable nutritious feedstuff with high production ability. It also gives starch, protein concentrates, natural dye, a good source of squalen and antioxidants. It is being intensively tested as one of the interesting energy crops *(Anon.,* 2015a). Squalene is a natural 30-carbon organic compound originally obtained for commercial purposes primarily from shark liver oil (hence its name), although plant sources (primarily vegetable oils) are now used as well, including amaranth seed, rice bran, wheat germ` and olives.

Nutritional Value of Amaranth Leaves

The major attributes of amaranths are their adaptability to a wide range of climatic and soil conditions, superior nutritional quality of grain with high protein (12-19%) and complementary amino acid profiles (lysine 5-7%), easily digestible starch, presence of cholesterol lowering fractions in the seed oil and high carotene (pro-vitamin A) contents in the leaves (Ebert *et al.,* 2011). Amaranth is a highly nutritious leafy vegetable, both in raw and cooked form. Its nutritional value is comparable to spinach, but much higher than cabbage and Chinese cabbage. Amaranth is well adapted to high temperatures in the tropics, while spinach lacks heat tolerance and is not an option for tropical climates. Amaranth is low in saturated fats and sodium; cholesterol is absent. It is a good source of calcium, iron, magnesium, phosphorus, potassium, zinc, copper and manganese. It is a very good source of high quality protein with well-balanced amino acids. Many vitamins are found in high levels in both raw and cooked amaranth leaves and stems: vitamin A, vitamin C, riboflavin (vitamin B2), vitamin B6, folate (vitamin B9) and niacin (vitamin B3). Amaranth also contains betacyanines—amaranthine and iso-amaranthine—responsible for the red-violet colors of leaves, stems, flowers and seeds. These pigments dissolve in cooking water, which is poured off. Leaves and stems contain the antinutrients nitrate (mostly in stems) and oxalate at a level similar to other leaf vegetables such as spinach and spinach beet. Consumption of 100-200 g of fresh produce per day is safe. Cooking in water removes the toxic components (Ebert et al., 2011). According to Nath and Swamy (2015) there are a number of different types of amaranth. All are grown during the summer and the rainy season. This is the most common leafy vegetable grown during the summer in India. The leaves and tender stems are rich in protein, minerals and vitamins A and C. Amranth is a common leafy vegetable grown in most parts of India. The fresh, tender leaves and stems give delicious preparation on cooking as in case of other fresh leafy vegetables. Cooked similar to spinach or spinach beet, it is a cheap vegetable for the common people and is highly rich in vitamin A and C. Besides having the highest protein contents among leafy vegetables, it also contains carbohydrates, calcium and iron. The vitamin A content in different species varies from 23,000 - 54, 110 IU. The leaves contain 130-173 mg vitamin C, 100-130 mg vitamin B, 4 g protein, 397 mg calcium, 83 mg phosphorus, 25.5 mg iron, 341 mg potassium, 247 mg magnesium and 9200 IU of vitamin A.

Health Benefits of Amaranth Grain and Leaves

Amaranth was cultivated by the Aztecs and in other tropical climates, but is now experiencing a resurgence in popularity as a gluten-free protein. Though amaranth is derived from the fruit of a flowering plant, it is often referred to as a grain. Very little research has been conducted on amaranth's beneficial properties, but the studies that have focused on amaranth's role in a healthy diet have revealed three very important reasons to add it to our diet. i) It's a protein powerhouse: At about 13-14%, it easily trumps the protein content of most other grains. We may hear the protein in amaranth referred to as "complete" because it contains lysine, an amino acid missing or negligible in many grains. ii) It is good for our heart: Amaranth has shown potential as a cholesterol-lowering whole grain. iii) It is naturally gluten-free: Gluten is the major protein in many grains and is responsible for the elasticity in dough, allows for leavening, and contributes chewiness to baked products. But more and more people are finding they cannot comfortably – or even safely – eat products containing gluten, often due to Celiac disease. When people with celiac disease eat gluten (a protein found in wheat, rye and barley), their body mounts an immune response that attacks the small intestine. These attacks lead to damage on the villi, small fingerlike projections that line the small intestine, that promote nutrient absorption. When the villi get damaged, nutrients cannot be absorbed properly into the body. It is an autoimmune digestive disease that damages the body's ability to absorb nutrients from food. This makes amaranth an important grain to take note Celiac Awareness (*Anon.*, 2015b). Mom (2013) reported the following health benefits of Amaranth:

1. **Amaranth is gluten-free:** Cook amaranth grain as a hot cereal to eat in the morning. Find it as flour and use if for baking. Some even pop it like popcorn.
2. **It has more protein than other grains:** One cup of amaranth grain has 28.1 g of protein compared to oats at 26.1 g. It is healthier to receive protein from plant-based sources rather than animals, because the latter often comes with fat and cholesterol.
3. **Amaranth provides essential lysine:** Amaranth has far more lysine, an essential amino acid that the body can't manufacture, than other grains. Lysine helps metabolize fatty acids into energy, absorb calcium, and even keep the hair on our head intact.
4. **Helps with hair loss and graying:** Eating amaranth helps with hair loss; juice the leaves and apply it after shampooing. It is said that it helps moisturize and flatten wiry grey hair.
5. **Lowers cholesterol and risk of cardiovascular disease:** Amaranth seeds and oil (found in the seed) have fiber which contributes to lower cholesterol and risk of constipation. It is also rich in phytosterols, which are known for lowering cholesterol.
6. **It is high in calcium:** Amaranth helps reduce risk of osteoporosis and other calcium deficiencies because it has twice the calcium as milk.
7. **Amaranth is full of antioxidants and minerals:** It is the only grain to have vitamin C, but it is high in vitamin E, iron, magnesium, phosphorus and potassium which are necessary for overall health. The leaves are high in vitamin C, vitamin A and folate.
8. **Works as an appetite suppressant:** Protein reduces insulin levels in the blood stream and releases a hormone that makes us feel less hungry. Since amaranth is roughly 15% protein, the fact that it aids in weight loss or maintaining weight is one of the health benefits.

9. **Improves eyesight:** Some cultures believe that amaranth greens are a natural way to improve eyesight.

10. **Amaranth is easy to digest:** Amaranth is traditionally given to patients recovering from illness or people coming off of fasts. It is the mix of amino acids that allows for very easy digestion.

Cytotaxonomic Background

Basic Botany of the Amaranth

Amaranth belongs to the family Amranthaceae and the genus *Amaranthus*. The *Amaranthus* genus (Magnoliophyta: Caryophyllidae) comprises 70 species grouped into three subgenera (Mosyakin and Robertson, 2003). *Amaranthus* shows a wide variety of morphological diversity among and even within certain species. Although the family Amaranthaceae is distinctive, the genus has a few distinguishing characters among the 70 species included. This complicates taxonomy and *Amaranthus* has generally been considered among systematists as a "difficult" genus. Formerly, Sauer (1955) classified the genus into two subgenera, differentiating only between monoecious and dioecious species: *Acnida* and *Amaranthus*. Although this classification was widely accepted, further infrageneric classification was (and still is) needed to differentiate this widely diverse group. Currently, *Amaranthus* includes three recognized subgenera and 70 species, although species numbers are questionable due to hybridization and species concepts. Infrageneric classification focuses on inflorescence, flower characters and whether a species is monoecious or dioecious, as in the Sauer (1955) suggested classification. A modified infrageneric classification of *Amaranthus* was published by Mosyakin & Robertson (1996) and includes three subgenera: *Acnida*, *Amaranthus*, and *Albersia*. The taxonomy is further differentiated by sections within each of the subgenera (WIKI, 2015). Sauer (1967) recognized two sections in *Amaranthus, viz.*, *Amaranthus* and *Blitopsis*, with equal number of species in each. Section *Amaranthus* consists of species having terminal flower clusters and includes grain types; whereas section *Blitopsis* consists of species having flower clusters in axils and includes the green types or vegetable types.

The most economically important is the subgenus *Amaranthus* proper, which includes the three species domesticated for grain production: *Amaranthus hypochondriacus*, *Amaranthus cruentus*, and *Amaranthus caudatus*. Other species of amaranths have been domesticated as leaf-vegetables, for fodder, as potherbs, or as ornamentals; among these species, *A. tricolor*, from South Asia, is probably the most important (Sauer, 1967). According to Ebert *et al.* (2011) the genus *Amaranthus* comprises about 70 species, 40 of which are native to the Americas. Among the 70 species, 17 are vegetable amaranths with edible leaves, and three are grain amaranths with edible seeds (*A. caudatus*– Inca wheat, love-lies-bleeding; *A. cruentus*– purple amaranth; *A. hypochondriacus*– Prince's feather).

There are several species of *Amranthus* used for leaves, grains or for both. There is a taxonomic confusion because species are quickly adapted in any environment, differences among various species are small, many specific and common names have been used throughout the world, almost interchangeably and also due to quick intergradation of a species in the region itself where it thrives well when cultivated and appears adapted.

In spite of this confusion, some species are sufficiently recognized to merit universal acceptance. The best of the species for grains is *A. hypochandriacus* L. and for edible leaves *A.gangeticus* L., *A. cruentus* L and *A. dubius* Mart ex. Thell. *A. hypochandriacus* was used as a grain in India and Sri Lanka in the 18th century. It became prevalent in the foot hills of the Himalayas during the 19th century, where it became a staple food. It is important now in Nepal, China, Manchuria, Uganda, *etc.* In India, *A. hypochandriacus* and *A. caudatus* (grains) and *A. gangeticus* (leaves) are of major importance. *A. mangostanus* (Syn. *A. tricolor* var *mangostanus), A. lividus* and *A. dubius* (a recognized tetraploid) are grown on a limited scale in Odisha and other states (Nath and Swamy, 2015).

The important species of leafy amaranth, *Amaranthus tricolor* L, occupies a predominant position in India with different morphological forms in colour and shape of leaves. *Amaranthus dubius, A. lividus, A. blitum, A tristis* L., *A. spinosus* L and *A. viridis* are other amaranth species, which are under cultivation. *Amaranthus* is an annual herb, erect or trailing, scarce to profuse branched, shallow to deep tap-rooted, stem green to purple, leaf simple, alternate or opposite, colour green to purple. Inflorescence terminal and axillary, branched spikes, flower small, regular, mostly unisexual, monoecious. In general the cultivated species are monoecious (Nath and Swamy, 2015).

Amranth *(Amaranthus* spp.) is a bushy plant growing 1 to 3 m, depending if it is wild or cultivated, a vegetable variety or a grain variety. Feral versions are green, sometimes with red stems, spindly and usually no more than 60 cm, rarely 90 cm. Cultivated versions can be all red, or all green, showy or dull. Grain varieties can be 2 m tall. The lens-shaped seeds are tiny — eye of a needle size — and can be gold to black. Plants produce on an average 50,000 seeds each. Amaranth will grow under a variety of conditions and climates (Green Deane, 2012).

Amaranth is an annual plant with C4 type of photosynthesis. Depending on species amaranth leaves vary in shape, size and colour (green, red, purple). This plant can grow up to 3 m. Its stem, sometimes branched, is terminated by branched inflorescence (panicle). Inflorescence is usually indeterminant and reaches different lengths. Basic unit in inflorescence is called glomerulus containing female, male or both flowerets. Seed has lenticular shape (1-2 mm). Amaranth shows a high coefficient of propagation *(Anon.,* 2015a).

Species of Genus *Amaranthus*

Although over 70 species of *Amaranthus* have been reported, a brief description for some of the species is furnished as under (WIKI, 2013):

1. ***Amaranthus acanthochiton* (Green stripe)**, is an annual plant species. It is native to the southwestern United States and northern Mexico, growing at altitudes of 1000-2000 m where it is uncommon. It is a dioecious plant growing to 10-80 cm tall. The leaves are slender, 2-8 cm long and 2-12 mm broad. The flowers are pale green, produced in dense terminal spikes. The seeds are brown, 1-1.3 mm diameter, contained in a 2-2.5 mm achene. The seeds and young leaves were used as a food source.

2. ***Amaranthus albus*** is an annual species. It is native to the tropical Americas but a widespread introduced species in other places, including Europe, Africa, and Australia. When it dries, it forms tumbleweeds. Common names include common tumbleweed, tumble pigweed, tumbleweed, pigweed amaranth, prostrate pigweed, white amaranth, and white pigweed.
3. ***Amaranthus anderssoni*** is endemic to Ecuador. *i.e.,* it is Native to Ecuador or limited to a certain region of Ecuador.
4. ***Amaranthus arenicola***, commonly called sand amaranth or sand hill amaranth, is a plant species found in many states of the contiguous United States. It is an annual species found in sandy areas, near riverbeds, lakes, and fields. It is native to the central or South Great Plains, extending from Texas to South Dakota, and was introduced to other areas. This flowering plant can grow up to 2 meters in height.
5. ***Amaranthus australis*** is also known as Southern amaranth or Southern water-hemp. The plant usually grows from 1 to 3 m in height, though some have been known to grow up to 9 m high. The stems can be up to 30 cm in diameter. It is a herbaceous annual. It is found in many Southern states of the USA, Mexico, the West Indies, and South America. They are frequently found in wetland areas. It is herbaceous, short lived perennial. The largest is 5 m tall. As with most plants used for "greens" young and tender leaves are usually the best, take them from the top. Leave the older, larger leaves to collect energy for the plant. If you want to collect the seeds after they form, take a large, non-porous bag, put it over the top of the plant, gently tip the plant to the side, and shake. The seeds will come loose.
6. ***Amaranthus bigelovii*** is a species commonly known as Bigelow's amaranth. It is an annual native to New Mexico, Texas, and Louisiana.
7. ***Amaranthus blitoides***, commonly called prostrate amaranth, prostrate pigweed, mat weed, or mat amaranth, is a glabrous annual plants species. It usually grows up to 0.6 m, though it may grow up to 1 m. It flowers in the summer to fall. It is believed to have been a native of the central or Eastern United States, but it has naturalized in almost all of temperate North America. It has also naturalized in South America and Eurasia. Some authorities list it as an invasive species.
8. ***Amaranthus blitum***, commonly called purple amaranth is an annual plant species. Native to the Mediterranean region, it is naturalized in other parts of the world, including much of eastern North America. Although weedy, it is eaten in many parts of the world.
9. ***Amaranthus brownii*** is an annual herb. The plant is found only on the small island of Nihoa in the Northwestern Hawaiian Islands. *A. brownii* differs from other Hawaiian species of *Amaranthus* with its spineless leaf axils, linear leaves, and indehiscent fruits.
10. ***Amaranthus californicus*** is a species of flowering plant known by the common name California amaranth, California pigweed. It is a glabrous monoecious annual that is native to most of the Western United states and Canada. The plant grows from 10 to 50 cm tall. It is found in moist flats or near bodies of water, and blooms from summer to fall.

11. ***Amaranthus cannabinus*** is also known as salt marsh water hemp or salt marsh pigweed, tidal marsh amaranth. It is a herbaceous perennial found in most of the eastern United States. It grows from 1 to 3 m in height. It is often mistaken for *A. australis*.

12. ***Amaranthus caudatus*** is a species of annual flowering plant. It goes by common names such as love-lies-bleeding, pendant amaranth, tassel flower, velvet flower, foxtail amaranth, and quilete. Many parts of the plants, including the leaves and seeds, are edible, and are frequently used as a source of food in India and South America. The exact origin is unknown, as *A. caudatus* is believed to be a wild *A.hybridus* aggregate. The red color of the inflorescences is due to a high content of betacyanins. Ornamental garden varieties sold under the latter name are either *A. cruentus* or a hybrid between *A. cruentus* and *A. powelli*.

13. ***Amaranthus chihuahuensis*** is known as Chihuahuan amaranth. It is not native to the United States. It is found in Mexico. Some reports have suggested that it is present in lower Texas, but further evidence is necessary. Its taxonomic identity is considered to be unsure.

14. ***Amaranthus crassipes***, also known as spreading amaranth, is a glabrous annual plant that is both native and introduced in the United States. It is also found in Mexico, the West Indies, and South America. The plant can grow up to 60 cm in height. It flowers in the summer and fall. It is usually found near wet habitats or disturbed areas.

15. ***Amaranthus crispus*** is also known as crisp leaf amaranth. It is a herbaceous, sparsely pubescent annual. It can grow up to 0.5 m in height. It flowers in summer to fall. It usually grows in waste places, disturbed habitats, or near water. It is found in the U.S. It is native to Argentina, and it has also been introduced into parts of Eurasia.

16. ***Amaranthus cruentus*** is a flowering plant species that yields the nutritious staple amaranth grains. It is one of three *Amaranthus* species cultivated as a grain source, the other two being *A. hypochondriacus* and *A. caudatus*. In English it has several common names, including blood amaranth, red amaranth, purple amaranth, prince's feather and Mexican grain amaranth. In Maharashtra, it is called as *shravani maath* or *rajgira*.

17. ***Amaranthus deflexus*** is also known by the common names large-fruit amaranth, low amaranth, Argentina amaranth, and perennial pigweed. It is native to South America. It is a short-lived perennial or annual plant. The plant can grow up to 0.5 m in height. It flowers in the summer to fall, and was introduced into many states. It has been introduced into many warm or temperate regions of the globe. It grows the best in weedy areas or in disturbed habitats.

18. ***Amaranthus dubius*** is known as spleen amaranth, *khada sag*. This plant is native to South America and was introduced to Asia, Europe and Africa. In Tamil Nadu the plant is known as "Araikeerai". In general 'Keerai' means 'greens' in Tamil. In Telugu it is called 'Yerra thotakura'. In Kerala, it is called "Cheera".Usually it grows to a size of 80–120 cm. It has both green and red varieties, as well as some with

mixed colors. The green variety is practically indistinguishable from *A. viridis.* It flowers from summer to fall in the tropics, but can flower throughout the year in subtropical conditions. It is a ruderal species, usually found in waste places or disturbed habitats. A ruderal species is a plant species that is first to colonize disturbed lands. The disturbance may be natural – for example, wildfires or avalanches – or a consequence of human activity, such as constructionof building, mining, *etc.* or agriculture. The word *ruderal* comes from the Latin *rudus* rubble. *Amaranthus dubius* is considered to be a morphologically deviant allopolyploid. It is very close genetically to *A. spinosus* and other *Amaranthus* species.

19. ***Amaranthus fimbriatus*** is a species of glabrous flowering plant. It is also known by common names such as fringed amaranth or fringed pigweed. The plant can often grow up to 0.7 m in height. The flower is greenish to maroon. It is found in the Southwestern U.S. and in Mexico. It often grows on sandy, gravelly slopes or in disturbed habitats. It usually blooms after the summer rains in these arid regions. It is considered to be an invasive weed.

20. ***Amaranthus floridanus*** is a flowering plant that can grow up to 1.5 m in height. It flowers from late spring to fall, and is found in Florida. It usually grows in moist places, near dunes, swamps, marshes, or in disturbed habitats. It is known as Florida water hemp, Florida amaranth.

21. ***Amaranthus furcatus*** is a species of plant endemic to Ecuador (Native to or limited to Ecuador).

22. ***Amaranthus graecizans*** is an African species naturalized in North America. Common names include tumble weed and pigweed. The edible leaves are used as a vegetable throughout Africa.

23. ***Amaranthus grandiflorus*** is a species found in Australia. It is an annual plant, reaching up to 40 cm tall. The leaves are ovate to lanceolate, and up to 5 cm, with an acute tip. The flowers are clustered into inflorences, borne in the axils. The petals are 5–7 mm long.

24. ***Amaranthus greggii*** is also known as Gregg's amaranth or Josiah amaranth. It is a glabrous annual flowering plant native to USA and Mexico. The plant can grow up to 1 m in height. It is found in sand dunes and near sea beaches. The species name *greggii* honors Josiah Gregg (1806 – 1850), a merchant, explorer, naturalist, and author of the American Southwest and Northern Mexico.

25. ***Amaranthus hybridus***, commonly called smooth amaranth, smooth pigweed, red amaranth, green amaranth or slim amaranth, is a species of annual flowering plant. It is a weedy species found now over much of North America and introduced into Europe and Eurasia.

26. ***Amaranthus hypochondriacus*** is an ornamental plant commonly known as prince-of Wales feather or prince's feather. Originally endemic to Mexico, it is called *quelite, blero* and *quintonil* in Spanish. In Africa, it is valued as source of food.

27. ***Amaranthus mitchellii*** is commonly known as boggadri weed. It is a generally useful plant and is said to be "edible".

28. ***Amaranthus palmeri*** is a species of edible flowering plant. It has several common names, including Palmer's amaranth, Palmer amaranth, Palmer's pigweed, and careless weed. It is native to most of the Southern half of North America. Populations in the Eastern United States are probably naturalized. It has also been introduced to Europe, Australia, and other areas. The plant is fast-growing and highly competitive. The leaves, stems and seeds of Palmer amaranth, like those of other amaranths, are edible and highly nutritious. Palmer amaranth was once widely cultivated and eaten by Native Americans across North America, both for its abundant seeds and as a cooked or dried green vegetable. The plant can be toxic to livestock animals due to the presence of nitrates in the leaves. Palmeri amaranth has a tendency to absorb excess soil nitrogen, and if grown in overly fertilized soils, it can contain excessive levels of nitrates, even for humans. Like spinach and many other leafy greens, amaranth leaves also contain oxalic acid, which can be harmful to individuals with kidney problems if consumed in excess.

29. ***Amaranthus polygonoides*** is a species of flowering plant. It goes by the common name of Tropical Amaranth.

30. ***Amaranthus powellii*** is a species known by the common names Powell's amaranth, Powell's pigweed and green amaranth. It is native to the Southwestern United States and Northern Mexico, but it is common throughout most of the rest of the temperate Americas as a naturalized species. It has also been introduced to other continents, including Australia and Europe. This is an erect annual herb growing to a maximum height of 2 m. It has leaves up to 9 centimeters long, those on the upper part of the plant lance-shaped and lower on the stem diamond or roughly oval in shape. The inflorence holds several long, narrow clusters of both male and female flowers interspersed with spiny green bracts. The fruit is a smooth dehiscent capsule about 3 mm long containing shiny reddish black seeds.

31. ***Amaranthus pumilus***, the seaside amaranth or sea beach amaranth, is an annual plant. This annual plant is now a threatened species, although it was formerly scattered along the eastern coast of the United States, its native range. The plant consists of many low and prostrate stems with fleshy leaves. Larger plants with hundreds of stems may cover an area of about a meter. Yellow flowers are obscure, but many seeds are produced in July. The lengthy viability of these seeds may account for the reappearance of *Amaranthus pumilus* in places where it had formerly vanished.

32. ***Amaranthus retroflexus*** is a species of flowering plant with several common names, including red-root amaranth, redroot pigweed, red-rooted pigweed, common amaranth, pigweed amaranth, and common tumble weed. This plant is eaten as a vegetable in different places of the world. No species of genus *Amaranthus* is known to be poisonous, but the leaves contain oxalic acid and may contain nitrates if grown in nitrate-rich soils, so the water should be discarded after boiling. *A. retroflexus* was used for a multitude of food and medicinal purposes by many Native American groups. It is used in Kerala to prepare a popular dish known as *thoran*. The seeds are edible raw or toasted, and can be ground into flour and used for bread, hot cereal, or as a thickener.

33. ***Amaranthus sclerantoides/ Amaranthus scleranthoides*** is a species endemic to Ecuador. These tiny plants (with the smallest individuals about the size of a thumbnail) are an example of phenotypic plasticity in the endemic *Galapagos* species *Amaranthus sclerantoides.* In dry or salt-stressed environments, these plants tend to strongly express betalain pigments, which gives the entire plant a dull orange, yellow, or pink color. A cluster of plants resembles seaweed washed up on the beach. *Amaranthus sclerantoides* also starts flowering at much smaller sizes in such exposed environments. The plants can grow up to half a meter tall.

34. ***Amaranthus spinosus***, commonly known as the spiny amaranth, prickly amaranth or thorny amaranth is native to the tropical Americas, but it is present on most continents as an introduced species and sometimes a noxious weed. It can be a serious weed of rice cultivation in Asia. Like several related species, *Amaranthus spinosus* is a valued food plant in Africa. In Tamil it is called *mullik keerai.* In Sanskrit it is called *tanduliyaka.* Spiny amaranth has edible leaves and might be a medicine and sex aid (Green Deane, 2012).

35. ***Amaranthus thunbergii***, commonly known as Thunberg's amaranth, is found in Africa. The leaves are used as a flavouring or leafy vegetable.

36. ***Amaranthus torreyi*** is a species of flowering plant that is sometimes considered to be a synonym of *A. watsonii.* Its common name is Torrey's Amaranth. It is native to the Southwestern United States and Northern Mexico.

37. ***Amaranthus tricolor*** is an ornamental plant known as tandaljo or tandalja bhaji in India, and Joseph's coat after the Biblical figure Joseph, who is said to have worn a coat of many colors. Although it is native to South America, many varieties of amaranth can be found across the world in a myriad of different climates due to it being a C4 carbon fixation plant, which allows it to convert carbon dioxide into biomass at an extremely efficient rate when compared to other plants. Cultivars have striking yellow, red and green foliage. The leaves may be eaten as a salad vegetable as well as the stems. In Africa, it is usually cooked as a leafy vegetable. It is usually steamed as a side dish in both China and Japan.

38. ***Amaranthus gangeticus*** is considered a synonym of *A. tricolor*, but has been recognized as a separate species in the past. *Amaranthus gangeticus* is also known as elephant-head amaranth. It is an annual flowering plant with deep purple flowers. It can grow from 60-90 cm in height. It has been used as a leafy vegetable.

39. ***Amaranthus tuberculatus***, commonly known as tall water hemp or rough fruit amaranth, is a species of flowering plant. It is a summer annual plant with broadleaf and germination period that lasts several months· Tall water hemp has been reported as a weed in 40 of 50 U.S. States. Tall water hemp is a dioecious plant. The seed head branches in the female are numerous, short, and smooth. The male seed head branches are fewer, longer, and more slender than those of the female. The species has terminal spike inflorescences and very short bracts with simple to highly branched flowers. Seed produced is reddish to black in colour and less than 1mm in diameter.

40. ***Amaranthus viridis*** is a cosmopolitan species and is commonly known as slender amaranth or green amaranth. It is also eaten as a vegetable in parts of Africa. The leaves of this plant have been used in the diet of the Maldives for centuries in dishes. *A. viridis* is used as a medicinal herb in traditional Ayurvedic medicine, under the Sanskrit name *tanduliya.*

41. ***Amaranthus watsonii*** is a species known by the common name Watson's amaranth. It is native to the Southwestern United States and Northern Mexico, where it grows in sandy places such as deserts and beaches, and disturbed areas. It is also known as a rare introduced species in parts of Europe. This is an erect annual herb producing a glandular hairy stem to a maximum height of about a meter. The leaves are generally oval-shaped and up to 8 cm long, with a petiole of up to 9 cm. The species is dioecious, with male and female plants producing different types of flowers. The inflorescence is a long spike cluster of flowers interspersed with spiny green glandular bracts. The fruit is a smooth capsule about 2 mm long that snaps in half to reveal a small shiny reddish black seed.

42. ***Amaranthus wrightii*** is a species of flowering plant. It goes by the common name of Wright's Amaranth.

43. ***Amaranthus acutilobus*** is synonym of *A. viridis* L.

44. ***Amaranthus interruptus,*** is a species of flowering plant. The common names are Australian amaranth, native amaranth. Decumbent to erect herb 20-60 cm tall. Leaf petioles grooved on the upper surface, 5-10 mm long. Leaf blades 10-27 x 7-16 mm, smaller towards the apex of the plant; underside of the leaf blade sparsely clothed in dark brown hairs. Inflorescence 6-8 mm long. Flowers small, about 1 x 0.5 mm, each subtended by a bract about 1.25 cm long. Perianth consists of 3 tepals fused only near the base. Anthers about 1.5 mm long, filaments 1-1.25 mm long. Pollen cream, Olvules solitary in each ovary. Fruits about 2 mm long with the styles persistent at the apex and perianth lobes persistent at the base. Fruits split along the equator. Seeds about 1 mm diameter.

45. ***Amaranthus minimus*** is a very rare and critically endangered Cuban endemic sea beach amaranth. *A. minimus* emerged as a pioneer species extremely adapted to its littoral (situated on the shore of the sea or a lake) and seasonal habitat, with a specialized dispersal strategy, a rich seed bank and variable germination rate which are key life history traits.

46. ***A. muricatus*** is commonly known as muricate amaranth, African amaranth. Plants- annual or short-lived perennial, glabrous or slightly pubescent near tips. Stems- ascending or prostrate, much-branched from stout rootstock, 0.1-0.4 m. Leaves: petiole to $^1/_2$ as long as blade; blade linear to narrowly lanceolate, 1.5-8 × 0.2-0.5(-1) mm, base tapering, margins entire, plane to undulate, apex obtuse and often emarginate. Inflorescences- terminal, compact pyramidal panicles and axillary glomerules, erect or reflexed, green, leafless at least distally. Bracts of pistillate flowers linear, 0.7-1.2 mm, $^1/_2$-$^2/_3$ as long as tepals. Pistillate flowers: tepals 5, narrowly oblanceolate, not clawed, equal, 1.5-2 mm, apex obtuse or subacute; style branches erect; stigmas 3. Staminate flowers intermixed with pistillate or

at tips of inflorescences; tepals 5; stamens 5. Seeds black, lenticular, 1-1.2 mm diam., semiglossy. The vernacular name "African amaranth" is sometimes used for this species which is a misnomer; the species is native to South America and naturalized in Africa.

47. ***Amaranthus obcordatus.*** Plants- glabrous. Stems- erect or ascending, branched, 0.1-0.5 m. Leaves: petiole $^1/_2$ as long as blade; blade oblong-lanceolate to lanceolate-linear, 1-3 × 0.2-1 cm, base cuneate to narrowly cuneate, margins entire, plane or sometimes undulate, apex rounded or obtuse-truncate, mucronate. Inflorescences mostly axillary, but at apex flowers also condensed in terminal spikes or spicate panicles, usually leafy proximally or nearly leafless distally. Bracts broadly ovate, 1 mm or shorter, $^1/_2$ or less length of petals, apex acute. Pistillate flowers: tepals 5, spreading at maturity, spatulate, clawed, subequal, 2 mm, margins fimbriate, apex rounded or shallowly emarginate; stigmas 3. Staminate flowers: tepals (3-)5; stamens 3. Seeds dark reddish brown to nearly black, lenticular or broadly lenticular, 0.6-0.8 mm diam., smooth. Flowering summer-fall. *Amaranthus obcordatus* has been reported only from Southern Arizona.

48. ***Amaranthus pringlei*** belongs to the group of annual and biennial plants. The common name is Pringle's amaranth. *A. pringlei* is deciduous. The simple leaves are alternate. They are entire and petiolate. It is native to Mexico.

49. ***Amaranthus scleropoides*** is called as bone bract amaranth, bone bract pigweed. Plants annual, glabrous. Stems ascending to prostrate, erect when young, or main stems ± erect, branched proximally, 0.1-0.6 m. Leaves: petiole equaling or $^1/_2$ as long as blade; blade elliptic, oblanceolate to lanceolate, (0.5-)1-3(-3.5) × 0.3-2 cm, base tapering, margins entire, plane to slightly undulate, apex broadly rounded or emarginate. Inflorescences axillary clusters borne from base to top, axes thickened and inflated, becoming indurate at maturity. Bracts of pistillate flowers keeled, ovate-triangular, minute. Pistillate flowers: tepals 5, narrowly spatulate, slightly clawed, with expanded blade, equal or subequal, (1.2-)1.5-2.5 mm, apex acute to apiculate; claws indurate at maturity; style branches spreading; stigmas 2-3. Staminate flowers intermixed with pistillate; tepals 5, membranaceous; stamens 3. Seeds dark brownish black to black, compressed-ovoid to broadly lenticular, 0.9-1.1 mm diam., shiny.

50. ***Amaranthus standleyanus*** is an annual plant growing to 0.7 m. It is frost tender. The flowers are monoecious (individual flowers are either male or female, but both sexes can be found on the same plant) and are pollinated by wind, self. The plant is self-fertile. Edible parts are leaves and seeds.

51. ***Amaranthus tamariscinus*** is commonly called as tall water hemp, prostrate water hemp. It is native of U.S. Its habitat is disturbed moist sandy soil. It flowers from August to October.

Cultivated Species of Vegetable Amaranth

There are six vegetable amaranth species, *viz.*, *A. blitum* (syn. *A. blitum* var. *oleraceus* and *A. lividus*), *A. cruentus* (syn. *A. paniculatus*), *A. dubius*, *A. spinosus*, *A. tricolor* (syn. *A. melancholicus*, *A. tristis*, *A. mangostanus* and *A. gangeticus*), *A. viridis (A. gracillis)*. All these species are used as leafy vegetables (Ebert *et al.*, 2011).

1. ***Amaranthus blitum:*** Common names are livid amaranth, slender amaranth. The probable origin of *A. blitum* is the Mediterranean region; it is found worldwide from the tropics to temperate climates. It is a popular cultivated vegetable in India and is also cultivated in East and Central Africa. It is grown in home gardens in southeastern Europe (Greece), where it is used as a spinach substitute during the hot and dry summer months. It is not suitable for fresh consumption due to relatively high levels of hydrocyanic acid and oxalic acid. It is also a cosmopolitan weed.
2. ***A. cruentus:*** Common names are purple amaranth, red amaranth, red shank, bush greens, African spinach, Indian spinach. This amaranth species was domesticated as a grain in Mesoamerica and found its way to the tropics and subtropics of the Old World during colonial times. It is used for grain production (pseudo cereal), as a leafy vegetable, and for ornamental purposes. In tropical Africa it is a traditional, highly productive, nutritious and economically important leafy vegetable. It is also widely grown as a leafy vegetable throughout Southeast Asia and South Asia. In Indonesia it is mostly grown in mountainous areas where it is too cold for the popular amaranth species *A. tricolor*. The leaves and tender stems are cut and cooked or fried in oil and served as a side dish. The high calcium oxalate content limits its use as fodder. *Grain amaranth* types of *A. cruentus* are common in Central America and northern South America. Grain amaranth is also popular in India and Nepal and is commercially grown in hot and dry areas of the United States, Argentina, and China. *Ornamental types* of *A. cruentus* have large bright-red inflorescences and are often found in the tropics and subtropics. The inflorescences can be used to produce a red dye.
3. ***A. dubius:*** Common name is spleen amaranth. Cultivated types of *A. dubius* may have been derived from the weedy ancestor in tropical Asia (Indonesia, India) and may have been introduced to Africa and Central America by immigrants. It is usually grown from sea level to 1200 m elevation and reaches a yield of 25 t/ha in eight weeks. The plants are used as a cooked leafy vegetable. *A. dubius* is more susceptible to drought than *A. cruentus.*
4. ***A. spinosus:*** Common names are spiny amaranth, thorny pigweed. This species probably originated from the lowland tropics in South and Central America. It is now found in most tropical and subtropical regions, including Africa, but is rarely cultivated due to the spines and the poor taste. It is usually collected for home consumption and eaten cooked, steamed or fried; it can help bridge periods of drought. To some extent, it is also used as forage. The species has multiple medicinal uses. The root has diuretic effects; the plant sap is used as an eye wash. The plant is also used as an expectorant and to relieve breathing of patients suffering from acute bronchitis.
5. ***A. tricolor*:** Common names are Chinese amaranth, Chinese spinach, Joseph's coat. The probable origin of this amaranth species is tropical Asia. In South and Southeast Asia, *A. tricolor* is a major leafy vegetable species and is the dominant cultivated amaranth species, followed by *A. dubius* and *A. cruentus*. It is occasionally cultivated in East, West, and Southern Africa. It is generally used as a cooked vegetable, but occasionally also eaten raw in salads. In India, the soft stems are

eaten like asparagus. Leaves and stems contain the antinutrients nitrate (mostly in stems) and oxalate, but adverse nutritional effects are unlikely if consumption does not exceed 200 g per day. Cooking in ample water removes the toxic components. Types with red, yellow and green leaves are widely cultivated as ornamentals. *A. tricolor* is also used as a diuretic and to treat inflammations. *A. tricolor* is an easy-to-grow, productive, tasty, and nutritious leafy vegetable.

6. ***A. viridis*:** Common names are green amaranth, pigweed, slender amaranth. *A. viridis* is possibly of Asian origin. It is considered a pan-tropical weed and has penetrated into warm temperate regions worldwide. It is also a widespread and common weed in Africa and occasionally cultivated. Although mostly growing as a weed, its nutritional value is high. Leaves and young plants are eaten as a cooked vegetable. The plant also serves as fodder for cattle and green manure. The leaves have diuretic and purgative properties and are used in traditional medicine to cure many different ailments.

Cultivated Species of Grain Amaranth

Grain Amaranth species include *A. hypochondriacus, A. cruentus A. caudatus, A. retroflexus, A. albus* and *A. graecizans* (syn. *A. blitoides, A. angustifolius*). However, the popular *Amaranthus* species that are used as grain are *A. caudatus, A. cruentus* and *A. hypochondriacus.*

1. ***Amaranthus caudatus*** is a species of annual flowering plant. It goes by common names such as love-lies-bleeding, pendant amaranth, tassel flower, velvet flower, foxtail amaranth, and quilete. Many parts of the plants, including the leaves and seeds, are edible, and are frequently used as a source of food in India and South America. The exact origin is unknown, as *A. caudatus* is believed to be a wild *A.hybridus* aggregate. The red color of the inflorescences is due to a high content of betacyanins. Ornamental garden varieties sold under the latter name are either *A. cruentus* or a hybrid between *A. cruentus* and *A. powelli.*

2. ***Amaranthus cruentus*** is a flowering plant species that yields the nutritious staple amaranth grains. It is one of three *Amaranthus* species cultivated as a grain source, the other two being *A. hypochondriacus* and *A. caudatus.* In English it has several common names, including blood amaranth, red amaranth, purple amaranth, prince's feather and Mexican grain amaranth. In Maharashtra, it is called as *shravani maath* or *rajgira.*

3. ***Amaranthus hypochondriacus*** is an ornamental plant commonly known as prince-of Wales feather or prince's feather. Originally endemic to Mexico, it is called *quelite, blero* and *quintonil* in Spanish. In Africa, it is valued as source of food. **(Fig. 12-13).**

Cytogeneticalcal Studies on *Amaranthus* Species

Madhusoodanan and Pal (1981) studied cytology of 5 species of *Amaranthus* (section *Blitopsis*), viz., *A. tricolor, A. lividus, A. graecizans, A. viridis,* and *A. albus.* They reported that all these species were diploids with X=17, or X=16, the former being more common. Mallika (1987) conducted cytogenetical studies on 8 species of *Amaranthus, Viz., A. tricolor,*

A. lividus, A. viridis, A. spinosus, A. dubius, A. hypochondriacus, A. cruentus and *A. caudatus* and their hybrids. The first 3 belong to the Section *Blitopsis* and the remaining 5 belong to the section *Amaranthus.* Meiosis in both the sections was normal with regular formation of bivalents. Of 8 species, 7 were diploids with n=16 or 17 and one species, *A. dubius* was a polyploidy with n=32. 7 intraspecic hybrids were obtained which showed normal growth and flowering. The chromosome counts indicated *A. tricolor* to be 2n=2X=34, *A.cruentus* to be 2n=32 and *A. dubius* to be 2n=64.

Queiros (1989) reported the following chromosome numbers in 8 *Amaranthus* species in Portugal: 2*n* = 32 for *A. hybridus, A. blitoides, A. albus* and *A. graecizans*; and 2*n* = 34 for *A. paniculatus* [*A. cruentus*], *A. retroflexus, A. deflexus* and *A. lividus.* Bao-Hua *et al.* (2002) from China reported chromosome numbers of 14 species of the genus *Amaranthus* : *A. retroflexus* 2n=34, *A. caudatus* 2n=32, *A. hybridus* 2n=32, *A. spinosus* 2n =34, *A. cruentus* 2n=34, *A. hypochendriacus* 2n=32, *A. paniculdatus* 2n=32, *A. roxburghianus* 2n=34, *A. blitoides* 2n=32, *A. polygonoides* 2n=34, *A. albus* 2n=32, *A. viridis* 2n = 34, *A. lividus* 2n = 34, *A. tricolor* 2n = 34. The number in *A. roxburghianus* is reported here for the first time. The basic chromosome numbers in this genus are x = 16 and x = 17 and both numbers are found in sect. *Amaranthus* and sect. *Blitopsis.* The chromosomes in this genus are very small in size, hampering a detailed karyotype analysis.

Cytological observations were made by Grant (1959) for four species of *Amaranthus, viz., A. arenicola, A. palmeri, A. tamariscinus,* and *A. tuberculatus.* Three of these species have a diploid chromosome number of 32, whereas *A. palmeri* has a somatic chromosome number of 34. A single spontaneous triploid (2*n*=48) female plant was found in collections of *A. tamariscinus* and a tetraploid (2*n*=64) male plant in collections of *A. tuberculatus.* A fifth species, *A. australis,* has previously been reported as having 32 somatic chromosomes. The chromosome numbers for half the dioecious species of *Amaranthus* have now been determined. Detailed observations on the dividing chromosomes of these species in mitosis, in meiosis, and in the first division of the nucleus in the pollen grain have been made and have failed to distinguish heteromorphic chromosomes which might be associated with sex determination. The small size of the chromosomes has made detailed morphological studies impractical and there is no marked difference in absolute size of the chromosomes between species. Since haploid numbers of 16 and 17 are found in both monoecious and dioecious species, it would seem that the aneuploid condition in *Amaranthus* arose early and hybridization within the genus has resulted in promoting the genie condition which has been necessary for the expression of the dioecious condition. Behra and Patnaik (1974) studied the cytology of 23 taxa belonging to 9 genera of the family Amaranthaceae. The cytological analysis indicated that the genus *Amaranthus* L. was characterised by greater amount of homogeneity in its chromosome numbers than other genera of the family. Basing on Grant's (1959) view that the basic chromosome number of x=17 is a probable derivation from x=16, two lines of evolution from this basic number in the cytological level was visualised; one through euploidy resulting n=32 in few species and the other through aneuploidy resulting n=17 in larger number of species.

Karyotypical studies in the genus are scarce, probably due to the small size of the chromosomes, which makes morphological analysis difficult. Updated data have indicated that there are two basic chromosome numbers, x = 16 and x = 17, and, in some cases, both numbers were cited for the same species suggested that the gametic number n = 17originates from n = 16 through primary trisomy. Greizerstein and Poggio (1992) supported this hypothesis through the analysis of meiotic behavior of species and interspecific hybrids. Studies carried out on chromosome morphology of some species of the genus have indicated variation in number of chromosome pairs with satellites. Greizerstein and Poggio (1994) proposed karyotypic formulae of various accessions of cultivated species (*Amaranthus cruentus, Amaranthus hypochondriacus, Amaranthus mantegazzianus* and *Amaranthus caudatus*). In all studied species, only one pair of chromosomes with a satellite was found (Greizerstein and Poggio, 1994). In the study conducted by Bonasora *et al.* (2013), the chromosome numbers were confirmed, 2n = 34 for *Amaranthus cruentus* L., and 2n = 32 for *Amaranthus hypochondriacus* L., *Amaranthus mantegazzianus* Passer, and *Amaranthus caudatus* L.

Pandey (1999) carried out meiotic studies in four accessions of three grain species, *viz. Amaranthus cruentus, A. powellii* and *A. retroflexus* and their F1 hybrids to elucidate the genome relationships between the cultivated and wild types and the cytogenetical mechanisms involved in speciation. All the three species were morphologically distinct and cytologically uniform with 17 bivalents at metaphase I. Morphologically the interspecific hybrids were either intermediate or had an overall dominance of wild parents. Chromosome analysis at meiotic metaphase I in the F1 interspecific hybrids of A. powellii with the Indian and Mexican accessions of *A. cruentus* showed an average of chromosome association of 1.0 IV + 0.10 III + 14.78 II + 0.14 I and 1.0 IV + 0.45 III + 14.20 II + 0.25 I and 5.81% in the former and 8.44% in the later pollen grain fertility, respectively and that of *A. retroflexus* with *A. cruentus* (Indian and Mexican) showed almost similar chromosomal associations. These studies have shown close genomic homology amongst all these three species involving certain chromosomal aberrations resulting in their evolution.

Early History

Origin and Distribution of Amaranth

The grain amaranth ('ramdana', 'marcha', 'ganhar', lathe') considered by many as the crop of the future has been associated with man since prehistoric times (4,800 BC). The genus *Amrnthus* is notable mainly because of the success of many of its members as fellow-travelers of mankind. Sometimes deliberately cultivated as grain crops, green vegetables, dye plants, or ornamentals, more often unintentionally encouraged weeds, many amaranth species have spread and diversified with the advance of artificial habitats on every continent (Sauer, 1957). *Amaranthus*, collectively known as amaranth, is a cosmopolitan genus of annual or short-lived perennial plants. Some amaranth species are cultivated as leaf vegetables, cereals, and ornamental plants. Most of the species from *Amaranthus* are summer annual weeds and are commonly referred to as pigweeds. Catkin-like cymes of densely packed flowers grow in summer or autumn. Approximately 60 species are recognized, with inflorescences and foliage ranging from purple and red to green or gold. Members of this genus share many characteristics and uses with members of the closely related genus *Celosia* (WIKI, 2015). Amaranth is a grain with high nutrition value, comparable to those of maize and rice. Amaranth has been a staple in Mesoamerica for thousands of years, first collected as a wild food, and then domesticated at least as early as 4000 BC. The edible parts are the seeds, which are consumed whole toasted or milled into flour. Other uses of amaranth include dye, forage and ornamental purposes. Amaranth is a plant of the family of *Amaranthaceae*. About 60 species are native to the Americas, whereas less numerous are the species originally from Europe, Africa and Asia. The most widespread species are native of North, Central and South America, and these are *A. cruentus*, *A. caudatus*, and *A. hypochondriacus*. *Amaranthus cruentus*, and *A. hypochondriacus* are native of Mexico and Guatemala. The first one is used in Mexico to produce typical sweets called *alegría*, in which the amaranth grains are toasted and mixed with honey or chocolate. *Amaranthus caudatus* is a widely distributed staple food both in South America and in India.

This species originated as one of the staple foods for the ancient inhabitants of the Andean region (Sauer, 1967; Maestri, 2015). Some species of green amaranth (*Amaranthus* spp.), especially, *Amaranthus gangeticus*, *A. mangostanus*, *A. paniculatus*, *A. angustifolius*, are supposed to have originated in India or Indo-Chinese region. Other spiecies have originated in various other centres like North America, Central America, Mexico, South America and Mediterranean region (Nath and Swamy, 2015).

The main cultivated species of vegetable type Amaranthus, *A. tricolor* L., has been originated in South or Southeast Asia, particularly in India. Buddhist monks and Muslim invaders took the crop to neighboring countries. Another vegetable type, *A. dubius* L., shows diversity in Central America, Indonesia, India and in Africa. *A. lividus* seems to have been a popular vegetable in Southern and Central Europe. As a hot season vegetable, it is cultivated throughout the year in the tropics (Varalakshmi, 2015).

The cultivated species of grain type *Amaranthus* and their probable native regions are as follows (Sauer, 1986):

1. *Anacardium hypochondriacus* (= *A. frumentaceus*, *A. leucocarpus*, *etc.*) of Northwestern and Central Mexico.

2. *A. cruentus* (= *A. paniculatus*, *etc.*) of Southern Mexico and Central America.

3. *A. caudatus* of the Andes. In the Argentine Andes, the typical form is grown together with a conspicuous mutant that produces club-shaped inflorescence branches with determinate growth, a trait unknown in wild Amaranths. This mutant has commonly been given specific rank (as *A. edulis*) but may better be treated as *A. caudatus* ssp. *mantegazzianus*.

The wild species that appear most closely related to the above grain type species are as follows (Sauer, 1986):

1. *A. powellii*, a pioneer of canyons, desert washes and other open habitats in the Western Cordillera of the Americas. An aberrant form with indehiscent utricles has sometimes been given specific rank as *A. bouchonii*.

2. *A. hybridus* (= *A. chlorostachys*, *A. patulus*, *etc.*), a riverbank pioneer of moister regions of Eastern North America and the mild highlands of Central America.

3. *A. quitensis*, a riverbank pioneer of highland and subtropical South America.

Most interspecific hybrids, both spontaneous and artificial, have been reported among the grain species and their wild relatives. Some experimental hybrids show heterosis and nearly normal meiosis but are partially sterile; others have abnormal growth and are totally sterile, including *A. caudatus x A. hypochondriacus.* No spontaneous polyploids are known among the grain Amaranths but colchicines-induced autotetraploids and amphiploids have been bred from some of them. Seed weight in the tetraploids is about double that of the diploids, suggesting agro-economic potential (Sauer, 1986).

Amaranth was probably widely used among hunter-gatherers in both North and South America. The wild seeds, even if small in size, are produced in abundance by the plant and are easy to collect. Evidence of domesticated amaranth seeds comes from the Coxcatlan cave in the Tehuacan valley of Mexico and dates as early as 4000 BC. Coxcatlan Cave is a rockshelter in the Tehuacan Valley of Mexico, and it was occupied by humans for nearly 10,000 years. Later evidence, like caches with charred amaranth seeds, has been found throughout the US Southwest and the Hopewell culture of the US Midwest. Domesticated species are usually larger and have shorter and weaker leaves which make the collection of the grains simpler. As other grains, seeds are collected through rubbing the inflorescences between the hands (Sauer, 1967; Maestri, 2015). Amaranth grain has a long and colorful history in Mexico and is considered a native crop in Peru. It was a major food crop of the Aztecs, and some have estimated amaranth was domesticated between 6,000 and 8,000 years ago. During the pre-Columbian period, the Aztecs cultivated amaranth as a staple grain crop. Annual grain tributes of amaranth to the Aztec emperor were roughly equal to corn tributes. The Aztecs didn't just grow and eat amaranth, they also used the grains as part of their religious practices. Many ceremonies would include the creation of a deity's image that had been made from a combination of amaranth grains and honey. Once formed, the images were worshipped before being broken into pieces and distributed for people to eat. But things changed when the Spanish conquistadors (adventurers or conquerors, especially one of the Spanish conquerors of the New World in the 16th century) arrived. The *conquistadors* were Spanish explorers and warriors who successfully conquered much of America in the 16th century. When Cortez and his Spaniards landed in the New World in the sixteenth century, they immediately began fervent and often forceful attempts to convert the Aztecs to Christianity. One of their first moves was to outlaw foods involved in "heathen" festivals and religious ceremonies, amaranth included. Although severe punishment was handed to anyone found growing or possessing amaranth, complete eradication of this culturally important, fast-growing, and very prevalent plant proved to be impossible (*Anon.*, 2015b).

According to Howard (2013) "Native folks would pop the seeds and mix them with sacrificial human blood. They would form the seeds into sculptures and then eat them in religious ceremonies. This was seen as pagan [by the Spanish], so it was outlawed." Pagan is a volcanic island in the Mariana Islands archipelago in the Pacific Ocean, belonging to the Commonwealth of the Northern Mariana Islands. Formerly inhabited, the inhabitants were evacuated due to volcanic eruptions in 1981. Amaranth crops were seized, fields were burned, and those who tried to grow the plant were punished. The locals replaced their former staple by eating more corn. But amaranth cultivation did survive in a few isolated pockets. The grain lived on in a traditional treat called *alegria* (joy), in which popped, whole-grain amaranth is made into bars with honey and sunflower and pumpkin seeds. The bars are often enjoyed during Day of the Dead and other festivals. It has high levels of micro- and macro-nutrients that are lacking in the Oaxacan diet (*Oaxaca* is a city in Mexico well-known for its exceptional food), and it is a source of vitamins and nutrients that can help combat malnutrition" (Howard, 2013). Amaranth was preserved on hard to reach places of mountainous Central and South America. Amaranth was first introduced as an ornamental plant in Europe in the 16th century. Different species of amaranth spread throughout the world during 17th, 18th and 19th century. In India, China and under the harsh conditions of Himalayas this plant became important grain and/or vegetable crop (*Anon.*, 2015a). Because the plant has continued to grow as a weed since that time, its genetic base has been largely maintained. Research on grain amaranth began in the U.S. in the 1970s. By the end of the 1970s, a few thousand acres were being cultivated. Much of the grain currently grown is sold in health food shops. Grain amaranth is also grown as a food crop in limited amounts in Mexico, where it is used to make a candy called *alegría* (Spanish for happiness) at festival times. Amaranth species that are still used as a grain are *Amaranthus caudatus, A. cruentus,* and *A. hypochondriacus.* The grain is popped and mixed with honey. In North India, it is called "rājgīrā" . The popped grain is mixed with melted jaggery in proper proportion to make iron and energy rich "laddus," a popular food provided at the Mid-day Meal Programme in municipal schools. Amaranth grain can also be used to extract amaranth oil - a particularly valued pressed seed oil with many commercial uses (WIKI, 2013).

Use of Amaranth in Ancient Mesoamerica

In ancient Mesoamerica, amaranth seeds were commonly used. The term Mesoamerica, literally Middle America, was introduced by anthropologist Paul Kirchhoff in 1943 to define a huge geographic and cultural area that included the central and southern portion of Mexico, all of Guatemala, Belize and part of El Salvador, Honduras, and Nicaragua. Mesoamerica is home to many different cultures that flourished here before and after the contact with European people.The Aztec/Mexica cultivated large quantities of amaranth and it was also used as form of tribute payment. Its name in Nahuatl was *huauhtli.* Among the Aztecs, amaranth flour was used to make baked images of their patron deity, Huitzilopochtli (one of the most important Aztec gods), especially during the festival called *Panquetzaliztli,* which means "raising banners". During these ceremonies, amaranth dough figurines of Huitzilopochtli were carried around in processions and then divided up among the population. The Mixtecs of Oaxaca also recognized a great importance to this plant. The Mixtecs are an indigenous group of Mexico. In pre-Hispanic times, they lived in the western region of the state of Oaxaca and part of the states of Puebla and Guerrero and they were one of the most important groups of Mesoamerica.The precious postclassic turquoise mosaic covering the skull encountered within Tomb 7 at Monte Alban was actually kept together by a sticky amaranth paste. Cultivation of amaranth decreased and almost disappeared in Colonial times, under the Spanish rule. The Spanish banished the crop because of its religious importance and use in ceremonies that the newcomers were trying to extirpate/ eradicate (Maestri, 2015; Sauer, 1967). According to Green Deane (2012) a book could be written about amaranth, and probably has, if not several. A grain, a green, a cultural icon, a religious symbol... amaranth is colorful plant with a colorful history. It's also nutritious. Amaranth was a staple food of pre-Colombian Aztecs, who imbued it with supernatural powers and made it part of their religious ceremonies. Pre-Colombian Aztecs (1325–1521 AD) is related to the history and cultures of the Americas before the arrival of Columbus in 1492. They would mix amaranth flour and human blood then shaped the dough into idols that were eaten in their well-known sacrificial ceremonies. Human sacrifices were extremely common among equatorial jungle people around the world. The idea they shared was as decaying plants nourish living plants, decaying flesh, preferably the enemy's, would nourish the people. Occasionally, however, only a young maiden would assure good crops. Spanish conquistadors, who saw no religious parallel with their own communion beliefs, thought eliminating amaranth would stop the sacrifices. The blood ceremonies did stop but for more reasons than the outlawing of amaranth. The plant's use declined and was forgotten except in remote villages where it was still raised for food.

Evolution and Domestication of Amaranth

Evolution is change in heritable traits of biological populations over successive generations. Evolutionary processes give rise to diversity at every level of biological organisation, including the level of species, individual organisms, and at the level of molecular evolution. Evolution can be defined as a process of gradual and relatively continuous change from a lower, simpler, or worse to a higher, more complex, or better state/growth. It is the historical development of a biological group (as a race or species (phylogeny). It is a theory that the various types of animals and plants have their origin in other preexisting types and that the distinguishable differences are due to modifications in successive generations (WIKI, 2015a). Whatever their origin, the grain *Amaranthus* species arose from a wild progenitor or progenitors by domestication. Grain types are not found in the wild condition anywhere in the world. The principles involved during a long selection history resulted, firstly, in the development of rather short and weak bracts, in order to make the inflorescence less prickly when rubbed between the hands while extracting grain. The perfectly dehiscent utricle has been an added advantage in this direction. Secondly, selection has been made for large plant body, particularly large compound inflorescences, so as to give enormous grain yield without an increase in grain size. Thirdly, there has been a decided preference for white seeds with good popping qualities and flavor. During these selection processes, there has been inadvertent selection for the right proportions of protein, carbohydrate and oil components to make it a balanced food with higher calorific value. These features also distinguish the grain types from their wild ancestors (Pal and Khoshoo, 1973a; Sauer, 1967).

Sauer (1957) studied a group of ten related North American species, sharply set off from all other amaranths of the world by their dioecious habit. He made an attempt to reconstruct some of their history. This group is more open to such reconstruction than most of the genus because its distribution and heredity were less radically reshaped by ancient man. None of the dioecious amaranths have become domesticated plants or pandemic weeds. Yet, they have also been increasing in geographic and genetic ranges in response to civilization. Tracing their comparatively plain case histories may be a fair approach to understanding the behavior of the whole genus. He reported that the ten dioecious species of *Amaranthus*, all originally pioneer plants of naturally disturbed habitats, have behaved quite differently during increasing human disturbance of their native continent, North America. Those at home in coastal marshes, beaches, and dunes have remained static while those from river floodplains and desert washes have been expanding and diversifying as weeds in artificially opened habitats. Direct historical evidence, mainly from a hundred year record of herbarium collections, permits partial reconstruction of the geographic and genetic changes. The amount of geographic advance has been very unequal even along different borders of a single species. In some cases expansion was local and completed before the present century; elsewhere migration has extended across whole states and is still active. However, unlike the common weedy monoecious amaranths, even the weediest dioecious species have remained geographically cohesive. With geographic expansion, former limited overlap between species has greatly deepened and exceptional hybridization has become common. Apparent hybrids derived from a large number of species pairs, involving not only all the dioecious but also some monoecious species. Crosses between dioecious and monoecious species run into a blind alley of sterility but crosses among dioecious species have yielded fertile off-spring, in some cases so abundantly that the taxonomic distinctness of the parent species is threatened. Hybrid populations are almost invariably concentrated in the newer, artificial habitats, perhaps because of less rigorous selection in such places. In natural habitats the older types of populations have remained relatively immune to invasion and hybridization.

According to Sauer (1986) wild amaranth seeds were commonly gathered by many prehistoric American Indian people. The wild seeds are as nutritious and as large as those of the cultivated species. Archaeological proof of domestication comes with the appearance of pale white seeds, contrasting starkly with the dark brown wild type; the mutation producing this change has never been recorded historically. Associated with the change in colour are improved popping quality and flavor. A small proportion of dark seeds is generally present in the grain crops. Where selection for pale seed colour is relaxed, as when the plants are grown as ornamentals, the dark seeds became predominant. The earlier record of the pale-seeded crop is from Tehuacan, Puebla, Mexico, where *A. cruentus* appeared about 4,000 BC and was joined by *A. hypochondriacus* about AD 500. By the 14th century AD, pale-seeded *A. hypochondriacus* was also cultivated by Arizona cliff-dwellers. The earliest record of *A. caudatus* is from 2,000-year-old tombs in Northwestern Argentina, where its pale seeds were found mixed with those of *Chenopodium quinoa* and with dark seeds of weed amaranths and chenopods. The three grain type species may have been independently domesticated but there is an alternative possibility, namely that there was a single primary domestication of *A. cruentus* from *A. hybridus*, with the other two domesticates evolving secondarily by repeated crossing of *A. cruentus* with weedy *A. powellii* and *A. quitensis* as the crop spread into their respective territories. Evolution of all three domesticates has involved increased size of the whole plant and particularly of the inflorescence, resulting in greatly increased seed yield. All three domesticates also display the effects of selection for striking anthocyanin pigmentation of leaves, stem and inflorescences. Presumably, the intense red colour had ceremonial meaning. At the time of the Spanish Conquest, grain amaranths were important in rituals of the Aztecs and other Mexican peoples. Judging by later ethnographic evidence, ceremonial use of red amaranths extended from the Pueblo region of the Southwestern United States to the Andes and was more widespread than use as a grain crop. The ceremonial dye amaranths are generally extremely deep red forms of *A. cruentus* with dark seeds; in the Andes some may be *A. cruentus x A. quitensis* hybrids

Putnan (1992) stated that crops are truly artifacts of human history. We can ask ourselves this: "What would Africans do without peanuts, the Indians do without chillies, the Irish or Poles do without potatoes, the Italians do without tomatoes or the whole world do without maize?". All were new crops from Americas (as is amaranth). Each of these species found an important role in the culture of the society to which they were transported, and filled an important need or niche not previously met by an 'Old' crop. Crops such as amaranth became only a curiosities in Europe largely because those cultures did not discover ways to easily incorporate the crop into their diet; the grains of amaranth were too small to be easily ground by primitive flour mills. Initial mistrust may have also played a role, as it did with the tomato. Such was not the case in the introduction of amaranth to India centuries ago. Not only amaranth is grown by farmers at the highest habitable regions of the Himalayas, but it is grown in arid regions of the plains and throughout the Indian subcontinent. It is eaten as a leafy vegetable or popped as a candy, "chikki". Amaranth is widely adapted and reasonably high yielding. Amaranth is cited in ancient religious texts and has many Indian names. It is hypothesized that amaranth found its way to Surat and South India via Portuguese traders, and eventually made its way to the Himalayas. In the Indian subcontinent, amaranth is still very much an underutilized crop. The tradition here dictates eating habits, and amaranth is not widely utilized in daily diets. In many Indian Institutes, research on amaranth is being carried out for decades.

Recent History

In Spanish America or Hispanic America (the region comprising the Spanish-speaking nations in the Americas) after the 16th century, grain Amaranth cultivation was regarded as a symbol of paganism and repressed; thus the crop nearly disappeared from history. However, by the early 19th century, grain Amaranths had appeared as a staple food crop in the Nilgiri Hills of South India and in the Himalaya; they have since been noted over an increasingly wide area of India as well as across the interior of China to Manchuria and Eastern Siberia. Pale-seeded *Amaranthus hypochondriacus* constitutes the bulk of Asiatic crop; dark-seeded *A. hypochondriacus* and pale-seeded *A. caudatus* are minor components. In the 1940s, cultivation of *A. hypochondriacus* was begun in East Africa to supply grain to the local Indian population. The wide latitudinal spread of these species in the Old World presumably required evolutionary changes, because their flowering is controlled by photoperiod. *A. cruentus* has not become established as a grain crop in the Old World. However, dark-seeded, deep red forms of this species have been widely planted in tropical Africa and Asia for over a century as ornamentals, dye plants, fetishes and potherbs. All these three of the domesticated species may have been introduced to the Old World via Europe; they have been grown in European gardens as ornamentals and curiosities for at least 250 years. Only dark-seeded forms of *A. hypochondriacus* were known to have present in Europe (Sauer, 1986).

Amaranth is also making a comeback as a popular super food. The seed, commonly referred to as grain, is gluten-free and a good source of protein. It may also have a political advantage over quinoa, another healthy grain that is growing in popularity. Quinoa has drawn some controversy recently because the price of the grain has increased so dramatically that the indigenous populations in Bolivia, Ecuador, and Peru (where the grain is grown) can no longer afford to eat it. Some argue that the increased demand creates income for otherwise impoverished areas. Amaranth, on the other hand, is inexpensive in comparison to quinoa, and easily cultivated in a wider variety of conditions. It can survive in arid climates and is currently grown everywhere from Long Island and Iowa to India and

Fiji. As a bonus, the plant's stems and leaves are also edible and vitamin-rich. In Caribbean cultures, certain varieties of amaranth leaves are known as *callaloo* (Wang, 2012). In true "never-fading" fashion, seeds from the amaranth plant spread around the world and both leaves and grain became important food sources in areas of Africa, India, and Nepal. In the past two decades, amaranth has reached a much larger number of farmers and can now be found in many non-native regions such as China, Russia, Thailand, and Nigeria, as well as Mexico and parts of South America. It prefers high elevation to low but is impressively adaptive and can grow well in moist, loose soil with good drainage at almost any elevation and in just about any temperate climate. Once established, amaranth can continue to thrive in low-water conditions, making it especially valuable in sub-Sahara Africa where water sources are few, especially in the dry season. Looking a little closer to home, amaranth received renewed interested as a food source in the United States back in the 1970s. Today, you can find it growing in small amounts in some pretty surprising locations, including Iowa, Nebraska, Missouri, North Dakota, and even Long Island, NY! At present amaranth is grown in the USA, South America, India, China and Russia. The Czech Republic is the most important grower in Europe (*Anon.*, 2015b).

Studies on Cytogenetic Relationships in Amaranth

Pal and Khoshoo (1973a) studied the cytogenetic relationships in grain amaranths. This group of amaranths was studied using four domesticated species (*Amaranthus hypochondriacus, A. cruentus, A. caudatus, A. caudatus* var. *atropurpureus* and *A. edulis*), two ancestral weedy species (*A. hybridus, A. powellii*) and eight hybrids, namely *A. edulis, A. hypochondriacus, A. edulis* x *A. caudatus, A. edulis* x *A. caudatus* var. *atropurpureus, A. caudatus* x *A. hybridus, A. edulis* x *A. hybridus, A. caudatus* x *A. hypochondriacus, A. hybridus* x *A. hypochondriacus* and *A. powellii* x *A. hypochondriacus*. They reported that the parents have perfectly normal meiosis and pollen and seed fertility. Except for *A. powellii* and *A. cruentus* ($n = 17$), the species have $n = 16$. However, the hybrids may be divided into three groups. The first group contains *A. edulis* x *A. cruentus*, involving parents with $n = 16$ and 17, which failed totally, although, under the same conditions, crosses between *A. powellii* ($x = 17$) and *A. hypochondriacus* ($n = 16$) and those between species with $n = 16$ succeeded with ease. The second group is made up of *A. edulis* x *A. hypochondriacus, A. caudatus* x *A. hypochondriacus, A. caudatus* x *A. hybridus, A. edulis* x *A. hybridus* and probably also *A. powellii* x *A. hypochondriacus*. Of these, the two combinations, *A. caudatus* x *A. hybridus* and *A. edulis* x *A. hybridus*, did not proceed beyond the two-leaf stage. At pachytene, the other hybrids showed unmistakable evidence of structural hybridity, with deletions, long or short differentiated segments and inversions. Although bivalents were formed, they possessed a chiasma frequency lower than that of either parent. There was total pollen and seed sterility. The third group comprises *A. edulis* x *A. caudatus, A. edulis* x *A. caudatus* var. *atropurpureus* and *A. hybridus* x *A. hypochondriacus*, which did not show serious developmental defects, the F_1 being vigorous, with good meiotic pairing associated with a reasonable amount of differentiation in the chromosomes leading to 25– 55% fertile pollen and 49 to 66% threshable seed. In the F_2 there were 11–18% unthrifty plants, which disturb the ratios of gene combinations controlling the different characters in the two parents. Plants very near one or both parental phenotypes were recovered, and also those showing different degrees of recombination of characters. Amphidiploids from the F_1 hybrids showed the typical autoploid or segmental alloploid type of meiosis indicating that the parental chromosomes are quite homologous. They concluded that in view of the present experimental evidence and possible parallel mutations in different grains and weed amaranths, it is not certain whether the cases of natural hybridization and, in particular, of introgression can be taken as evidence for or against the two hypotheses proposed by Sauer (1967) on the basis of his brilliant ecogeographical, morphological, ethnobotanical and archaeological studies of this group of amaranths. The only point that can be stated categorically is that *A. caudatus* has given rise to *A. edulis*. The dominance of the characters of *A. caudatus* over those of *A. edulis* strengthens such a view, but the latter is sufficiently differentiated morphologically and genetically to deserve independent status. *A. caudatus* var. *atropurpureus* is a fertile but unstabilized hybrid segregate between *A. caudatus* and *A. edulis*. This is borne out by its morphological, cytogenetic and breeding behaviour, and its hybrids with *A. edulis*, and, above all, by the recovery of plants identical with this variety from the F_2 progeny of *A. edulis* x *A. caudatus*. Whatever the origin of grain types, at present they exist only in cultivation and appear to have a long history, having been selected for large plant body, huge compound inflorescences, large number of female flowers per glomerule, small and soft bracts and pale coloured seed in a dehiscent utricle. At the same time, there has also been inadvertent selection for higher and correctly balanced amounts of protein, carbohydrate and fat.

Pal and Khoshoo (1973b) studied the cytogenetic relationships in vegetable amaranths. Of the four interspecific hybrids, three (*A. graecizans* var. *graecizans* x *A. tricolor* cv. 'Purple leaf', *A. lividus* var. *lividus* x *A. tricolor* var. *viridis* and *A. gracilis* x *A. tricolor* cv. 'Purple leaf') were studied cytologically. In all the three, differentiation between the parents is chiefly a result of interchanges and paracentric inversions. The interchange complexes may involve from four (*A. gracilis* x *A. tricolor* cv. 'Purple leaf') to fourteen (*A. lividus* var. *lividus* x *A. tricolor* var. *viridis*) chromosomes, indicating that the parents differ from each other in 1 to 6 interchanges. Because of the small size of the chromosomes, it is possible that crossing-over in interchange for small segments is restricted. The particular parental species representing the ancestral condition from which others were derived or compounded is difficult to pin-point. With preferential pairing and the restoration of fertility in the amphididiploid *A. lividus-tricolor*, it became clear that it is very likely that the interchanged segments are small and sterility in the hybrid is entirely chromosomal.

Studies on Interspecies Hybridization

Detailed basic studies, have provided a very clear picture of evolutionary dynamics of this group of plants. This was possible by undertaking a systematic programme involving breeding systems, intra-interspecific hybridization, diallel analysis on a very large amount of plant material secured from all over the world, in order to decipher not only the genetic relationships but also the size of the gene pool available for their improvement. These studies have revealed a very interesting series of reproductive barriers like unidirectional incompatability, male sterility, hybrid sterility and 'virus' like syndromes, the last being unique and unravelled for the first time in this group of plants. The cytogenetic investigation of F_1 and F_2 progenies of the dibasic interspecific crosses involving *A. hybridus* (2n=16), *A. hypochondriacus* (2n=16), *A. cruentus* (2n=17) and *A. retroflexus* (n=17) have shown that x=16 is ancestral and x=17 derived through primary trisomy. A close genetic homology has been found between the progenitors like *A. hybridus* and *A. quitensis* on the one hand and between them and their respective domesticates, namely, *A. hypochondriacus, A. cruentus* and *A. caudatus edulis.* The domesticates and the progenitor species thus constitute essentially a single 'gene pool' in which despite be achieved through the wild progenitor species. The discovery for the first time of the existence of a wild species *viz. Amaranthus retroflexus* L. in the Ladakh and adjacent regions of India and evaluation of its cytogenetic relationship with the cultivated grain amaranth species viz. *A. cruentus* L. (n=17) is of special significance in as much as the fertility shown by the hybrid could be used for evolving a grain amaranth crop for cold dry desert regions of India and other adjacent countries. Earlier *Amaranthus retroflexus* was not considered a close relative of grain species and thus the study has also helped in broadening the 'gene pool' of grain amaranths (Pal, 1999).

Hybridization studies have been very important in establishing evolutionary relations and gene pools accessible for conventional breeding programs. Murray (1940) was one of the first to systematically assess interspecies hybridization within the genus, in a study to elucidate the mechanisms involved with sex determination in Amaranthaceae. Murray classified monoecious species according to the arrangement pattern shown by male flowers in inflorescences. He identified two types of species, with type I plants having male flowers interspersed with female flowers, whereas type II plants have male flowers clustered at the terminal ends of inflorescences. Murray performed a number of different crosses between and among type I monoecious species (including A. *caudatus,* A. *hybridus,* A. *retrojiexus,* and A. *powellii),* type II monoecious species (A. *spinosus),*and dioecious taxa. Crosses between monoecious species produced hybrids with different ease, with type I by type II crosses showing the most difficulty at hybrid production. Hybrids were readily obtained among species of the type I floral arrangement and between type I species and dioecious taxa, suggesting evolutionary proximity between these species. A. *hybridus* by A. *caudatus* crosses were among the most prolific, consistent with the weak pre-zygotic isolation expected of closely related taxa. Interestingly, similarly prolific were crosses between A. *hybridus* and A. *caudatus* with A. *tuberculatus,* insinuating an evolutionary relationship that is closer than is morphologically apparent (Trucco and Tranel, 2011).

Studies on Herbicide Resistance in *Amaranthus* Weeds

Amaranthus is a large and diverse genus, containing over sixty member species including locally adapted leaf vegetable and grain crops, horticultural varieties, and noxious weeds. Most species in this genus are herbaceous annuals. *Amaranthus* species are generally wind pollinated and can produce as many as two hundred thousand seeds per female plant per season. Starting in 2002, reports indicated that amaranth species, mainly *Amaranthus palmeri* and *Amaranthus rudis* (or *A. tuberculatus* subs. *rudis*), were becoming increasingly resistant to standard herbicide applications. An example of the potential economic impact of this resistance is in *Zea mays*, with studies indicating that as few as 0.5 *A. palmeri* plants per row of corn can reduce yield by more than 10%. To understand the mechanisms underlying the rapid evolution of herbicide resistance in the *Amaranthus* genus, testing alternative models of divergence population genetics interactions among and between populations and species in this genus is being carried out in Clemson University (Rauh, 2003).

A primary characteristic contributing to the infamy of *Amaranthus* species as weeds of modem agriculture is their demonstrated ability to evolve herbicide resistance. *Amaranthus* weeds comprise over 5% of worldwide cases of herbicide-resistant weeds and have evolved resistances to diverse herbicide modes of action. For example, *A. tuberculatus* has evolved resistance to single population (or even within a single plant), making control of *A. tuberculatus* a significant practical problem. In the southeastern US, resistance to glyphosate (which inhibits EPSPS/ 5-enolpyruvylshikimate-3-phosphate synthase) has become widespread in *A. palmeri* in recent years and is posing a very significant weed management challenge. The frequent occurrence of herbicide resistance in *Amaranthus* weeds suggests it should be possible to select the same traits in cultivated *Amaranthus* crops. Alternatively, it should be possible to transfer the resistance traits from the weeds to the crops via hybridization. For example, it should be straightforward to cross grain amaranth with *A. hybridus* containing resistance to ALS inhibitors, and then obtain the herbicide-resistant crop by recurrent backcrossing along with selection for the resistance. Unfortunately, however, that these herbicide resistances are widespread in many of the *Amaranthus* weeds would limit their utillty in the crop. Nevertheless, the only *Amaranthus* species thus far to have evolved resistance to PPO (Protoporphyrinogen oxidase) inhibitors is *A. tuberculatus*, and to EPSPS inhibitors are *A. tuberculatus* and *A. palmei*.. Thus, resistance to one or both of these herbicides in cultivated amaranth may have value, particularly in regions where these two weeds are not present. The mechanism conferring resistance to PPO inhibitors in *A. tuberculatus* was determined to be a deletion of a glycine residue in a conserved region of the *PPX2* gene. The gene was predicted to encode both mitochondria- and chloroplast-'targeted PPO, thereby resulting in herbicide-insensitive enzymes in both organelles. Through genetic transformation, one could insert the *A. tuberculatus* herbicide resistant *PPX2* into the crop species. The homologous gene was obtained from *A. hypochondriacus* and also shown to contain the dual-targeting signal sequences. Site-directed mutagenesis of the native *A. hypochondriacus PPX2* to obtain the glycine codon deletion followed by transformation would be another route to obtain resistance to PPO inhibitors. This latter approach might be met with greater public acceptance since the crop would not be carrying a gene from a weed species (although the encoded proteins from the two genes are over 97% identical. A major challenge beyond the development of a herbicide-resistant amaranth crop would be maintaining the utility of the trait by preventing its escape into coexisting *Amaranthus* weeds. In this regard, the body of work on interspecific hybridization should provide the framework for the development of adequate protocols for technology stewardship (Trucco and Tranel, 2011).

Development of Nutritionally Superior Lines/Varieties

The National Botanical Research Institute of India (NBRI) has built up perhaps, one of the best qualitative collections-of amaranth 'germplasm' in the world, comprising nearly 400 accessions, referable to 20 species, of which nearly half belong to the grain types. The most precious amaranth 'germplasm' in NBRI's collection is that of wild progenitor species of both the grain and vegetable domesticates as determined through an extensive hybridization programme. This could prove to be of immense value in developing varieties with desirable attributes for use under different agroclimatic regions in the developing countries (Pal, 1999). The grain amaranths constitute a group of pseudocereals which have a long history of domestication and cultivation (7,000 years). These are important source form of subsidiary food especially in the Himalayan Valleys because of their high nutritive value and excellent amino acid composition. The vegetable amaranths are used all over India as pot herbs and are rich in Vitamin A (2,000 to 11,0000 IU/ 100 g) and leaf protein

(2-3%). The foliage of sixty one lines comprising both the grain and vegetable amaranths referable to 10 species were evaluated for carotenoid, protein, nitrate and oxalate contents (fresh weight). Carotenoids varied from 9.0 to 20.0 mg/100 g in vegetable and 6.0 to 20.0 mg/100 g in the leaves of grain type. Variation with leaf protein was found to be 1.4 to 3.0%, 1.5 to 4.3%, nitrate 0.18 to 0.80%, 0.41 to 0.92% and oxalate 0.51 to 1.92% and 0.3 to 1.65% in vegetable and grain types respectively. The results were compared with the other cereals and leaf vegetables. Protein and amino acid composition was analysed in 19 lines of *Amaranthus hypochondriacus* cultivated solely for grains in India. Variation of protein was from 8.9 to 15.7% and lysine 3.8 to 5.5%. Seed protein and amino acid composition over a ten year period, in grain species *A. hypochondriacus* revealed stability for these features. Considerable variation as revealed by the present studies is of significance for developing nutritionally superior lines both in the vegetable and grain amaranths (Pal, 1999).

Prospects

Amaranths have been a staple crop of pre-Columbian cultures, and they have received interest in the last two to three decades as an alternative crop. Much of the recent interest in amaranths is based on the exceptional nutritional profile of the grain proteins, which are rich in amino acids that are usually deficient in other crops. Additional interest is generated by the oil and carbohydrate profiles of amaranth seeds, which present opportunities for different industrial applications, from the use of amaranth squalene (a natural 30-carbon organic compound) as a cosmetic oil to that of micro-sized starch in the formulation of foods. Studies are to be conducted to develop and optimize technologies aimed at exploiting these properties of amaranth (Trucco and Tranel, 2011).

From an agronomic perspective, drought tolerance and environmental plasticity are attractive traits promoting amaranth adoption in areas where traditional crops face greater challenges. Yet, modem amaranth cultivars still face several difficulties, which have been overcome in most major crops. Recent breeding efforts to try to solve some of these difficulties have been modest, and very few cultivars have been registered over the last decade. In an unusual contrast, *Amaranthus* weeds have been the subject of leading weed science research over the same timeframe. In fact, weedy amaranths have been proposed as a model system for the study of plant weediness and valuable genomic resources are being generated with these species. The path to improving important crop traits may be realized through the judicious exploitation of the wealth found in weedy amaranth resources (Trucco and Tranel, 2011).

The patterns of gene exchange among the different taxonomic groups constitute a first and incomplete attempt at drafting a roadmap for the exchange of adaptations among species. Simllarly, the herbicide resistance covers a number of possibilities yet to be explored by amaranth breeders. It is important to note that the great success of amaranths as weeds is not found in any one adaptation but in their ability to adapt quickly to changing weed management practices. Infamy due to the evolution of numerous herbicide resistant populations is a reflection of their adaptability, or from a different perspective a reflection of their ability to successfully respond to selection. Interestingly, what constitutes a threat to farm economies at one level may be the most valuable asset for the development of competitive cultivars at another. It is up to us to transform this serious challenge into a beneficial force (Trucco and Tranel, 2011).

The basic studies carried out at National Botanical Research Institute (NBRI) constitute important steps in the genetic upgrading of vegetable and grain amaranths to meet the dietary requirements of the populations in the developing countries. However, for optimum utilization of the multidimentional potential of amaranths efforts in many areas especially in food processing, commodity/product development and marketing are needed (Pal, 1999).

"Amaranth may have some environmental advantages over corn. The plant needs less water to grow, which is particularly important in water-stressed areas like much of Oaxaca. Amaranth can exist up to 40 days without rain and still produce seeds, unlike corn. Amaranth also grows fast and is easy to harvest, and can help reduce reliance on imported food. Farmers can get three to four times more money for a bushel (25-27kg) of amaranth than a bushel of other grains. Even so, adoption of the forgotten plant has been slow. It is advised that consumers may consider choosing products that incorporate amaranth into processed foods like bread, chips, pasta, and even desserts like marzipan (marzipan is made of almonds; it smooth, sweet, and often dyed and molded into shapes) and ice cream. It is also better to promote consumption of amaranth in as pure a form as possible. Health-conscious consumers in the U.S. and other developed countries would likely gobble up (eat a large amount of amaranth quickly) Oaxacan amaranth" (Howard, 2013).

The positive attributes of amaranth grain will remain academic unless we develop products or applications which consumers like and will be widely accepted. There are several unique and underexploited characteristics of amaranth. At CSK Himachal Pradesh Agricultural University, Palampur, scientists have produced many different dishes of amaranth, including a pancake made with chickpea, spinach-like preparations of the greens, and a cereal made from the popped grain with cardamom. More of this type of imaginative thinking is required to incorporate amaranth in the diet. We should not underestimate the potential worldwide value of amaranth as a crop. By the turn of the next century, India's population is likely to exceed one billion, a distinction that China has already attained. Amaranth's drought resistance, high growth rate, and high protein and energy content of the seed, and potential for multiple use, may enable a future role for this crop in helping to feed future populations. With research and experimentation, amaranth could be readily incorporated into flours, chapaties, cereal products and other commonly used foods to increase quality (Putnan, 1992).

References

Anonymous, 2015a. Amaranth (*Amaranthus* sp. L.). http://www.vurv.cz/altercrop/amaranth.htm

Anonymous, 2015b. Amaranth - May Grain of the Month.Whole Grains Council. http://wholegrainscouncil.org/whole-grains-101/amaranth-may-grain-of-the-month-0

Bao-Hua, Song., Xue-Jie, Zhang., Fa-Zeng, Li and Peng, Wan. 2002. Chromosome numbers of 14 species in Amaranthus from China. *Acta Phytotaxonomica Sinica,* 40(5): 428-432.

Behra,Bharati and Patnaik, S. 1974. Cytotaxonomic Studies in the Family Amaranthaceae. *Cytologia,* 39 (1): 121-131.

Bonasora, Marisa Graciela., Poggio, Lidia. and Greizerstein, Eduardo José. 2013. Cytogenetic studies in four cultivated *Amaranthus* (Amaranthaceae) species. *Comp. Cytogenet.*, 7(1): 53–61.

Ebert, Andreas W., Wu, Tien-hor and Wang, San-tai. 2011. Vegetable amaranth (*Amaranthus* L.). International Cooperator's Guide. AVRDC Publication Number: 11-754.

Grant, William F. 1959. Cytogenetic studies in *Amaranthus*: I. Cytological aspects of sex determination in dioecious species. *Canadian Journal of Botany*, 37(3): 413-417.

Green Deane. 2012. Amaranth, the forgotten food. Amaranth: Grain, Vegetable, Icon. Eat the weeds and other things too. http://www.eattheweeds.com/amaranth-grain-vegetable-icon/

Greizerstein EJ, and Poggio L. 1992. Estudios citogenéticos en seis híbridos del género *Amaranthus* (Amaranthaceae). *Darwiniana*, 31: 159-165.

Greizerstein EJ, and Poggio L. 1994. Karyological studies in grain Amaranths. *Cytology*, 59: 25-30.

Howard, Brian Clark. 2013. Amaranth: Another Ancient Wonder Food, But Who Will Eat It?. National Geographic. Published August 12, (2013). http://news.nationalgeographic.com/news/2013/08/130812-amaranth-oaxaca-mexico-obesity-puente-food/?google_editors_picks=true

Madhusoodanan, K.J. and Pal, M. 1981. Cytology of vegetable amaranths. Botanical Journal of the Linnean Society, London, 82: 61-68.

Maestri, Nicoletta. 2015. Amaranth-The Origin and Use of Amaranth in Ancient Mesoamerica. http://archaeology.about.com/od/amthroughanterms/a/Amaranth.htm

Mallika, V.K. 1987. Genome analysis in the genus amaranthus. PhD thesis, Kerala Agricultural University, Trichur, Kerala.

Mom, La Jolla. 2013. 10 health benefits of amaranth grain and leaves. In Food and Drink. http://lajollamom.com/amaranth-grain-leaves-health-benefits/

Mosyakin and Robertson 1996. "New infrageneric taxa and combinations in *Amaranthus* (Amaranthaceae)". *Ann. Bot. Fennici* 33: 275–281.

Mosyakin SL, Robertson KR 2003. Amaranthus. In: Flora of North America. North of Mexico. Oxford University Press, New York, USA.

Murray MJ 1940. The genetics of sex determination in the family *Amaranthaceae.* Genetics 25:409431.

Nath, Prem and Swamy, K.R.M. 2015. Vegetables for the Tropical Region (in Press). ICAR, New Delhi. pp. 450.

Pal, M. 1999. Amaranthus : Evolution, Genetic Resources and Utilization. *EnviroNews*, 5(4): reproduced from the archives of *EnviroNews* - Newsletter of ISEB India.

Pal, M. and Khoshoo, T. N. 1973a. Evolution and improvement of cultivated amaranths. VI. Cytogenetic relationships in grain types. *Theor. Appl. Gen.*, 43: 242–251.

Pal, M. and Khoshoo, T.N.1973b. Evolution and improvements of cultivated Amaranths VII. Cytogenetic relationships in vegetable amaranths. *Theor. Appl. Gen.*, 43: 343-350.

Pandey, R.M. 1999. Evolution and improvement of cultivated amaranths with reference to genome relationship among *A. cruentus, A. powellii* and *A. retroflexus. Genetic Resources and Crop Evaluation,* 46(3): 219-224.

Putnan, Dan. 1992. Amaranth and the currents of history. Legacy, 5(1):1-4.

Queiros, M. 1989. Cytotaxonomic studies of *Amaranthus* in Portugal. *Lazaroa*, 11: 9-17.

Rauh, Amy Lawton. 2003. Research activities: North Carolina State University. http://www.clemson.edu/genbiochem/people/alawton_rauh.php

Sauer, J.D. 1955. Revision of dioecious amaranths. *Madrono,* 13:5-46.

Sauer, J. 1957. Recent migration and evolution of the dioecious amaranths. *Evolution, 2(1): 11-31.* Society for the Study of Evolution. http://www.jstor.org/stable/2405808
Sauer, J. D. 1967. The grain amaranths and their relatives: A revised taxonomic and geographic. *Annals of the Missouri Botanical Garden*, 54(2): 103-137.

Sauer, J.D. 1986. Grain amaranths (*Amaranthus* spp.—Amaranthaceae). (In:) (Ed. N.W. Simmonds) Evolution of crop plants. Longman Scientific & Technical, England. pp. 4-7.

Trucco, Federico and Tranel, Patrick J. 2011. Amaranthus. (In:) (Ed. C. Kole) *Wild Crop Relatives: Genomic and Breeding Resources, Vegetables,*Springer-Verlag Berlin Heidelberg. (pp. 11-21) 282 pp.

Varalakshmi, B. 2015. Vegetable Amaranth. (In:) (Eds. K.V. Peter and Pranab Hazra) *Handbook of Vegetables* Vol.3. pp. 487-520. Studium Press LLC, USA.

Wang, Joy Y. 2012. Last Chance Foods: Amaranth's Ancient History. http://www.wnyc.org/story/225970-last-chance-foods-amaranths-ancient-history/

WIKI. 2013. Category:Amaranthus. From Wikipedia, the free encyclopedia. http://en.wikipedia.org/wiki/Category:Amaranthus

WIKI. 2015. Amaranth. From Wikipedia, the free encyclopedia. http://en.wikipedia.org/wiki/Amaranth

WIKI. 2015a. Evolution. From Wikipedia, the free encyclopedia. *http://en.wikipedia.org/wiki/Evolution*

Fig. 6: *A. blitum*

Fig. 7: *A. cruentus*

Fig. 8: *A. dubius*

Fig. 9: *A. spinosus*

Fig. 10: *A. tricolor*

Fig. 11: *A. viridi*

Fig. 12: *A.caudatus*

Fig. 13: *A. hypochondriacus*

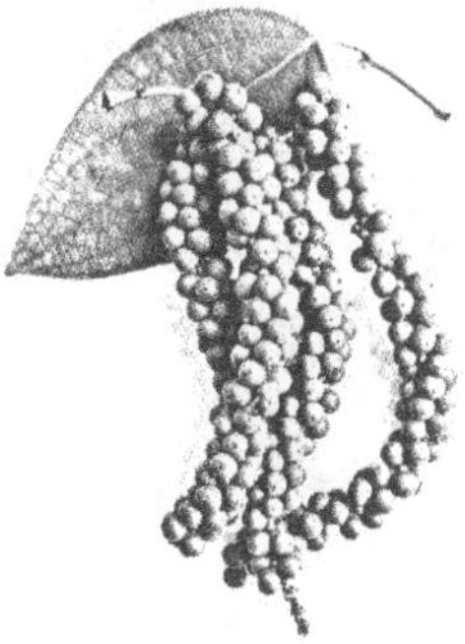

2

Black pepper

V P Neema and P M Ajith

Black pepper (*Piper nigrum* L.) known as "The King of Spices", is one of the oldest and the most popular spices in the world. It is a perennial climber in the family Piperaceae, , a native of Western Ghats of India which is cultivated for its dried mature berries and used as a spice . Pepper fruit is botanically a drupe but known as berry and is dark red when fully mature and like all drupes contains a single seed. Black pepper is the whole dried fruit (berry). White pepper is the dried berry after removal of mesocarp . The name pepper comes from the Sanskrit word *pippali* meaning berry.

Black pepper is native to South India, and is extensively cultivated there and elsewhere in tropical regions. The humid tropical evergreen forests bordering the Malabar Coast, the Western Ghats is the centre of origin and diversity of Black Pepper. The Malabar Coast was involved in the cultivation and trade of pepper from very early times from where it was taken to Indonesia, Malaysia and to other pepper growing countries.

At present apart from India, black pepper is widely cultivated throughout Indonesia, Malaysia, Thailand, Africa, Brazil, Sri Lanka, Vietnam and China also. Kerala is the original home of pepper and Indonesia is the second most ancient pepper growing country. Currently pepper is grown in about 26 countries.

The cultivars of black pepper might have originated from the wild ones through domestication and selection. In Kerala cultivar diversity is the richest and contributed much in the development of present day high yielding varieties.

Black Pepper is valued for its pungency and aroma. Dried ground pepper has been used since antiquity for both its flavour and as a traditional medicine. Black pepper is the world's most traded spice. It is one of the most common spices added to European cuisine and its descendants. The spiciness of black pepper is due to the alkaloid piperine and flavour is the result of volatile oil .It is ubiquitous in the modern world as a seasoning, and is often paired with salt.

Taxonomy

Piper nigrum L. (Piperaceae)

Common name:	Black Pepper
Kingdom:	Plantae – Plants
Subkingdom:	Tracheobionta – Vascular plants
Superdivision:	Spermatophyta – Seed plants
Division:	Magnoliophyta – Flowering plants
Class:	Magnoliopsida – Dicotyledons
Subclass:	Magnoliidae
Order:	Piperales
Family:	Piperaceae – Pepper family
Genus:	Piper L. – pepper
Species:	*Piper nigrum* L. – black pepper

History

Pepper is native to South Asia and Southeast Asia and has been known to Indian cooking since at least 2 BCE. J. Innes Miller notes that while pepper was grown in Southern Thailand and in Malaysia, its most important source was India, particularly the Malabar Coast of Kerala .

Before the 16th century, pepper was being grown in Java, Sunda,Sumatra, Madagascar, Malaysia, and everywhere in Southeast Asia. These areas traded mainly with China.

Ancient Times

Black peppercorns were found stuffed in the nostrils of Ramesses II, placed there as part of the mummification rituals shortly after his death in 1213 BCE. Little else is known about the use of pepper in ancient Egypt and how it reached the Nile from South Asia.

Pepper (both long and black) was known in Greece at least as early as the 4th century BCE, though it was probably an uncommon and expensive item that only the very rich could afford. Trade routes of the time were by land, or in ships which hugged the coastlines of the Arabian Sea. Long pepper, growing in the north-western part of India, was more accessible than the black pepper from further south; this trade advantage, plus long pepper's greater spiciness, probably made black pepper less popular at the time.

By the time of the early Roman Empire, especially after Rome's conquest of Egypt in 30 BCE, open-ocean crossing of the Arabian Sea direct to southern India's Malabar Coast was near routine. Details of this trading across the Indian Ocean have been passed down in the *Periplus of the Erythraean Sea*. According to the Roman geographer Strabo, the early Empire sent a fleet of around 120 ships on an annual one-year trip to China, Southeast Asia, India and back. The fleet timed its travel across the Arabian Sea to take advantage of the predictable monsoon winds. Returning from India, the ships travelled up the Red Sea, from where the cargo was carried overland or via the Nile-Red Sea canal to the Nile River, barged to Alexandria, and shipped from there to Italy and Rome. The rough geographical outlines of this same trade route would dominate the pepper trade into Europe for a millennium and a half to come.

With ships sailing directly to the Malabar coast, black pepper was now travelling a shorter trade route than long pepper, and the prices reflected it. Pliny the Elder's *Natural History* tells us the prices in Rome around 77 CE: "Long pepper ... is fifteen denarii per pound, while that of white pepper is seven, and of black, four." Pliny also complains "there is no year in which India does not drain the Roman Empire of fifty million sesterces," and further moralizes on pepper.

It is quite surprising that the use of pepper has come so much into fashion, seeing that in other substances which we use, it is sometimes their sweetness, and sometimes their appearance that has attracted our notice; whereas, pepper has nothing in it that can plead as a recommendation to either fruit or berry, its only desirable quality being a certain pungency; and yet it is for this that we import it all the way from India.

Black pepper was a well-known and widespread, if expensive, seasoning in the Roman Empire. Apicius' De re coquinaria, a 3rd-century cookbook probably based at least partly on one from the 1st century CE, includes pepper in a majority of its recipes. Edward Gibbon wrote, in *The History of the Decline and Fall of the Roman Empire*, that pepper was "a favorite ingredient of the most expensive Roman cookery".

Post Classical Europe

Pepper was so valuable that it was often used as collateral or even currency. In the Dutch language, "pepper expensive" (*peperduur*) is an expression for something very expensive. The taste for pepper (or the appreciation of its monetary value) was passed on to those who would see Rome fall. Alaric the Visigoth included 3,000 pounds of pepper as part of the ransom he demanded from Rome when he besieged the city in 5th century. After the fall of Rome, others took over the middle legs of the spice trade, first the Persians and then the Arabs; Innes Miller cites the account of Cosmas Indicopleustes, who travelled east to India, as proof that "pepper was still being exported from India in the sixth century". By the end of the Early Middle Ages, the central portions of the spice trade were firmly under Islamic control. Once into the Mediterranean, the trade was largely monopolized by Italian powers, especially Venice and Genoa. The rise of these city-states was funded in large part by the spice trade.

It is commonly believed that during the Middle Ages, pepper was used to conceal the taste of partially rotten meat. There is no evidence to support this claim, and historians view it as highly unlikely: in the Middle Ages, pepper was a luxury item, affordable only to the wealthy, who certainly had unspoiled meat available as well. In addition, people of the time certainly knew that eating spoiled food would make them sick. Similarly, the belief that pepper was widely used as a preservative is questionable: it is true that piperine, the compound that gives pepper its spiciness, has some antimicrobial properties, but at the concentrations present when pepper is used as a spice, the effect is small.Salt is a much more effective preservative, and salt-cured meats were common fare, especially in winter. However, pepper and other spices certainly played a role in improving the taste of long-preserved meats.

Its exorbitant price during the Middle Ages—and the monopoly on the trade held by Italy—was one of the inducements which led the Portuguese to seek a sea route to India. In 1498, Vasco da Gama became the first person to reach India by sailing around Africa asked by Arabs in Calicut why they had come, his representative replied, "we seek Christians and spices". Though this first trip to India by way of the southern tip of Africa was only a modest success, the Portuguese quickly returned in greater numbers and eventually gained much greater control of trade on the Arabian sea. It was given additional legitimacy by the 1494 Treaty of Tordesillas, which granted Portugal exclusive rights to the half of the world where black pepper originated.

The Portuguese proved unable to maintain their stranglehold on the spice trade for long. The old Arab and Venetian trade networks successfully 'smuggled' enormous quantities of spices through the patchy Portuguese blockade, and pepper once again flowed through Alexandria and Italy, as well as around Africa. In the 17th century, the Portuguese lost almost all of their valuable Indian Ocean trade to the Dutch and the English who, taking advantage from the Spanish ruling over Portugal (1580–1640), occupied by force almost all Portuguese dominations in the area. The pepper ports of Malabar began to trade increasingly with the Dutch in the period 1661–1663.

As pepper supplies into Europe increased, the price of pepper declined (though the total value of the import trade generally did not). Pepper, which in the early Middle Ages had been an item exclusively for the rich, started to become more of an everyday seasoning among those of more average means. Today, pepper accounts for one-fifth of the world's spice trade.

China

It is possible that black pepper was known in China in the 2nd century BCE, if poetic reports regarding an explorer named Tang Meng are correct. Sent by Emperor Wu to what is now south-west China, Tang Meng is said to have come across something called *jujiang* or "sauce-betel". He was told it came from the markets of Shu, an area in what is now the Sichuan province. The traditional view among historians is that "sauce-betel" is a sauce made from betel leaves, but arguments have been made that it actually refers to pepper, either long or black.

In the 3rd century CE, black pepper made its first definite appearance in Chinese texts, as *hujiao* or "foreign pepper". It does not appear to have been widely known at the time, failing to appear in a 4th-century work describing a wide variety of spices from beyond China's southern border, including long pepper. By the 12th century, however, black pepper had become a popular ingredient in the cuisine of the wealthy and powerful, sometimes taking the place of China's native Sichuan pepper (the tongue-numbing dried fruit of an unrelated plant).

Marco Polo testifies to pepper's popularity in 13th-century China when he relates what he is told of its consumption in the city of Kinsay .Messer Marco heard it stated by one of the Great Kaan's officers of customs that the quantity of pepper introduced daily for consumption into the city of Kinsay amounted to 43 loads, each load being equal to 223 lbs. Marco Polo is not considered a very reliable source regarding China, and this second-hand data may be even more suspect, but if this estimated 10,000 pounds (4,500 kg) a day for one city is anywhere near the truth, China's pepper imports may have dwarfed Europe's.

During the course of the treasure voyages in the early 15th century, Admiral Zheng He and his expeditionary fleets returned with such a large amount of black pepper that the once-costly luxury became a common commodity.

Taxonomic History of Indian *Piper*

The genus *Piper* was established by Linnaeus (1753) in his *Species Plantarum* and 17 species were included in *Piper* family. The name *Piper* was derived from *Peperi,* Greek name for Black pepper. According to Rosengarten (1973) European names for black pepper were derived from the Sanskrit root *Pippali* , the name for long pepper (*Piper longum*).The family name Piperaceae was first used by L.C. Rich in 1815.

Rheede (1678) described five types of *Piper* including black pepper, long pepper and wild peppers. Linnaeaus (1753) included seventeen species in *Species Plantarum* . Roxburgh (1832) described seven species of *Piper* and Miquel (1848) included seven wild species from India in his monograph on *Piper.* De Candolle in 1869 recognized more than 1000species in the two genera, *Piper* and *Peperomia* and also prepared the key to the family Piperaceae. William Trelease made extensive collection of American Piperaceae which led to the revision of the Piperaceae of the Northern South America by Trelease and Yuncker (1950)

Hooker (1886) in an elaborate study of Piper species divided the genus into six sections and black pepper was included in the section Eupiper. Rama Rao (1914) enumerated Piper species of Western Ghats and listed twelve species. The taxonomic keys for thirteen species were prepared by Gamble (1925) in his Flora of Presidency of Madras. Ravindran *et.al.* (1987) reported *Piper silentvalleyensis,* the only bisexual wild species. *Piper pseudonigrum* and *Piper sugandhi* were reported in 1990's.

Ravindran (1991) suggested a taxonomic key for *Piper* species and subdivided the genus into two sections namely Pippali (long pepper) and Maricha (pepper)based on the orientation of spikes , erect or pendent.

Piper species closely related to Black Pepper from the centre of origin were listed with brief descriptions by Ravindran *et al* (2000).

1. *Piper* Linn. *Sect.* Pippali – Spikes erect

   Piper longum

   Piper hapnium

   Piper mullesua

   Piper silentvalleyensis

2. *Piper* Linn. Sect. Maricha – Spikes pendent, almost always filiform, rarely cylindrical

   Piper argyrophyllum

   Piper attenuatum

   Piper galeatum

   Piper hymenophyllum

   Piper nigrum (common black pepper)

   Piper pseudonigrum

   Piper sugandhi

   Piper trichostachyon

   Piper schmidtii

   Piper wightii

   Piper barberi

Morphology

Piper nigrum L. species are perennial woody climbers growing up to 10 metres in height on supporting trees, poles, or trellises. The mature vine has a columnar appearance of 1.5 m diameter and 4 m height. It is a spreading vine, rooting readily where trailing stems touch the ground. Pepper plants are perennial vines with dimorphic branching . Leaves are alternate, petiolate and simple.

Pepper has an adventitious root system. There are about 10-20 adventitious roots at the base of the main stem ,3-4 m long and penetrate to a depth of 1-2 m with an extensive mat of surface feeding roots. Both aerial roots and underground roots are adventitious and differ in anatomical features. Pepper plants also produce adventitious runner shoots from the base which are used for propagation.

The young shoot tips are protected by the sheathing petiole of the leaf in runners and orthotrops. In plagiotrops shoot tip and spike emerge from a cap like structure- prophyll.

The shape and size of leaf lamina are variable-ovate, ovate-elliptic, ovate-lanceolate, elliptic- lanceolate and cordate . The leaf bases are round, cordate, acute or oblique. Leaf tip is acute or acuminate and leaf margins are even or wavy. Leaf texture are glabrous coriaceous or membraneous. Wax glands/ Pearl glands are present on the upper or lower surfaces of the leaf. Petiole is grooved on the upper surface.

Pepper exhibits dimorphic branching – straight , upward growing, monopodial, orthotropic branches and the sympodial laterally growing plagiotropic fruiting branches. The orthotropic vegetative climbing stem give the frame work of the plant. The stem become stout and woody. Each node has clinging roots which helps in anchoring and climbing. From the axils of orthotropic shoots lateral branches arise which are plagiotropic. The terminal bud gets modified into spike. Lateral fruiting branches have no roots.

Pepper inflorescence is a spike and is leaf opposite, produced on the lateral plagiotropic branches. These flowering lateral branches have sympodial mode of growth, the apical bud develops into the inflorescence and the growth is continued by the activity of the axillary bud. As the axillary bud develops the inflorescence is pushed out so that it becomes leaf opposed. The flowering starts usually in May- June following the South West monsoon in India. At the early stage the developing spikes and the leaf opposite are enclosed in a prophyll and it takes 3-4 weeks for the full spike emergence.

Pepper flowers are sessile, bracteate, achlamydeous, uni or bisexual. Cultivated types have bisexual flowers. Bracts oblong and develop into shallow cups. Stamens 2, anthers two celled, carpel single, ovary spherical, style absent, stigma 3-5 lobed, pappilate , white when receptive and later turning black

The stigma emergence takes place in 10-20 days. The receptivity of stigma varies from 3-9 days. The flowers in the basal portion open first. Protogyny exists in most of the cultivars. Anthers emerge on either side of the ovary and 2-4 days are required for anthesis. Pepper is predominantly self pollinated and the pollen dispersal is aided by rain or dew and also by gravitational descending of pollen.

Fruit is a drupe and referred as berry. Fruit is single seeded having fleshy pericarp and hard endocarp. Seeds are globose and glabrous . Berries are bold, medium or small .The berries are spherical in shape and obovate and oblong types also exists. The colour change of berries on ripening also varies with species. In most of the cultivars the colour change is from green to yellow, orange or red on ripening and green to black in wild types. The fruits are pungent in cultivars while it is bitter in wild types. Pepper fruit takes 6-8 months for full maturity from flowering. The berries at the top of the spike turns yellow and then red . Harvesting can be initiated when few spikes in a vine starts ripening.

Pepper seed is recalcitrant and takes longer time for germination, needs 25 to 40 days on an average.

The genus in general is characterized by very small flowers and most of the South Indian taxa are dioecious but cultivated pepper is bisexual. The cultivars of pepper might have originated from the wild ones through domestication and selection. Cultivar diversity is richest in Kerala followed by Karnataka. Most of them are bisexual .The migration process and human activities influenced the selective spread of high yielding cultivars and they became popular in all pepper tracts. The cultivar diversity and hybrid populations in Western Ghats also supports centre of origin of this species.

Cytology

The cultivated pepper is having the somatic chromosome number of 2n = 52 while 2n=104 in wild types. The chromosome length ranged from 1.0 to 3.0 μm. All the species studied from South India and Sri Lanka have a common basic number of x = 13, while the North Indian species seem to have a basic number of x = 12.Mathew (1958) suggested x=13 as the valid chromosome number of the genus and this might be a result of hybridization of types with x=6 and x=7. Jose and Sharma (1984) also suggest that the basic chromosome number of Piper is x=13 and 2n= 26 are diploids and 2n =52 are tetraploids.

Mathew (1972) carried out a karyotypic comparison of the different cultivated and wild varieties based on absolute chromosome size and got a positive correlation between spike length and chromatin content in general. Jose and Sharma (1984) distinguished six classes of chromosomes in Piper. Type A : Relatively long chromosomes with two constrictions, one median, and the other nearly sub median . Type B : Long chromosome with two constrictions, primary and secondary, nearly submedian at the opposite ends of the chromosome dividing it into two, outer short and middle larger segment. Type C : Short chromosomes with nearly median to nearly submedian primary constriction and a satellite at the distal end of shorter arm, joined by a SAT thread. Type D : Relatively long chromosome with nearly median to nearly submedian primary constrictions. Type E : Short chromosome with nearly median to nearly submedian primary constriction.

Nair *et al.* (1993) reported a natural triploid among pepper cultivars. The triploids could have originated under natural condition by hybridization between 2n = 104 and 2n = 52 forms, and because of the successful vegetative propagation the triploid have high survival value.

Origin and Interrelationships

Pepper has originated in the evergreen forests of Western Ghats of South India where the wild pepper plants are found extensively and cytological studies also suggest that *Piper nigrum* might have originated through hybridization between different species followed by polyploidization.

Ravindran (1991) suggested three species namely *Piper wightii, P. galeatum* and *P. trichostachyon* as the putative parents of *Piper nigrum* and all these species are woody climbers with similar leaf morphology and texture.Their spikes and fruits are more similar to *P.nigrum*.

Based on the study of morphology of bracts of *P. nigrum* and the putative parents Ravindran found the intermediate nature of bracts for *P. nigrum*.In the species *P. galeatum* the bracts are connate, fleshy and shoe shaped and in *P. wightii* bracts are fully adnate to the rachis with oblong shape. In *P. nigrum* the bracts form a shallow cup like structure which is typically intermediary of the two species.

Ravindran(1991) suggested the following scheme for origin of black pepper.

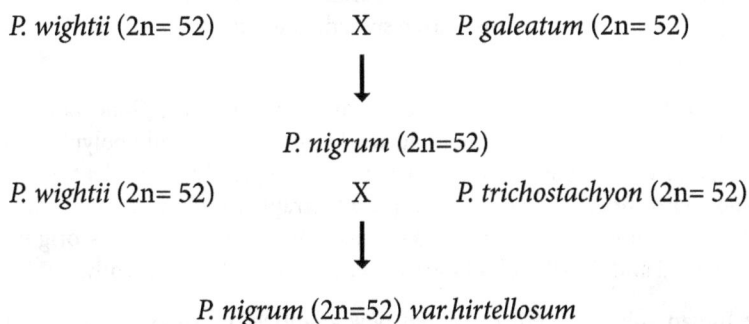

P. wightii (2n= 52) X *P. galeatum* (2n= 52)

↓

P. nigrum (2n=52)

P. wightii (2n= 52) X *P. trichostachyon* (2n= 52)

↓

P. nigrum (2n=52) *var.hirtellosum*

The wild parents *P. wightii* occurring in higher elevations and and *P. galeatum* in lower elevations are threatened species now. They might have coexisted in the past and sometimes climbed up the same support trees producing natural hybrids which in the absence of active pollen transfer mechanism might have isolated from their parents and evolved separately. The process might have continued producing fertile hybrids and segregants which largely constituted the present day cultivars.

The present day *Piper nigrum* are probably the descendents of a segregating population existed in the past and due to absence of active gene flow mechanism diverged subsequently finally leading to the present day wild and cultivated types of pepper.

Datta and Dasgupta (1979) studied the anatomical features of Piperaceae concluded that the evolution of *Piper* and *Peperomia* are related to adaptation, *Piper* to damper lowlands and *Peperomia* to drier high lands.

Directional Evolution of Cultivars

Successful vegetative propagation and lack of active gene flow have been the key factors for development of cultivars. From time immemorial black pepper was collected and exchanged as a forest produce. When brought under cultivation, farmers might undoubtedly have selected vines showing good spike length and good fruit set which are directly decided by the increasing degrees of bisexuality in the vines. This selection process can be assumed to be the main factor for directional evolution of bisexuality in cultivars.

New and Old World Piper

New World (South and CentralAmerica) and Old World (South Indian) species of *Piper* have evolved independently through two different lines from two different ancestral forms. The evolutionary development in the genus represents a major case of dichotomy. South and Central American pipers are usually shrubs or small trees and display bisexuality (Yuncker 1958).They usually have erect spikes. But Old World (South Indian) pipers usually are climbing vines with filiform spikes and are dioecious .

Northern South and Central America where more than 60 % of the species have been found, can be considered as the centre of diversity of the genus Piper where it originated during the mid Cretaceous period. When the old Gondwana land split due to plate tectonic activity, the genus became isolated by the oceans lying in between continents. Friis *et al.* (1987) suggested that angiosperms started spreading to different geographic regions during late Cretaceous period.

The present day forms might have reached India before the Gondwana land split up ie. before the late Cretaceous period. Old world pipers are usually polyploids while their original diploid counterparts have been reported frequently from Central America and Brazil. The South Indian species of Piper are all tetraploids while diploid forms are absent indicating that these ancestral forms have become extinct and in its original home in Central America and Brazil diploid forms have been reported frequently.

The Indian sub continent presented new adaptive pressures especially less sunlight penetration and competition from evergreen trees. This necessitated the highly sophisticated and self perpetuating vegetative reproductive mechanisms encountered in the species along with the highly competitive climbing habit which enables capture of whatever little sunlight available. The runners which help in natural vegetative propagation have no sexual tendencies unless they grow against gravity in a quest to produce a new plantlet. They then readily put forth plagiotrops and capture sunlight and start flower production. This indirectly suggested that seed propagation is the normal mode of reproduction in black pepper while vegetative propagation is only an adaptation which fortunately or unfortunately decided the course of evolution. There is structural and functional similarity between prop roots of *P.colubrinum* and supportive roots of *P. nigrum* which points to a common ancestry of the two lines. Seedlings and plantlets of *P. nigrum* produce supportive roots in an unsuccessful attempt to keep the plant stature steady under lack of sufficient support while mature plants produce supportive roots profusely under good moisture availability (Ajith and Neema, PRS, Panniyur, unpublished).

Chromosomes and Evolution

Somatic chromosome number of all cultivated black pepper is $2n = 52$. Mathew (1958) suggested a XX–XY female-male sex determination system in black pepper. The highest chromosome number so far reported for the species is $2n = 156$ in a line of *P. peepuloids* (Jose and Sharma,1984).

Decrease in chromatin size and spike length has been observed with progressive evolution of black pepper (Mathew 1972). *P. cubeba* having large chromosome size and total chromatin length is usually considered as primitive. This species is having basic chromosome number $n= 12$. Other species with $n= 13$ might have progressively developed from this basic number and might have sustained successfully due to selective advantage (Jose and Sharma, 1984).

Geitonogamy and Apomixis-Keys to Population Structure and Dynamics

Geitonogamy is a self pollination mechanism where pollination occurs due to gravity and rain water or dew .Though many cultivars of black pepper display protogyny as an adaptation for cross pollination, it becomes ineffective in preventing self pollination which frequently occurs in black pepper due to large number of pendent spikes at different stages of maturity.

Apomixis is found in many isolated wild populations of female black pepper plants that results in viable seeds. This is common in *P. longum* and *P. chaba*. In cultivars apomixis is greatly suppressed and not usually expressed except in predominantly female cultivars like Uthirankotta, Karivilanchy and Cholamundi. Among popular cultivars, Karimunda has been reported to express apomixis under controlled conditions (Sasikumar *et al.* 1992).

Summary

Black pepper (*Piper nigrum*), the king of spices, is one of the oldest and the most popular spice in the world. It is a perennial, climbing vine indigenous to the Malabar Coast of India. *Piper nigrum* might have originated as a hybrid. The overlapping of putative parents led to hybrids in many locations, separated in space and time. Because of the successful vegetative propagation, and the absence of active pollen transfer and gene flow an isolation barrier developed which led to the establishment of localized small populations and diversity of cultivars.

Because of selection pressure for better fruit set the bisexual forms might have selected which led to the directional evolution of bisexuality in the cultivated forms.

References

Datta, P.C. and Dasgupta, A. 1979. Comparison of vegetative anatomy of Piperales III. Vascular supplies of leaves . *Act Bot.India*, 7,39-46.

Friis, E.M; Chaloner, W.G. and Crane, P.R. 1987. Introduction to Angiosperms. (In). E.K.Friis, W.G. Chaloner and P.R.Crane (Eds.). The Origin of Angiosperms and their Biological Consequences, Cambridge Univ.Press, pp.1-15.

Gamble , J.S. 1925. *Flora of the Presidency of Madras. Vol. II.*

Hooker, J.D. 1886. *The Flora of British India.*Vol V, L. Reeve and Co. London (Rep.) pp. 78-95.

Jose, J.and Sharma, A.K. 1984. Chromosome studies in the genus *Piper* L. *J.Indian Bot.Soc.*, 63,313-319.

Linnaeus, C. 1753. *Species plantarum*, London.

Miquel, F.A.W. 1848. Piperaceae Reinwardtiiana. *Linnaea,*21,480-488.

Mathew, P.M. 1958. Studies on Piperaceae. *J.Indian Bot.Soc.*,37,155-171.

Mathew, P.M. 1972. Karyomorphological studies on *Piper nigrum. Proc. Natl. Symp. Plantation Crops.J.Plantation Crops(suppl.)*,1,15-18.

Nair , R.R., Sasikumar,B. and Ravindran, P.N. 1993. Polyploidy in a cultivar of black pepper (*Piper nigrum* L.) and its open pollinated progenies.*Cytologia*,58,27-31.

Sasikumar, B., George, J.K. and Ravindran, P.N. 1992. Breeding behavior of black pepper (*Piper nigrum* L.).*Indian J.Genet.*,52,17-21.

Rama Rao, M. 1914. *Flowering Plants of Travancore.* Govt. Press. Trivandrum, pp 336-338 (Reprint).

Ravindran, P.N., Nair, M.K. and Nair R.A. 1987. Two taxa of *Piper* (Piperaceae) from Silent Valley Forests, Kerala. *J.Eco.Tax.Bot.*,10,167-169.

Ravindran, P.N. 1991. Studies on Black Pepper and Some of its wild relatives. Ph.D thesis, Univ.Calicut.

Ravindran, P.N., Nirmal Babu, K., Sasikumar, B., and Krishnamurthy, K.S. 2000. Botany and crop improvement of Black Pepper, (In) Black Pepper, *Piper nigrum* (Ed.) Ravindran pp. 24-33.

Rheede, H.Van 1678. *Hortus Indicus Malabaricus,*Vol.7, Amstelodami, pp. 23-31, pl.12-16

Rich, L.C. 1815. Cited from Burger 1972.

Rosengarten, F.Jr. 1973. *The Book of Spices.* Pyramid books, Newyork.

Roxburgh, W. 1832. *Flora India* Vol. 1,pp.153-163.Serampore,India.

Trelease,W. and Yuncker, T.G. 1950. The Piperaceae of Northern South America. Univ. Illinois,USA.

Yuncker, T.G. 1958. The Piperaceae: A family profile. *Brittonia,* 10, 1-17.

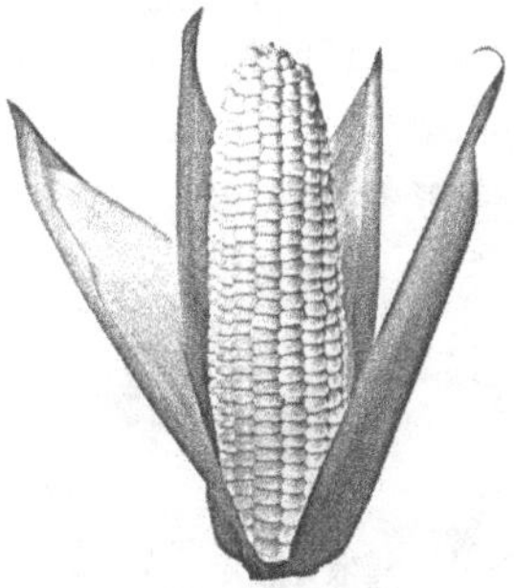

3

Baby corn

J P Shahi and Varsha Gayatonde

Maize is adaptable, high-yielding and fast-growing cereal crop adjusting to a wide range of environment. Maize is considered the 'Drosophila' of crop plant on account of extensive genetical, cytogenetical, breeding investigations and utility. These characteristics make it suitable for wider production and use in the developing world. The unique feature of corn is, having several subspecies which can be utilized for multiple purposes, the property which cannot be found in other cereals. Maize being a cross-pollinated (Fig. 1) crop there is wide scope for the development of stable hybrids and varieties. Evolutionary studies of inbred lines and hybrid production program may provide a better opportunity in increasing maize production, productivity and other properties.

Among the different utilities of corn, baby corn is the one which emerged before a half decade. Baby corn production was initiated in Taiwan in late 60's. However, the success of baby corn production occured in Thailand dates back to 1973 (Chutkaew *et al.*, 1994), where it has been continuously developing for more than 40 years. Thailand has achieved considerable success in growing maize as an alternative to rice, demonstrating that the young cobs, or baby corn, can be used as a nutritious vegetable and an export crop. Baby corn is used in two ways, fresh or processed consumption. Baby corn ears are popular as canned ears or with stir-fried vegetables in Chinese–American, European and Asiatic restaurants (Galinat, 1985). Recently, a market for fresh baby corn ears in trays has emerged in many countries, mainly for use as a decorative, crisp vegetable in salads. Another type of product is canned baby corn juices and drinks developed by Kulvadee *et al.*, (1988). Baby corn is a popular vegetable because of its high nutritive value and freedom from pesticides compared with other vegetables. Generally, there is no need to apply pesticides: the young cob is wrapped tightly in its husk.

Fig. 1: Nature of cross pollination in maize, silk and anthers at different positions

Baby corn (*Zea mays*) is a popular Asian vegetable that can be consumed cooked or raw due to its sweet and succulent taste. Many people presume the tiny ears come from dwarf corn plants. In fact, baby corn is the immature ear of fully grown standard cultivars; ears are harvested two or three days after silk emergence, but prior to fertilization.

The young and unfertilized ear of the corn (*Zea mays* L.) plant is harvested when the silks have either not emerged or just emerged (1 to 3 cm). The husked young ear in canned or fresh ear style is a more popular vegetable because of its sweetness, flavor, and crispness. Generally, the requirements of baby corn for the fresh market or processing are (Yodpet C, 1979) ear size of 4 to 9 cm length and 1.0 to 1.5 cm diameter and (Bar-Zur *et al.*, 1993) good quality, *i.e.*, yellow colour, straight ovary row arrangement, unfertilized and unbroken ear, and size within factory specifications

For baby corn production, varieties are grown under high plant densities (120,000 to 160,000 plants per ha), irrigation and high nitrogen application. Someone might use low plant density for picking the first ear for baby corn and the second ear as sweet or field corn.

Thailand is the world's largest exporter of baby corn. In 1973, it exported only 90 tons of *canned* baby corn, worth $31,000. By 1998, the volume and value had increased to 54,643 tons, worth $42.89 million. Thailand also exported 2,220 tons of fresh baby corn in 1988, worth $1.54 million, but exports increased dramatically to 11,924 tons in 1998, worth $1.87 million and in 2013 it is 20,224 tons. The greatest consumers of canned baby corn besides Thailand are the U.S., Netherlands, Japan, Germany, Canada, Australia, Hong Kong, U.K.,

France, Singapore, and South Korea, and recently China and India had been added to this list. In 1995, for example, the Food and Agriculture Organization of the United Nations (FAO) requested a leading Thai researcher to train scientists from 15 developing countries in the Asia-Pacific region in baby corn production.

The researchers realized the importance of finding a baby *corn* variety suited to industrial requirements in order to increase farmers' incomes and started a baby corn breeding program in 1976 in Asiatic countries. There were initial problems: the existence of diverse germplasm sources for yield and quality; baby corn was a new crop; lack of awareness of the economic importance of the baby corn crop; crops as baby corn were restricted to areas where irrigation water was available; resources for research and development were limited. The objective was to breed a composite variety with high yield, yellow color, good row arrangement, resistance to downy mildew and wide adaptation that would meet the specifications of the canning industry. Another benefit to baby corn production has been integrated with dairy farms is because only 13 to 20% of fresh ear weight is used as human food. The rest (husk and silk) can be used as green herbage for ruminants and pigs. Faungfupong *et al.*(1987), reported that it could produce 40 to 43 t/ ha of plant fresh weight. Out of this amount, 6 to 7% was attributed to the detasseling plant part, 8 to 10% to the husk and 83 to 86% to the remaining plant parts after ear harvesting. These plant materials are highly nutritive and can be used as roughage or silage for beef cattle and dairy cow. Cheva-Isarakul and Paripattananont, (1988), found that baby corn waste contained 86.4% moisture, 94.4% organic matter (OM), 10.6% crude protein, 55.1% neutral detergent fiber (NDF), 26.8% acid-detergent fiber (ADF), and 2.0% acid-detergent lignin. Digestibilities of DM, OM, nitrogen-free extract, NDF, and ADF, were over 70%. The nutritive value of baby corn residue obtained after the harvest for 1, 5, 10, and 15 days contained 18.9, 19.3, 20.1, and 21.3% dry matter; 9.9, 8.8, 8.7, and 8.4% protein; and 23.3, 26.6, 29.0, and 29.9% crude fibre, respectively. Chunjula indicated that the baby corn residue was good quality roughage for the beef cattle compared with Guinea grasses. Snitwong and Rungroung, (1991)showed a significant difference ($P < 0.05$) between para grass and baby corn husk; the value being 7.97 vs. 10.18 kg/head/day of 2,375.53 vs. 3,105.22 kg for the production of milk yield at 305 days. Both had nonsignificant differences in protein and lactose. They indicated that income from baby corn wastes (stover, husk, and stover and husk) was higher than from para grass.

Baby corn is a highly nutritious vegetable. Its nutritional value compared with other vegetables (cauliflower, cabbage, tomato, eggplant, and cucumber) is shown in Table 1, according to Yodpet, Bar-Zur and Schaffer, (1993), total sugar content of baby corn observed in all genotypes studied ranged from 20 to 30 mg/g fresh weight and the major sugars were the reducing sugars, glucose and fructose, in about equal amounts of 10 to 15 mg/g fresh weight. Sucrose was present in only small amount at silking. Moreover, the content of 17 amino acids and nine essential amino acids, including lysine, was found in baby corn (Yu *et al.*1993).

Baby corn provides benefits to people from every walk of life and all disciplines. Farmers can grow four crops in a year, and the production of baby corn generates employment amongst the rural poor's, from children to the elderly persons. Other sectors of society who are also benefited from the crop are the regional brokers who buy from farmers, canneries, wholesale merchants (for the local market), retail merchants and exporters.

The demand for baby corn is rapidly increasing in urban areas in India. Baby corn is not a separate type of corn like sweet corn or popcorn and any corn type can be used as baby corn. It is delicacy, which can be profitably used in prolific types of corn *i.e.* those types, which bear two or more ears per plant. The shank with unpollinated silk is baby corn. The economic product is harvested just after the silks emerge (1-2 cm long). Baby corn has immense potential as a salad and as cooked vegetable. It is used as an ingredient in Chop Suey (Chinese dish), soups, deep fried baby corn with meat, rice and other vegetables (Fig. 2). Large number of dishes may be prepared from baby corn as discussed subsequently.

Fig. 2: Different dishes prepared by baby corn: Baby corn candy, Garnished cornbits, Canned corn, Baby corn for raw consumption, Dried baby corn, Baby corn Manchurian, Baby corn noodles, Baby corn with rice (Dish), Fried corn and salad.(Source: Google images)

Since only immature cob is harvested as the economic produce, the crop meant as baby corn can be harvested within 50-55 days. Thus in the areas adjoining cities or other urban areas (peri-urban agriculture) multiple crop of baby corn can be raised which would fetch greater income to the farmers. Baby corn can be effectively used as both a nutritious vegetable and as an export crop to earn valuable foreign exchange. After harvest the still young plants may be used as fodder for cattle.

Study of baby corn evolution is a great concept to know its value and utility in the past, its breeding technique and related aspects which can boost the enhanced cultivation of present and future baby corn breeding and utilization program.

Table.1: Nutritional value of baby corn from analysis of 100 g compared with other vegetables

Components	Baby corn	Cabbage	Tomato	French bean	Lady's finger	Radish	Brinjal	Spinach
Moisture (%)	89.10	91.90	93.10	91.40	89.60	94.40	92.70	92.10
Fat (g)	0.20	0.20	0.20	-	0.20	-	0.20	0.70
Protein (g)	1.90	1.80	1.90	1.70	1.90	0.70	1.40	2.00
CHOs(g)	8.20	4.60	3.60	4.50	6.40	3.40	4.00	2.90
Ash (g)	0.06	0.70	1.60	-	-	-	0.60	-
Calcium (mg)	28.00	18.00	20.00	50.00	66.00	50.00	18.00	73.00
Phosphorus (mg)	86.00	47.00	36.00	28.00	56.00	22.00	47.00	21.00
Iron (mg)	0.10	0.90	1.80	1.70	1.50	0.40	0.90	10.90
Vitamin (iu)	64.00	75.00	735.00	-	-	-	130.0	-
Thiamine (mg)	0.50	0.04	0.07	0.08	0.07	0.06	0.04	0.03
Riboflavin (mg)	0.08	0.11	0.01	0.06	0.01	0.02	0.11	0.07
Ascorbic acid (mg)	11.00	12.00	31.00	11.00	13.00	15.00	12.00	28.00
Niacin (mg)	0.03	0.30	0.60	-	-	-	0.60	-

Source: Zebong M, Studies on sweet corn and its biochemical analysis as a potential young corn (*Zea mays* L.) PhD thesis, University of Philippines, Philippines, 2009.

Transfer of Baby corn Technology

About 40 years ago, the idea of producing and consuming corn as a vegetable was new to most of the countries. As in most developing countries, corn was seen as grain for the poor or animal feed. Baby corn production needed a higher investment in terms of inputs and labor than farmers were used to providing for maize. A public relations campaign was launched to promote the image and benefits of baby corn through newspaper and magazine articles, radio and television broadcasts and advertising. Interest was awakened in the baby corn industry, with consumers and farmers motivated to sample baby corn products and become involved in production, preparation and consumption.

Baby corn fairs, workshops and exhibitions were held across the country to spread the message, with cooking demonstrations and hands-on training by government and private-sector home economists. Farmers were alerted to the fact that producing corn offered a profitable alternative to rice. They were taken to research and demonstration plots and visited model farms. This was followed by hands-on training for each step of production, from land preparation to harvest, conducted by government and the public relations and education campaign have successfully fostered sustained growth of the baby corn industry. Domestic markets continue to expand and the number of farmers producing baby corn is still increasing.

The success of baby corn development in a country has depended upon various factors: The government policies have facilitated spectacular growth of the economy and the food processing industry in particular country for better development. A cooperation between the government and the baby corn industry in extensive public relations has spread the message about baby corn. Collaboration with FAO, the International Maize and Wheat Improvement Centre, the Japanese International Cooperation Agency, the Rockefeller Foundation and the United States Agency for International Development (USAID) have resulted in provision of breeding materials, research support and human resource development. New sources of germplasm have been introduced. Appropriate germplasm and breeding methods have been applied to evolve suitable varieties and hybrids and develop production and processing techniques. This has involved sustained cooperation among researchers and development officials for testing technologies. Progress in establishing baby corn research and development has been fostered by transferring technology through public awareness campaigns.

Research and Development of Baby corn

Some researchers suggested that hybrid corn was more suitable **for** future baby corn production because it produced ears of better quality and more uniform size. Since research related to improvement of yield and quality of baby corn was limited, an important strategy for improvement was the use of heterosis from diverse breeding sources to make baby corn hybrids. A practical measure of heterosis for baby corn is the percentage of yield and other agronomic characteristics of the hybrid or variety over the standard variety. Economic yield may be separated into three components: husk yield, young ear yield, and standard ear yield.

Fig. 3b: Ears quickly grow to be too large for baby corn; a harvest delay of 4 days can make an ear unmarketable.

Fig. 3a: Dwarf corn variety Tom Thumb grown in Montesano, Washington in 1997.

Selection of corn variety for baby corn production depends on the factors like its silking interval, kernel softness, color, sweetness, days to mature as a consumable baby corn etc.

e.g. By adopting Tom Thumb as a baby corn by Monsanto it did not give significant results (Fig.3a) and there may be some genotypes which increase their seed filling rate within a short period (Fig.3b). This is an opportunity to take baby corn as a catch crop. It helps the crop escape many of those biotic and abiotic stresses which appear after the flowering stage.

A. Requirements for the Baby corn Production

An ideal plant should bear at least three ears per plant without losing quality, size and shape of young ears (Fig. 1 and Fig. 4). In corn, Prolificacy is highly influenced and negatively correlated with the planting density. For Baby corn, the variety should not only bear more cobs but also it should be tolerance to high density planting. Selection of small tassel may be the one criterion to select genotypes tolerant to high density planting.

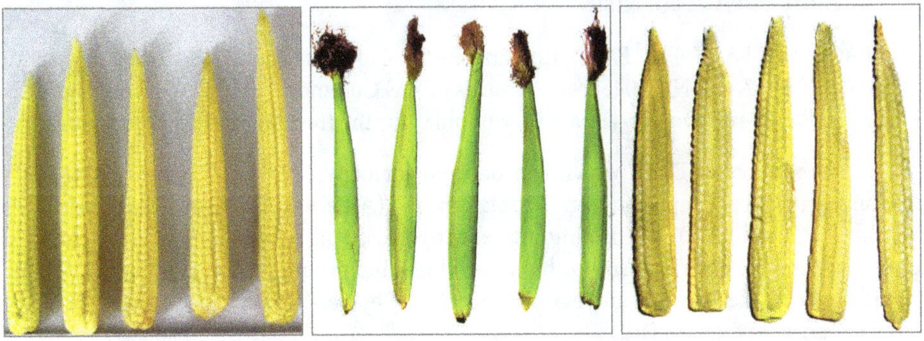

Fig. 4: Standard dehusked and husked cobs ideal for baby corn

B. Types of Germplasm Used in Breeding

Sweet corn, field corn, popcorn, silage corn, etc. grown as food or feed can be used as germplasm sources for baby corn breeding. However, only specific varieties will be used for baby corn production for the fresh market or processing. Bar-Zur and Schaffer, (1993), showed that there was no significant increase in sugar content of baby corn ears attributable to the *su*, *se*, or *sh2* genes compared with *Su*. Types of germplasm used in breeding considered for traits as follows: (1) high yields of unhusked, husked, and standard or good ear weights, (2) high husked and unhusked ear weight ratio, (3) yellow cob with beautiful shape, straight ovary row arrangement, and small ovary, (4) prolificacy, (5) earlyness (6) tolerance to higher plant density, (7) good root and stalk strength, (8) tolerance to leaf diseases and insects at vegetative growth, and (9) response to Nitrogenous fertilizer.

From reviewed literature, most baby corn breeding programs have been extensively developed in Thailand, and source materials were developed by Kasetsart University, Department of Agriculture, and private seed companies as follows:

1. Open-pollinated Varieties:

a) **Field corn varieties:** Suwan 1, Suwan 2, and Suwan 3, These are resistant to corn downy mildew (DMR) caused by *Peronosclerospora sorghi* (Sriwatanapongse *et al.*, 1993).

b) **Sweet corn variety:** Thai Supersweet Composite 1 DMR (Lavapaurya *et al.*, 1990).

c) **Baby corn varieties:** Rangsit, Baby Corn Thai Composite 1 DMR (Promson and Puddhanon, 1990), Chiangmai 90 and Kasetsart 1.

2. Field corn Inbreds:

Ki3, Ki11, Ki20, Ki21, Ki27, Ki28, Ki36, and Ki37. These Kasetsart inbreds are DMR inbreds, and they were released between 1982 and 1997.

3. Baby corn Inbreds:

Ki39 (Rangsit 1(H) C1-S_8-5) and Ki40 (Suwan 2(S) C7-S_8-2-4). Both were released in 1992.

4. Baby corn Hybrids:

PACB 421, PACB 116, PACB 129, and PACB 444 from Pacific Seeds Co., Thailand; G-5414, NTB 017, and NTB 048 from Novartis (USA) Co.; and Uniseed B-50 from Uniseeds Co., Asia, PACB 421 and G-5414 are very popular for the fresh market and processing.

Other sources of DMR which can be used for baby corn breeding are CIMMYT's lowland tropical germplasm; e.g. Population 28 (Amarillo Dentado), Population 31 (Amarillo Cristalino-2), etc. Prolific inbreds, hybrids, and populations of field corn are potential sources of germplasm for baby corn breeding programs; for example, baby corn prolific hybrids (i.e., NY 569 and NY 573) developed by Bar-Zur and Saadi, (1990) at Newe Ya'ar Experiment Station, Israel.

Another interesting source for baby corn breeding developed by Galinat, 1985, University of Massachusetts, U.S. is silkless baby corn, which cannot be pollinated and remains soft and edible. Galinat, (1985), used two different recessive genes for the tassel-seed trait (*ts2*and *ts1*) on chromosomes 1 and 2, respectively. These genes are restorers for the silkless gene, *sk*, also on chromosome 2. The double mutant, *sksk*, *ts2ts2*, and *sksk*,, *ts1ts1* with selection for normal sexual balance, function as normal corn. The double mutant hybrid *ts2Ts2 ts1Ts1 sksk* is completely silkless.

Baby corn breeding can use many of the techniques and theories developed by field corn breeders (Hallauer *et al.*, 1988). However, it is different in practice because of the different end use of the final product which is highly perishable. Crispness of baby corn declines with time after silking within 4 days depending on varieties (Bar-Zur *et al.*, 1990). The other consideration is size, which is determined by the length of the cob of baby corn, in accordance with the Table 2.

Table.2: Standard length for babycorn with specific code.

Size Code	Length (cm)
A	5.0-7.0
B	7.0-9.0
C	9.0-12.0

For all sizes, the minimum width should not be less than 1.0 cm and the maximum width not more than 2.0 cm.

C. Inbred and Hybrid Development in Baby corn

1. Inbred Development

Pedigree selection is the most commonly used method for inbred development from the germplasm sources (e.g. OPVs, synthetic varieties, and single cross, three-way cross, and double cross hybrids), and from related line crosses of field corn or sweet corn considered to have high potential for prolificacy. These germplasm sources have to be selected under higher population density to put more stress. Visual selection is effective during inbreeding under the stress, especially for important traits; e.g., stalk and root strength, early silk emergence prolificacy, and shorter plant height. Other characters should also be considered, such as seedling vigor, insect and disease resistance, pollen shedding, and plant and ear aspects, as is practiced in field corn.(Bauman,1981). The effectiveness of selection among and within progenies depends on the relative heritability of traits selected. Screening techniques for the important pests should be developed and used for selecting the tolerant or resistant lines; for example, the success in selection for DMR in artificial blocks in Suwan 1, Suwan 2 and Suwan 3 varieties in (Sriwatanapongse *et al.,* 1993). Pedigree selection can be modified with a conventional backcrossing program to add or recover some interested traits; e.g., prolificacy, DMR, etc. In Line X tester method for evaluation, Inbreds HUZM-221, HUZM-217, HUZM-329 and testers CM-119, HKI-323 exhibited good GCA effects for earliness, yield and quality traits, whereas, lines HUZM-91-1, HUZM-210-2 were good general combiner for early maturity and yield (Sarla Yadav *et al.,* 2009).

Prolificacy in field corn (Motto *et al.,* 1983) can be applied to baby corn breeding. It is not clear that prolificacy is controlled by partial to complete dominant (Robinson and Liable, 1955, 1968) or recessive (Collins, 1963) genes. Furthermore, opinions differ as to whether number of ears is polygenetically controlled (Early (*et al.,* 1974) or is more simply inherited.(Harris *et al.,*1976). However, most authors agree that gene action for controlling prolificacy is mainly of the additive type (Liable *et al.,*1968). Hallauer (1974) suggested that prolificacy fits the description of a threshold trait in that the inheritance is quantitative but expression is qualitative. Duvick (1974) demonstrated that multi-earedness traits can be transferred from a prolific to a nonprolific inbred line by continuous backcrossing. This result can be taken as evidence that a small number of genes can exert control over prolific potential in corn. Harris *et al.,*(1976), provided evidence that genotypic differences between one-and two-eared hybrids are affected by two loci with major effects.

2. Hybrid Development

Selected lines during inbreeding in baby corn should be evaluated in testcrosses in the S_3 or S_4 generations, as is practiced in field corn (Bauman,1981) However, there is limited information available on choice of testers, heterosis, and heterotic patterns for baby corn, compared with field corn. (Goodman and Hallauer, 1988) Practically, the use of tester depends on the breeding objectives. If the final hybrids are single crosses and three-way crosses, the testers should be an inbred and single cross, respectively. Heterosis in baby corn should be considered for prolificacy and other important traits, except for the fast development of young ears that is desirable for high grain yield in field corn. Hybrid development of baby corn has been conducted at Kasetsart University,(Promson and Aekatasanawan, 1994) the Department of Agriculture and Maejo University (Puddhanon *et al.*,1990) Thailand; Newe Ya'ar Experiment Station,(Bar-Zur,1993) Israel; Instititute of Agricultural Science Research, Hebei Province (Zao *et al.*,1991) and Yantai City Institute of Agricultural Science, (Yu *et al.*,1993) China. Other hybrid research has been done in Hungary, Vietnam, India, and Philippines. Bar-Zur and Saadi,(1990) showed that prolific baby corn hybrids, NY 569 and NY 573, gave the first baby corn ear on the stalk to reach the maximum size acceptable by consumers, and the successive ears were also within the size limits acceptable by the market. The ears picked 3 to 4 days after silking of the first ear had a similar taste and texture to those of sweet corn or field corn.

Trakoontiwakorn *et al.*, (1993), studied heterosis in baby corn using nine inbred lines from five different sources of germplasm (i.e., Suwan, CIMMYT, Hawiian Sugar Super Sweet, Tropical Corn Belt Stiff Stalk, and Tropical Corn Belt Lancaster) for making diallel cross. They found that inbred lines from Suwan, CIMMYT, and Tropical Corn Belt Stiff Stalk crossed with Hawiaan Sugar Super Sweet gave heterosis for unhusked yield range to 88% over Suwan 2(S) C7, an OPV check. Inbred lines from Suwan, CIMMYT, and Tropical Corn Belt Lancaster crossed with Hawiian Sugar Super Sweet had heterosis for husked yield range 78 to 130%. Suwan inbreds, however, showed good heterosis (260 to 335%) for standard ear yield, when mated with inbreds from CIMMYT and Tropical Corn Belt Stiff Stalk.

Line *per se* and topcross methods were compared for evaluating inbred lines for making baby corn hybrids by Sangtong *et al.*, (1990), using 32 lines of Rangsit 1 (RS 1(H) C1-S2) and 16 lines of Suwan 2 (SW 2 (S) C7-F2-S6). Their results showed that the two evaluation methods were not correlated for the unhusked, husked, and poor ear weights, but the good ear weight had significant correlation ($r = 0.33$, $p = 0.05$) between the two methods. They indicated that the topcross method was more effective than the line *per se* method by using the good ear weight as criteria for selecting inbreds for making hybrids. They also reported that unhusked ear weight was positively correlated to husked and poor ear weights ($r = 0.64$ and 0.50, respectively, $p = 0.01$) but unhusked ear weight was not correlated with good ear weight.

The prediction of three-way and double crosses of baby corn using single crosses data from a diallel cross of 10 inbred lines extracted from SW 2(S)C7, (RS 1(H)C1), and Kasetsart inbreds (Ki) and was investigated by Sittichoke *et al.*, (1990). They found positive correlation coefficients between predicted and actual three-way and double crosses when using data of the 1988 vs. 1989 early rainy seasons, within 1989 dry season, and 1989 dry season vs. average of three seasons, but not for the 1988 early rainy vs. 1989 dry seasons. They also showed that good ear weight was controlled by additive more than non-additive gene action.

Sonwai *et al.*, (1994), reported the breeding of baby corn hybrids using 10 inbred lines from three sources of germplasm (*i.e.*, four from SW 2(S) C7, RS 1(H) C1, and three from Kasetsart inbreds) for producing diallel crosses. Their result was in good agreement with Sittichoke *et al.*, (1990) that additive gene action was more important than nonadditive gene action for good ear weight. They also reported that SW 2(S) C7-F2-S6-28 and Ki28 gave high general combining ability (GCA). The single cross from both inbreds showed the highest yield for good ear weight of 354% over Suwan 2(S) C7. They also demonstrated that the highest yielders of three-way and double crosses produced good ear weights over the check by 418 and 332%, respectively. Field corn inbreds for baby corn production were evaluated by Aekatasanawan *et al.*, (1994). They used 10 Kasetsart inbreds for making diallel crosses. Results indicated that Ki20, Ki21, Ki32, and Ki44 gave significantly positive GCA for husked ear weight and Ki20 and Ki36 for good ear weight. The highest yielders for good ear weight were Ki20 × Ki37 and Ki36 × Ki37.

Population improvement in Baby Corn Composite 1 DMR by S_1 and full-sib selections for yield and quality of young ear corn and DMR was conducted by Promson *et al.* during 1987 to 1988 at the National Corn and Sorghum Research Center (Suwan Farm), Kasetsart University. Results revealed that all newly improved and original populations were not significantly improved in yield. After one cycle, S_1 recurrent selection showed higher potential for improving unhusked, husked, and good ear weights than full-sib selection. At the 10 and 20% selection intensities, the unhusked ear weight in the S_1 improved population was 9 and 16% higher than the initial population, while the husked ear weight was 7 and 10% higher, and the good ear weight was 24% higher. The 10 and 20% selection intensities for full-sib selection produced good ear weight 7% higher and 4% lower than the original population, while the husked ear weight was 2 and 0% higher, and the good ear weight was 10 and 7% higher. The average Downey mildew (DM) infection of the newly improved population was generally slightly higher than that of the original population. However, the DM infection of the improved population by the S_1 selection at 20% selection intensity was reduced by 6%. They also found that, over both S_1 and FS families, highly significant correlations (r) among unhusked, husked, and good ear weights, ranged from 0.36 to 0.90 ($p = 0.01$), and between the three traits and the total number of young ears, ranged from 0.15 to 0.74 ($p = 0.01$), except for the *r* value between the good ear weight and total number of ears for S_1 lines.

US Agency of International Development (USAID), the Rockefeller Foundation, the Japanese International Cooperation Agency (JICA), International Maize and Wheat Improvement Centre (CIMMYT) and the Food and Agriculture Organization (FAO), were necessitated for providing the breeding materials, research support, and human resource development to the countries which are interested in baby corn research. New sources of germplasm have been introduced. Appropriate germplasm and breeding methods have subsequently been supplied to evolve suitable varieties and hybrids and also to develop suitable production and processing techniques. This has involved sustained cooperation among researchers and development officials for the testing of technologies over locations, seasons and years. Finally, the progress in establishing sound baby corn research and development has been fostered by transferring technology through effective public awareness campaigns.

Present Status and Future Strategies of Breeding

A. Genetics of Baby corn

Two different recessive genes for the tassel-seed trait *ts2* and *ts1* on *chromosomes 1* and *2* respectively and silk restorers for the silkless gene *sk* also on *chromosome 2* governed traits for development of baby corn. Double mutants *sk sk, ts2 ts2* and *sk sk, ts1 ts1* with selection for a normal sexual balance, function as normal corn. Double hybrid *ts2 Ts2, ts1 Ts1, sk sk* between these two double mutants is 100% silkless because each parent carries the normal dominant allele that masks the recessive tassel-seed gene in the other parent. (Galinat, 1966) Baby ear production is further enhanced by the silkless condition because all of the energy becomes devoted to producing more cobs instead of going first into silk growth and then into kernel development.

At present, the target varieties of most baby corn breeding programs are superior, stable, high-yielding hybrids *i.e.,* single cross, three-way cross, and double-cross hybrids (Promson *et al.,* 1990-1991). Most breeders have developed semi-prolific or prolific inbreds from various germplasm sources of field corn and sweet corn for making prolific baby corn hybrids. Pedigree breeding with visual selection during line development is effective because of the relatively high heritability and additive effects of important traits in baby corn, especially for prolificacy, DMR, and resistance to root and stalk lodging. Baby corn hybrids have higher yield potential, better uniformity of silk emerged, shorter harvest, cob size, plant and ear heights, etc., compared with open-pollinated baby corn varieties (OPVs). Consequently, they have been used for the fresh market and processing and replaced OPVs very rapidly, particularly in this decade in Thailand.

B. Utilization of Male Sterility for Baby Corn Improvement

In baby corn production, the removal of tassels or detasseling is necessary for the stimulation of earlier harvest date, the enhancement of number of ears per plant or prolificacy, the increase of high yield, and the prevention of pollination. However, the detasseling results in higher cost and perhaps in yield loss affected by some leaf loss (Aekatasanawan *et al.,* 1994)

1. Development and Evaluation of Male-sterile Baby corn Varieties

Aekatasanawan *et al.,* (1994), solved these problems by using C cytoplasmic male sterility to improve non-detasseled baby corn. They reported that during 1988 to 1991, six selected male-sterile (no anther exserted) lines from IITA (Nigeria) and Guatemala were used as females in crossing with two baby corn varieties: Suwan 2(S) C7 (SW2) and Thai Supersweet Composite 1 DMR (TSC 1 DMR). Both SW2 and TSC 1 DMR were backcrossed to Nigeria and Guatemala lines for five and two times, respectively. Suwan 2 had the same ancestor (Thai Composite #1) as Suwan 1. A breeding scheme for Suwan 1 and Suwan 2 varieties and germplasm assembled in Thai Composite #1 respectively.

The results of the detasseling and non-detasseling in the normal baby corn or the fertile baby corn supported the study of Grogan, (1956), on detasseling responses in corn as affected by climate, soil, and competitive conditions. He concluded that under conditions of moisture and/or nutrient stress, yield increases associated with detasseling were due to the elimination of competition for nutrients between the ear and the tassel. He predicted that similar results might be expected if male-sterile cytoplasm were substituted for normal cytoplasm. Duncan *et al.,* (1967) and Hunter *et al.,* (1969), reported that the increase of grain yield from detasseling was larger and more consistent in higher plant densities resulting from the elimination of tassel light interception.

The results of detasseling and non-detasseling in the male-sterile baby corn supported the conclusions of Chinwuba *et al.,* (1961) that the male-sterile single cross outyielded their fertile counterparts at higher plant densities. And they reacted similarly to the the detasseled fertile single crosses at different plant densities when reductions in yield from mechanical injury and the initial stages of pollen development before the tassels were removed were considered. Sanford *et al.*, (1969) reported considerably more N accumulated in fertile tassels than in sterile tassels. After pollen shedding, there was no difference in the Nitrogen (N) content of fertile and sterile tassels. Sterile plants accumulated more N in ears and husks than did fertile plants due to the greater yield and number of ears produced by sterile plants. There were only slight differences between fertile and sterile plants in N content of leaves and stalks. They suggested that the comparatively fewer ears per plant produced by the fertile versions were due to competition for N between the ear primordia and the pollen.

2. Development and Evaluation of the Non-detasseled Baby corn Variety Developed Using Male Sterility

In the 1992 early rainy season, MS(1,2,6,7 × SW2) BC5 and (CU88A(18 × 19) × SW2) BC2 lines having 100% of male sterility, two ears per plant, and good agronomic characters were selected to form Male-Sterile Suwan 2 or MS-Suwan 2 by using bulk seeds. In the 1993 dry season, a maintainer was developed from the selected 11 S1 lines of Suwan 2(S) C7. During 1993 to 1994, the female parent (MS-Suwan 2) and the maintainer were improved for baby corn in isolation blocks. In 1995, the National Corn and Sorghum Research Center released the non-detasseled baby corn variety, MS-Suwan 2, to farmers and processing plants in the name of Kasetsart 1.

3. The Development of the Non-detasseled Baby corn Hybrids

Aekatasanawan *et al.*, (1997) evaluated topcross baby corn hybrids for male sterility using Kasetsart 1 topcrossed with 15 baby corn inbreds of Suwan 2(S)C7, 16 field corn inbreds of various germplasm sources (e.g., Suwan 1, second-cycle recovered lines of Ki21 and Ki27, etc.), and five TSC 1 DMR inbreds. They found that male-sterile topcrosses of baby corn, field corn, and sweet corn inbreds had 8 (53.33%), 6 (37.50%), and 3 (60.00%) varieties, respectively. Most male-sterile topcrosses of sweet corn inbreds had nonsignificant unhusked, husked, and good ear weights higher than those of Kasetsart 1. The results demonstrated that some male-sterile topcross hybrids can be used as the non-detasseled baby corn hybrid. They also have been developing the non-detasseled baby corn single crosses. And they found that Kasetsart 2, which was developed from Ki28cms crossed with KSei 14004 or [(*sh2* Syn 29 × KS 1) × Suwan 3(S) C4-F4-S8-24-2-2-4], gave higher unhusked, husked, and good ear weights and better quality. The National Corn and Sorghum Reseach Center released Kasetsart 2, the non-detasseled baby corn single cross, in 1999.

Brief Review of Baby corn Research in India

Baby corn cultivation started in India late 80's. Initially it was used as a cattle feed in off seasons in dairy farming. The major work was focused by the institutes like GB Pant University of Agriculture and Technology, Pantnagar, Uttarakhand, Godhra Research Station, Gujarat and ICAR-Indian Institute of Maize Research, New Delhi and other major stations like Almora, Bajauva, Coimbatore, BHU, Varanasi and Udaipur.

Some of the popular Indian baby corn varieties are D 901, D 921, D 903, D931, D 933, D 965, Tarun, Azad Uttam, Navin, Shweta, Kanchan, Surya, KH-101, KH-501, HIM-129,

Yey-705, VL-42, DEH-20194 and DEH-21696. Among these genotypes HIM-29 is having the highest dehusked cob yield *ie*, 17.22q/ha, (Tiwari and Verma, 1999, Pantnagar).Other better yielders are given in the Table.3 (Sharma *et al*, 2010).

Table 3: Important varieties and hybrids developed for baby corn production

Varieties	Hybrids
VL- 239, VL-42, HIM -129, Makka 16, Makka-2, Chandan Makka-1, Chandan Makka-3, Navin, Diara-3, D-765, Farm Sumeri, Kanchan, MCU-508, Makka-41, Kiran (J-660), VL Makka-88, Pusa Comp-1, Surya, Gujarat Makki-1, Madhuri(sweet corn), Panch, Ganga, Megha, Gurat Makai 2, Birsa Makai-1, C-14, JM.8, Shakti, Amar(D-941), Aravalimakka-1, Gujarat Makkai-4, Tarun, Narmada, Moti, Priya sweet etc.	Him 128, DEH-20194. Ganga, Deccan-1, DHM-107, Pusa early hybrid Makka-1, Pusa early hybrid Makka-2, Jh-3459, Pusa early hybrid Makka-3, Vivek Hybrid-9, Parkash etc.

Corn experiment in these stations resulted into the highest phenotypic variability which was obtained from corn yield and fodder yield traits. In baby corn yield is a better heritable character.

According to Thakur *et al.*, (2000), hybrids have significantly higher yield than composites, whereas fodder yield was more in early composites than hybrids. Removal of tassel just after its emergence gave 18% higher marketable baby corn yield than no detasseling.

According to Lolita *et al.*, (1991), the optimal stage for harvesting, the acceptable top ears is two days after silking for both wet and dry season trials which gave the largest acceptable cob, highest cob weight and biggest ear diameter, three days after silking in the wet season trial gave acceptable second ear.

Problems Associated with Breeding of Baby corn

Some problems associated with baby corn breeding can be summarized as follows:

1. There is limited information on baby corn breeding because a few baby corn breeding programs have been actively working. Research and development on baby corn breeding were initiated in the late 1970s, compared with field corn and sweet corn breeding programs developed about 100 years ago.
2. There are limited germplasm sources developed for specific purposes of baby corn.
3. There is a lack of breeders in the limited baby corn breeding programs. Young staff or new breeders should be developed and encouraged in each breeding program.
4. It is time-consuming and labor intensive, because it requires detasseling, harvesting many times, and dehusking of green ears.
5. Harvest time cannot be delayed as with field corn, because green ears must be harvested when silk emerges 1 to 3 cm or 1 to 2 days. And they are husked within the harvested day.

6. Heterosis in grain yield and heterotic pattern in field corn cannot be applied to baby corn breeding, because it needs more small young ears per plant and slow development of young ears for high yield and high quality.

Future Prospects of Baby corn

In India no cultivar has been exclusively bred for baby corn purpose. Prolific and early maturing cultivars have been mostly popularized as baby corn cultivar. In order to encourage uniformity in the material, more emphasis is to be given towards development of early maturing prolific hybrids. As baby corn with light yellow color and regular row arrangement fetches better market price, at the time of breeding for baby corn attention must be kept in this direction.

In recent past baby corn has gained popularity in regular vegetable markets in urban areas. However, keeping in mind the nutritive value of baby corn there is a need to popularize it further in other urban and rural areas. Though baby corn is being sold in domestic market, they are being sold without proper processing. As a result there is considerable reduction in qualities of the cobs. This is principally due to lack of awareness among the farmers and due to non-existence of proper storage facilities and location of the farms far away from the market. Thus there is a need to develop appropriate entrepreneurship and establishment of appropriate storage and marketing facilities and popularization of baby corn cultivation in peri-urban agriculture. However, this is dependent on organization of markets and support from government sectors. Where baby corn is being grown for further market and export, extra care is to be taken to process the cobs and can them within two to three hours of harvest. Otherwise they will lose their nutritive value.

It takes time for the output of inter-disciplinary research and development to be transferred to the persons involved, starting with the farmer who produces the raw materials, then to the canneries where the finished product is processed for the consumer who eventually prepares and eats the product. Nevertheless, this is essential for the success of the baby corn business. So, concerted efforts in this direction would be critical in the future.

Corn will remain one of the most important field crops in developing countries. Young people who are involved in this business will be able to make a contribution to society. Government in these countries should, therefore, concentrate on the development of young farmers, researchers and extension agents, cannery managers, and others involved in the private sector to find creative ways to sustain the baby corn industry. Thus, policies for required human resource development would go a long way in the proper adoption of such innovative policies.

Future use of baby corn breeding will be concentrated on single-cross hybrids to meet needs of farmers (e.g., high yield, resistance to root and stalk lodging, harvesting in a short period, etc.) and of factories (high uniformity and standard of raw materials, such as cob size, color, flavor, etc.). Thus, inbred and hybrid development will be selected for the desirable traits. The improvement or recycling of commercial or elite inbreds to correct some weak points will be emphasized for potential use as female parents for producing superior baby corn single crosses.

The required germplasm sources for future use in breeding should be more developed (i.e., genic male sterility or cytoplamic male sterility, silkless, etc.) for producing baby corn single-cross to reduce costs of production for the tasks of detasseling, husking, and removal of silks by hand.

Various types of molecular markers, such as restriction fragment length polymorphisms (RFLPs), randomly amplified polymorphic DNAs (RAPDs), amplified fragment length polymorphisms(AFLPs), and simple sequence repeats or microsatellites (SSR) will be exploited for helping breeders to select desirable genotypes resistant to important diseases and insects (for example, DMR, corn borer, etc.). To assess genetic diversity in germplasm of interest to breeders, molecular markers will be useful for identifying and classifying heterotic groups. However, heterosis in greater number of young ears per plant will be beneficial to baby corn breeders, because heterosis for fast ear development, as in field corn, results in cobs that are too large.

Conclusion

Considerable scope exists for promoting baby corn technology in the Asia-pacific region. The baby corn industry provides opportunities for higher income, generates employment for the rural poor and potential for export. Besides, its use as vegetable provides additional source of nutrition to the consumers. Baby corn is expected to catch the attention of more and more consumers and farmers because of its superior taste and texture. Using local produce gives fresh and nutritious food and keeps small farmers in business. Thus, help supports the local economy and conserving natural resources (raw materials for transportation and packaging are saved). In order to harness these benefits, research and development support and appropriate policies at the national level are required. Hence, the Governments should therefore, concentrate on framing policies and development of human resources. In addition, government policies should encompass motivating young farmers and finding creative ways to sustain baby corn industry by involving personnel involved in both private and public sectors. Further, for promoting baby corn industry, regional co-operation for exchange of information and germplasm, regional testing of selected hybrids and varieties, joint meetings and visits, human resource development, collaborative efforts for research and development and sensitization of policy makers for arriving at adoption of appropriate baby corn production and processing technology would be highly desirable.

References

Aekatasanawan, C., Jampatong, S., Aekatasanawan, C. and Chulchoho, N. 1994. Evaluation of field corn inbreds for baby corn production, *Kasetsart University Conf.*, 32, 478.

Aekatsanawan, C., Chulchoho, N. and Balla, C. 1997. Evaluation of topcross hybrids for male-sterility and for use in non-detasseled baby corn production, *National Vegetable Conf.*, 15, 293.

Bar-Zur, A. and Saadi, H. 1990. Prolific maize hybrids for baby corn, *J. Hortic. Sci.*, 65, 97.

Bar-Zur, A. and Schaffer, A. 1993. Size and carbohydrate content of ears of baby corn in relation to endosperm type (*Su, su, se, sh2*), *J. Am. Soc. Hortic. Sci.*, 118, 141.

Bauman, L. F. 1981. Review of methods used by breeders to develop superior inbreds, *Corn Sorghum Ind. Res. Conf.*, 36, 199.

Cheva-Isarakul, B. and Paripattananont, T. 1988. The nutritive value of fresh baby corn waste, *Ruminant Feeding Systems Utilizing Fibrous Agricutural Residues 1987,* International Development Program of Australian Universities and Colleges, Ltd., Canberra, 151.

Chinwuba, P. M., Grogan, C. O. and Zuber, M. S. 1961. Interacion of detasseling, sterility, and spacing on yield of maize hybrids, *Crop Sci.* 1, 279,

Chutkaew, C. and Paroda, R. S. 1994. *Baby Corn Production in Thailand — A Success Story,* APAARI Publication: 1994/1, FAO Regional Office for Asia and the Pacific, Bangkok.

Collins, W. K. 1963. Development of potential ears in Corn Belt *Zea mays* L., *Iowa State College J. Sci.*, 38, 187.

Duncan, W. G., Williams W. A. and Loomis, R. S. 1967. Tassels and productivity of maize, *Crop Sci.*7, 37.

Duvick, D. N. 1974.Continuous backcrossing to transfer prolificacy to a single-eared inbred line of maize, *Crop Sci.* 14, 69.

Early, E. B., Lyons, J. C., Inselberg, E., Maier, R. H. and Leng, E. R. 1974. Earshoot development of Midwest dent corn (*Zea mays* L.), *Illinois Exp. Sta. Bull.*

Faungfupong, S., Tungadulratana, R., Waitruardrok, Cholwiriyakul, W. and Samuttong, N. 1987. Corn and sorghum agronomic research in 1987, *Thailand National Corn and Sorghum Program Ann. Rep.*, 19, 148.

Galinat, W. C. and Lin, B.Y. 1988. Baby corn: Production in Taiwan and future outlook for production in the United States, *Econ. Bot.*, 42, 132.

Galinat, W. C. 1985. Silkless baby corn, seed production genetics, *Maize Genetics Coop. Newsl.*, 59, 102.

Galinat, W. C. 1985.Whole ear baby corn, a new way to eat corn, *Proceed. Northeast Corn Improvement Conf.*, 40, 22.

Goodman, M. M. and Brown, W. L. 1988. Races of corn, in *Corn and Corn Improvement,* Sprague, G. F. and Dudley, J. W. Eds., American Society of Agronomy, Madison, WI, 33.

Grogan, C. O. 1956. Detasseling responses in corn, *Agron. J.*, 48, 247.

Hallauer, A. R. and Miranda Fo, J. B. 1988. *Quantitative Genetics in Maize Breeding,* 2nd ed., Iowa State Univ. Press, Ames, IA.

Hallauer, A. R. 1974. Heritability of prolificacy in maize, *J. Heredity,* 65, 163.

Harris, R. E., Gardner, C. O. and Compton, W. A. 1972. Effects of mass selection and irradiation in corn measured by random S1 lines and their testcrosses, *Crop Sci.* 12, 594.

Hunter, R.B., Daynard, T.B., Hume, D.J., Tanner, J.W., Curtis J.D. and Kannenberg, L.W. 1969. Effect of tassel removal on grain yield of corn (*Zea mays* L.), *Crop Sci.* 9, 405.

Kulvadee, T., Supasri, R., and Maleehual, S. 1988. Canned vegetable juice from baby corn, Inst. of Food Research and Product Development, Kasetsart University, Bangkok, 19.

Lavapaurya, T., Saridniran, P., Chowchong, S., Juthawantana, P., Thongleung., S., Chuthatong, Y., and Trongpanich, K. 1990. Research and development of sweet corn and baby corn for fresh consumption and processing, *Kasetsart J.*, 24, 208.

Liable, C. A. and Dircks, V. A. 1968. Genetic variance and selection value of ear number in corn (*Zea mays* L.), *Crop Sci.* 8, 540.

Lolita T. Pino and Vivencio R. Mamaril, 1991 Summary of some important findings on Baby corn Rewiew, *PNAS:* 251-258.

Motto, M. and Moll, R. H. 1983. Prolificacy in maize: A review, *Maydica*, 28, 53.

Promson, S. 1987. Comparison on hybrid and open-pollinated varieties of baby corn, Special Problem, Graduate School, Kasetsart Univ., Bangkok.

Promson, S., Lavapaurya, T., Subhadrabandhu, S. and Chutkaew, C. 1990. Population improvement of baby corn Thai Composite 1 DMR by S1 and Full-sib selections, *Kasetsart J.*, 24, 417.

Puddhanon, P., Phetmanee, S., Nattriphob, P., Punsurintr, S., Kajornmalee, V., Vayupab, S. and Boonthum, A. 1990. Baby corn regional yield trial in 1990, *Thailand National Corn and Sorghum Program Ann. Rep.*, 22, 54.

Robinson, H. R., Comstock, R. E. and Harvey, P. H. 1955. Genetic variation in open-pollinated varieties of corn, *Genetics*, 40, 45.

Sangtong, V., Chutkaew, C., Pruksacheeva, S. and Pongtongkam, P. 1990. Evaluation of line per se versus topcross in baby corn, *Kasetsart University Ann. Conf.*, 28, 247.

Sarla Yadav., Shahi J.P., Singh P.K. and Srivastava K.(2009). Studies on combining ability for earliness, yield and quality traits in baby Corn (*Zea mays* L.*) International Journal of Plant Sciences,* Vol. 5 Issue 1: 1-5

Sharma G. and saikia R.B. 2010. Stability analysis for yield and yield attributing characters in baby corn (*Zea mays* L.), *Indian J. Hill Farming* 13: 30-34.

Sittichoke, J., Chutkaew, C., Pruksacheeva, S. and Ratisoontorn, P. 1990. Prediction of three-way and double crosses of baby corn by using single crosses data, *Kasetsart University Ann. Conf.*, 28, 241.

Snitwong, C. and Rungroung U. 1991 Stover and baby corn husk as based ration for milking cows, *J. Agric. (Thailand), Warasan Kaset,* 7, 96.

Sonwai, S., Chutkaew, C., Rojanaridpiched, C. and Ratisoontorn, P. 1994. Breeding of baby corn (*Zea mays* L.) hybrids (single, three-way and double crosses), *Kasetsart University Ann. Conf.*, 32, 458.

Sriwatanapongse, S., Jinahyon, S. and Vasal, S. K., 1993. *Suwan-1: Maize from Thailand to the World,* CIMMYT, Mexico, D.F.

Thakur D.R. 2000. Baby corn production technology, *Directorate of Maize Research; Indian Agricultural Research Institute,* New Delhi.

Trakoontiwakorn, P., Chutkaew, C., Kunta, T. and Chandhanamutta, P. 1993. Heterosis in baby corn hybrid, *Kasetsart University Ann. Conf.,* 31, 167.

Verma, S. S. 2001. Baby corn production technology, G B Pant University of Agriculture and Technology, Pantnagar, *Annual preceedings.*

Yodpet, C. 1979.Studies on sweet corn as potential young cob corn (*Zea mays* L.). Ph.D. thesis., University of the Philippines, Philippines.

Yu, W. Z., Sun, S. L., Zhou, D.Q. and Wang, Z. F. 1993. Lu Sun Yu 1, an elite maize cultivar for baby corn, *Chinese Vegetables,* 6, 52.

Zhao, W. Q. 1991. Maize cultivar Ji Te 3, *Crop Genetic Resource,* 1, 48.

4

Basil

Varughese George, Palpu Pushpangadan and T P Ijinu

Among all the herbs, *Ocimum basilicum* and *O. sanctum* are the only species venerated in Indian homes and places of worship. *Ocimum* leaves and inflorescence invariably form a part of the offerings to the deity during worship in Hindu Temples. Tulsi is called a plant without equal because of its multifarious uses. It is used as a spice to flavor food material and it is an important ingredient in a number of medicinal preparations. The curative, preventive and health promoting properties of *O. basilicum* was known for several millennia to ancient cultures and civilizations. Because of its aroma and medicinal properties, *O. basilicum* is an economically important herb and it is cultivated in many parts of the world.

Basil Down the Millennia

Basil, *Ocimum basilicum*, *Lamiaceae* is an annual spicy herb, indigenous to India. Several species of *Ocimum* are cultivated in India and their medicinal and aromatic uses were known for several millennia. Basil is considered as a plant without equal in ancient Indian writings. It is considered as a panacea for several ailments, the mother of all medicinal plants, pure and a provider of good health. Basil is invariably a constituent in all materials used for Worship in Hindu Temples. Just as in India, the ancient Greeks also used basil for the Worship of Greek gods and goddesses. The great Ayurveda Acharya Susruta classified Basil among vegetables. The ancient Ayurvedic texts such as *Charaka Samhita, Susrutha Samhita, Bhava Prakasa, Dhanwantari Nighandu, Ashtangahridayam etc.* also describe the medicinal properties of Basil. Basil is used as medicine in Africa, Arabia, Burma, Sri Lanka,

Philippines etc. Bhavamisra, the great Ayurvedic Acharya referred to *Ocimum basilicum* as 'barbari' (Pushpangadan *et al.*, 1993). It had been 'Herbe royale' to the French, a sign of love to Italians, and a sacred herb in India. The first century A.D. Roman naturalist Pliny reported that basil relieves flatulence, which had been subsequently proven true. In the Far East, the herb had been used as a cough medicine and in Africa it has been used to expel worms (Holistic online, 1December, 2010).

The first written history of basil goes back to 4000 years when it was grown in Egypt. The ancient Egyptians burned a mixture of basil and myrrh to appease their gods and embalmed their dead with it. In Persia and Malaysia, basil is planted on graves and in Egypt women scatter the flowers on the resting places of those belonging to them. To the ancient Greeks and Romans, the herb was a symbol of hostility and insanity. They painted poverty as a ragged woman with basil at her side. They believed that to grow truly fragrant basil, one had to shout and swear angrily while sowing its seeds. In French "sowing basil" (semer le basilic) means "ranting". Other folk traditions have associated the herb with love. During recent centuries, when an Italian woman placed a potted basil plant on her balcony, it signaled that she was ready to receive her lover. The French were introduced to it by Catherine de Medici in 1533 when she married King Henry II and brought with her Italian chefs and a taste for food well seasoned with basil. They dubbed it Herbe Royale. In northern Europe, lovers exchanged basil sprigs as signs of faithfulness. Haitians believe in basil's protective powers. Shopkeepers in Haiti sprinkle basil water around their stores to ward off evil spirits and to bring prosperity.

Etymology of *Ocimum basilicum*

Basil belongs to the genus *Ocimum*, derived from the Greek *Ozo* which means to smell, in reference to the strong odours of the species within the genus. The word Basil comes from the Greek word *Basileus* meaning King, as it has come to be associated with the Feast of the Cross commemorating the finding of the True Cross by St. Helena, Mother of the Emperor Constantine. The generic name may also come from the Greek *Okimum*, fragrant-lipped. Alternatively it was also called *Basilicon* relating to the Latin word *Basilisk* meaning scorpion since the herb is connected to Mars and under the scorpion. The Oxford English dictionary quotes speculation that basil may have been used in some royal unguent, bath or medicine. Basil is still considered the king of herbs by many cookery authors.

Geographical Distribution

Sweet basil is native to India or Iran. It arrived in Europe *via.* the Middle East in Italy (Bonar, 1985). From Italy it was taken to France in the 15th century. From France it was carried to England in the 17th century. The migrant British carried it across the Atlantic Ocean to America. It is cultivated commercially in southern Europe, Egypt, Morocco, Indonesia and California (Christman, 2010). It grows in tropical and sub-tropical regions (Ayurnepal, 1 December, 2010). It is an annual herb; cultivated extensively in France, Egypt, Hungary, Indonesia, Morocco, the United States (Arizona, California, New Mexico, North Carolina), Greece and Israel (Simon, 1998).

Distribution: Tropical Old World: Central Africa and Southeast Asia (Simon, 1998).

Other names: Tulsi (India); Sweet basil, St. Josephwort, Basilienkraut (German); Basilic (French); Basilico (Italian); Albahaca (Spanish); Basilkört (Swedish); Raihan (Arabic); Basilicum (Dutch); Manjericao (Portuguese); Bazilik (Russian); Meboki (Japanese); Lo-le, luole (Chinese); American Dittany, Alabahaca, Witches' Herb, Our Herb, Bazylia pospolita (Polish).

Description

Ocimum basilicum is an erect, almost glabrous herb, 30 to 90 cm high. Leaves are ovate, lanceolate, cucuminate, toothed or entire, glabrous on both surfaces and glandular. The flowers are white or pale purple in simple or many branched racemes that are often thyrsoid, the nutlets are ellipsoid black and pitted.

Mature leaves are ovate, reaching about 5 cm in length not including the petiole of about 2 cm. On the upper surface they are smooth and lustrous; on the lower surface along the midrib and on the petiole, short, stiff hairs occur sparingly.

The flowers are borne in long terminal racemose inflorescences. The greenish corolla is small and inconspicuous. The calyx partly grown together with branches enlarges itself post-florally and remains with the latter dry on the plant. The capitate hairs have commonly a two-celled head with a stalk so short as to appear sessile. Chief structural characters of the leaves are ovate, pointed, blunt toothed when young, smooth above, and sparingly hairy below and along the midrib (Prakash, 1990).

Polymorphism and cross-pollination under cultivation have given rise to a number of subspecies and varieties differing in height, habitat and growth, degree of hairiness and colour of stems, leaves, flowers and aroma ranging from lemon, camphor, cinnamon, clove, mint, licorice, rose and anise.

Chemical Composition

Extraction of Volatile Oils

Extraction of volatile oils from plants was known to almost all ancient civilizations. The extraction techniques have undergone constant refinements throughout the ages. The most common methods now employed for extraction of essential oil are hydro-distillation, steam distillation, super critical extraction and microwave extraction. The extraction technique is a careful choice based on the nature of the plant specimen and thermal stability of the target oil constituents (Baby and George, 2009).

As the quality of the essential oil, associated with color and aroma retention, is strongly influenced by postharvest handling, leaves and flowering tops are dried at low temperatures (<35^{0}C) to retain maximum color before grinding to marketable size or distilling for essential oil. Leaves should be washed and cleaned, with weeds and all extraneous materials removed. Basil is very sensitive to chilling injury. For essential oil, the cut basil is normally cut, then field dried for 1-3 days before the material is collected and distilled.

The basil oil contains a large number of terpenic hydrocarbons. Removal of these hydrocarobons renders the oil terpene less. These terpene less oils fetch better price in the market. In order to prepare terpene less oil, basil oil is subjected to fractional distillation under vacuum in fractionating columns. The fractionating columns are also used to isolate major components present in the oil.

Storage of Volatile Oil

Essential oils are affected by light, heat and air on longstanding. Oxidation, resinification, polymerization, hydrolysis of esters and interaction of functional groups cause darkening of oil and deterioration in oil quality. Prior to storage the oil must be made free from moisture and other impurities. Basil oil like any other essential oil must therefore be stored in opaque, air tight containers in a cool dark place.

Chemotypic Variation in Basil Oil

The flowers on an average yield 0.4% oil while the whole plant (Indian basil) contains 0.1 to 0.25% oil. By taking the initial 3 to 4 harvests of flowers (including main and sub inflorescence) and final harvest of the whole herb one can get about 3 to 4 tons of flowers and about 13 tons of whole herb per hectare corresponding to about 13 kg of the flower oil and about 27 kg of whole herb oil. In all, 40 kg of oil per hectare can be obtained. Oil of sweet basil produced both from herb as well as flowers has a potential market (Farrell, 1985; Pruthi, 1976). Oil of sweet basil is produced by the distillation of the herb.

Sweet basil oil possesses a clove like scent with an aromatic and somewhat saline taste. The basil oil is used as a flavoring agent and also as a perfume. Two types of basil oil are commonly recognized (Guenther, 1949; Anon., 1966). Pushpangadan and Bradu, (1995) reported that the essential oil composition of *O. basilicum* varied depending on the variety. They have reported methyl chavicol, linalool, methyl cinnamate and geraniol rich varieties of *Ocimum basilicum*. Based on TLC and gas chromatographic studies they reported that *O. basilicum* L. var. *minima* Benth. contained geraniol (45%) and eugenol (25 %) as the major compounds, *O. basilicum* L. var. *glabratum* Benth., Chemotype No.1 contained methyl chavicol (38%) and linalool (35%), *O. basilicum* L. var. *glabratum* Benth., Chemotype No.2 contained linalool (47%) and eugenol (20%) as the major components, *O. basilicum* L. var. *glabratum* Benth., Chemotype No.3 contained linalool (40%), eugenol (20%) and camphor (20%), *O. basilicum* L. var. *purpurascence* Benth. contained methyl cinnamate (20%) and linalool (60%), *O. basilicum* L. var. *tryrsiflora* Benth. contained methyl cinnamate (35%) and linalool (60%), *O. basilicum* L. var. *crispum* Benth. contained methyl chavicol (50%) and linalool (28%) and *O. basilicum* L. var. *darkapal* contained geraniol (35%), linalool (35%) and eugenol (25%).

Ji-Wen *et al.*, (2009) studied the composition of essential oil obtained from the aerial parts of *Ocimum basilicum* Linn. var. *pilosum* (Willd.) Benth., an endemic medicinal plant growing in China. They identified linalool (29.68%), (*Z*)- cinnamic acid methyl ester (21.49%), cyclohexene (4.41%), α-cadinol (3.99%), 2,4-diisopropenyl-1-methyl-1-vinylcyclohexane (2.27%), 3,5-pyridine-dicarboxylic acid, 2,6-dimethyl-diethyl ester (2.01%), β-cubebene (1.97%), guaia-1(10),11-diene (1.58%), cadinene (1.41%), (*E*)-cinnamic acid methyl ester (1.36%) and β-guaiene (1.30%) from this oil.

Koba *et al.*, (2009) reported five chemotypes of *Ocimum basilicum* (Lamiaceae) from Togo. They are estragole type; linalool/ estragole type; methyleugenol type; methyleugenol/ t-anethole type and t-anethole type.

In a study carried out by Klimankova *et al.*, (2008) in the aroma profiles of five basil cultivars grown under conventional and organic conditions it was found that linalool, methyl chavicol, eugenol, bergamotene, and methyl cinnamate were the dominant volatile components, the relative content of which was found to enable differentiating between the cultivars examined. The relative content of some sesquiterpenes, hydrocarbons benzenoid compounds and monoterpene hydrocarbons was lower in dried and frozen leaves as compared to fresh basil leaves.

Thus, the essential oil composition of *O. basilicum* varies depending on the variety, geographic origin, harvesting season *etc.* This considerably affects the perfumery and flavour characteristics of the oil. Literature search revealed that the oil constituents belonged to different classes of compounds such as mono and sesquiterpene hydrocarbons, oxygenated mono and sesquiterpenes, aliphatic alcohols, aldehydes, esters, ketones, acids, aromatic compounds etc.

Chemical Constituents of Basil Oil

1. **Monoterpene hydrocarbons:** α-Pinene, sabinene, myrcene, p-cymene, limonene, α-terpinene, (Z)-β-ocimene, (E)-cis-ocimene, γ-terpinene, terpinolene.

2. **Oxygenated monoterpenes:** 1,8-Cineole, linalool cis-furanoid, linalool oxide cis-furanoid, trans-sabinene hydrate, linalool trans-furanoid, linalool oxide trans-furanoid, ocimene oxide, camphor, 3,7-dimethyl-1,6-octadien-3-ol, linalool, linalyl acetate, trans-p-menth-2-en-1-ol, bornyl acetate, carvacryl methyl ether, exo-methylcamphenilol, 4-terpineol, cis-dihydrocarvone, hotrienol, terpinen-1-ol, L-menthol, trans-pinocarveol, d-terpineol, lavandulol, trans-verbenol, p-menth-1,8-dien-4-ol, terpinyl formate, α -terpineol, borneol, verbenone, exo-2-hydroxycineole acetate, dihydrocarveol, α -citral, exo-2-hydroxycineole, L-carvone, linalool oxide cis-pyranoid, trans-piperitol, linalool oxide trans-pyranoid, citronellol, yrtenol, nerol, trans-carveol, p-cymen-8-ol, geraniol, geranyl acetate, guaiacol, exo-2-hydroxycineole, piperitenone, L-perillyl alcohol, cuminyl alcohol, fenchone, estragole, t-anethole, carvacrol, thymol, bornyl acetate, methyleugenol, geranyl formate.

3. **Sesquiterpene hydrocarbons**: β-Cubebenec, d-cadinene, valencene, α-amorphene, WWWWW-selinene, dehydroaromadendrene, β-elemene, α-copaene, β-caryophyllene, α-humulene, (-)calamenene (E)-α-bergamotene, α-caryophyllene, β-selinene, α-zingiberene, bicyclogermacrene, α-muurolene, germacrene A, germacrene D, γ-cadinene.

4. **Oxygenated sesquiterpenes**: γ-Cadinol, spathulenol, caryophyllene oxide, α-humulene oxide, elemol, viridiflorol, spathulenol, α-cadinol, γ-murolol, β-bisabolol, β-bisabolol isomer, α-eudesmol, isospathulenol, β-eudesmol, caryophylla-4(12),8(13)-dien-5β-ol, dihydroactinidiolide, caryophylla-3,8(13)-dien-5α (or β)-ol.

5. **Aliphatic alcohols:** 1-Penten-3-ol, 3-methyl-3-buten-1-ol, (Z)-2-pentenol, 3-methyl-2-buten-1-ol, hexanol, (Z)-3-hexenol, 3-octanol, cyclohexanol, 1-octen-3-ol, octanol.

6. **Aliphatic aldehydes:** Hexanal, (E)-2-hexenal, (E,Z)-2,4-heptadienal, (E,E)-2,4-heptadienal.
7. **Aliphatic esters:** Methyl 2-methylbutyrate, (Z)-3-hexenyl acetate.
8. **Aliphatic ketones:** 3-Octanone, 3-hydroxy-2-butanone, 6-methyl-5-heptenone, 6-methyl-(E,E)-3,5-heptadien-2-one, β-ionone, cis-jasmone, trans- β-ionone-5,6-epoxide, methyl jasmine.
9. **Aliphatic acids:** Butanoic acid, octanoic acid, decanoic acid.
10. **Aromatic compounds:** Benzaldehyde, methyl benzoate, phenyl acetaldehyde, 1-methoxy-4-(2-propenyl) benzene, methyl salicylate, p-methylacetophenone, cuminaldehyde, anethol, safrole, benzyl alcohol, phenethyl alcohol, methyl cinnamate, methyl eugenol, α,α-dimethylphenylethyl alcohol, anisaldehyde, trans-cinnamaldehyde, methyl cinnamate, p-cresol, ethyl cinnnamate, eugenol, 2-isopropyl-5-methylphenol (thymol), 2-isopropyl-2-methylphenol (carvacrol), 5-isopropyl-3-methylphenol, 4-allylphenol, dillaiole, p-methoxycinnamaldehyde.
11. **Miscellaneous compounds:** 2,6-Dimethylpyrazine, c-butyrolactone, myristicin (Lee *et al.*, 2005).

The genetic variability within the species is responsible for the different types of basil aroma such as lemon, camphor, cinnamon, clove, mint, licorice, rose, anise etc. It is therefore important to identify the specific cultivars before *O. basilicum* is utilized for different end uses.

Cultivation

Soil and Climate Requirements (Pushpangadan and Bradu, 1995)

Ocimum species thrive well on a variety of soils and climatic conditions. Rich loam to poor laterite, saline and alkaline to moderately acidic soils are very suited for cultivation of *Ocimum* species. Well drained soil helps in better vegetative growth. Basil flourishes well under fairly to high rain fall and humid conditions. Long days and high temperature have been found favourable for plant growth and higher oil production. Tropical and subtropical climates are suited for cultivation.

Seed / Propagules

Since the *Ocimum* species, in general, are highly cross pollinated, a certain amount of heterozygosity is essential for vigours, high oil yield and better oil quality. These characters are mostly controlled by polygenes whose effect is mostly additive. The seeds are, therefore, likely to be deteriorated in future generations unless the selected lines are maintained, from which the polycross seed is produced and every time fresh seed is to be collected from the polycross lines. For fresh plantings, the growers have to take fresh seeds from the pedigree stock. Selected lines can be multiplied vegetatively by growing tender shoot tips. Good quality planting seeds can be obtained from reputed seed companies.

Nursery Practices

The plantations can be raised either by raising the seedlings in the nursery and then planting in the field or by direct sowing of seeds in the field.

Direct sowing: Experiments have shown that direct sowing is more efficient, economical and profitable, especially when the crop is to be raised from seeds when compared to raising separate nursery and then transplanting seedlings. The seeds (75-250 g/hectare) are mixed with dry sand to ensure an even distribution. Before sowing, the field is marked into long narrow furrows at a spacing of 50-60 cm by plough and the seed is sown in rows by hand or drilled. After the seeds are sown, the field is worked with light wooden plank to cover the seeds. The field must be flooded with water after 24h of sowing, if there is no rain.

The seeds germinate within 10-15 days. After 20-25 days, when the seedlings are sufficiently grown (15-20 cm high), first weeding, thinning/gap filling (depends on the situation), etc. if necessary, are done.

Nursery sowing: The plantation can be raised by raising seedlings in the nursery. Raised seed beds should be thoroughly prepared and well manured by addition of farmyard manure. The seeds should be sown in the nursery in the third week of February. About 200-300 g seeds are enough to raise the seedlings for planting one hectare land. After sowing the seeds in the nursery, a mixture of farmyard manure and soil is thinly spread over the seeds and irrigated with a sprinkler hose. The seeds take about 8-12 days to germinate and seedlings are ready for transplanting in about 6 weeks time. A spray of 2 per cent urea solution on the nursery plants 15 to 20 days before transplanting helps in getting very healthy plants for transplanting.

Transplanting: The seedlings at 4-5 leaf stage are ready for transplanting (6-7 weeks after sowing the seeds). Transplanting is generally started in the middle of April. This can be undertaken in the month of March if the seedlings are raised in the hot beds.

Spacing of 40 to 60 cm^2 is found suitable for *Ocimum* species. In *O. basilicum* the spacing of 60cm X60 cm was recommended by Singh *et al.,* 1970 for Assam and Gulati *et al.,* (1977) for Haldwani in UP, India.

Seedlings are established well by the time of second irrigation. It is proper time to get the gaps filled and replace the poor plants so that uniform stand is achieved. In summer, three irrigations per month are necessary whereas during the remaining period, it should be done as and when required except rainy season when no irrigation is required. About 12-15 irrigations are enough during the year.

Effect of manures and fertilizers: 20 to 25 kg of nitrogen and 10 to 15 kg P_2O_5 per hectare ensures good vegetative growth, herb and oil yield. Application of nitrogen up to 75 kg/ha has been found to be beneficial for optimum herb and oil yield.

Organic basil: The use of inorganic fertilizers should be avoided as far as possible since this may reduce the therapeutic property of basil. Organic materials such as farmyard manure, bio-gas slurry, compost, neem cake and other oil seed cakes, biofertilizers, green manure and cover crops can substitute the inorganic fertilizers. The use of organic manure will increase the organic matter of the soil, reducing bulk density and increasing the water holding capacity. These improve fertility of the soil, especially when soil is sandy in texture and low in organic matter. These should preferably be incorporated with the advent of summer as their decomposition rate increases with warm weather. Besides, use of organic manure is desired because they conserve nutrients against leaching losses and release them as a continual process. Organic manures also help in reducing soil erosion, while crop rotation and green manures involving legumes add to soil fertility. For control of pests, bioinsecticides such as tobacco extract, turmeric extract, garlic extract, garlic-neem extract, neem seed oil emulsion, neem seed kernel extract etc. can be used.

Basil as Therapeutic Agent

Basil in Traditional Medicine

In traditional medicine, infusions of *O. basilicum* are used to decrease plasma lipid content in some Mediterranean areas such as the Eastern Morocco (Zhang *et al.*, 2009). Basil has also been used as a medicinal plant in the treatment of diarrhea, constipation, warts, and worms (Simon *et al.*, 1999). Externally, basil can be used as an ointment for insect bites, and its oil is applied directly to the skin to treat acne (Waltz, 1996).

The leaves are used for inflammation, pain, bronchitis, colds, coughs, exhaustion, flatulence, flu, gout, insect bites, insect repellent, muscle aches, rheumatism and sinusitis. The juice expressed from the leaves is used against cold and cough. Aqueous extract of the seeds is used as a diuretic (Pushpangadan *et al.*, 1993).

The juice of leaves gives luster to eyes; good for toothache, ear-ache, headache and can be mixed with camphor to stop nasal haemorrhage. It also forms an excellent nostrum for the cure of ring worms, scorpion sting and snake bite. The washed and pounded seeds are used in poultices for unhealthy sores and sinuses. It is also used in sharbat in habitual constipation and in internal piles. The seeds are chewed in snake bite (Kirtikar and Basu, 1935). The juice of leaves warmed with honey is useful in the treatment of croup. The seeds steeped in water and eaten are said to be cooling and very nourishing. Its infusion is given in gonorrhea, diarrhea and chronic dysentery. The roots of plant are used for bowel complaints in children (Chopra *et al.*, 1956). A teaspoon full of seed infused in a tumbler of water with little sugar, when taken daily, acts as a demulcent in genito-urinary disease. A cold infusion of seeds is said to relieve after pains of child birth. An infusion of seed is also given in fever (Dastur, 1970). The Santhal tribe use sweet basil for headache, earache, cough, cold, inflammation, snake bite and rabies (Pushpangadan *et al.*, 1993).

In a palm leaf book entitled "Visha Retnamala" in Malayalam written by Raveendran Pillai of Kollamkode in 1882 a number of formulations containing *Ocimum basilicum*/ *O. tenuiflorum* for treatment of poisonous bites are given.

1. For snake bites, to slow down the spread of poison, it is recommended that 50 ml of the juice of Tulsi may be administered.

2. For snake bites 50 ml of Tulsi extract may be given internally and 2 drops each should be poured in to the eyes and ears. Further roots of Tulsi should be made in to a paste and applied on the wounded part.
3. To slow down the effect of poison 20 leaves of Tulsi and 10 fruit of Pepper may be administered in the form of a paste every 2 hrs.
4. If the patient looses consciousness after the snake bite it is recommended that a paste of Tulsi roots dispersed in human breast milk should be poured in the eyes and the nasal cavities.
5. For scorpion bites, equal amount of Tulsi leaves and Tulsi roots should be made in to a paste and applied on the affected part.
6. For scorpion bites Tulsi leaves may be eaten as such and a paste of leaves in cow's urine or lemon juice may be applied on the affected part.
7. For poisonous bites equal amount of Tulsi leaves, Tulsi flowers, Turmeric and *Boerhavia diffusa* should be made in to paste and applied on the affected part. More over 5-6 g of the paste should be administered to the patient thrice in day for seven days.

Basil in Ayurveda

Ocimum basilicum is described by the name barbari by Bhavamisra. According to Ayurveda the plant is used for diseases caused by aggravation of *Kapha* and *Vata* while the seeds are used for pacifying aggravation of *Vata* and *Pitta*. The medicinal properties of *O. basilicum* are described in the classical Ayurvedic texts such as Susruta Samhita, Charaka Samhita, Ashtangahridaya, Bhavaprakasham, Danwanthari Nighandu and Kaiyadeva Nighandu (Pushpangadan *et al*, 1993). In Ayurveda the whole plant is used for cough, asthma, bronchitis, ophthalmia, giddiness, intermittent and malarial fever, catarrh, otalgia, cephalalgia, dyspepsia and spasmodic affections (Udayan and Balachandran, 2009). The plant is stomachic, stimulant, carminative, diaphoretic, expectorant, diuretic, and antipyretic (Kirtikar and Basu, 1935).

Pharmacology of Basil

Basil for Heart Diseases

Hyperlipidaemia, atherosclerosis and related diseases are becoming a major health problem in developing countries. The alcoholic extract of *O. basilicum* exhibited a cardiotonic effect and the aqueous extract produced a ß-adrenergic effect in albino rats (Muralidharan and Dhananjayan, 2004). *O. basilicum* as medicinal plant is beneficial for cardiovascular system (Amrani *et al.*, 2009). *O. basilicum* may contain polar products able to lower plasma lipid concentrations and might be beneficial in preventing hyperlipidemia and related cardiovascular diseases (Harnafia *et al.*, 2009).

O. basilicum is one of the medicinal plants widely used in Morocco to reduce plasma cholesterol and to reduce the risk of atherosclerosis-related diseases. The hypolipidaemic effect exerted by *O. basilicum* extract was markedly stronger than the effect induced by fenofibrate treatments. Further it was demonstrated that *O. basilicum* aqueous extract displayed a very high antioxidant power. These results indicate that *O. basilicum* extract may contain hypolipidaemic and antioxidant substances and its use as a therapeutic tool in hyperlipidaemic subjects may be of benefit (Amrani *et al.*, 2006). Basil also exerts positive effects as a vasodilator and reduces plaque aggregation. Thus, it interferes positively in the mechanisms involved in the development of atherosclerotic disease (Benedec *et al.*, 2007; Amrani *et al.*, 2009).

Umar *et al.* (2011) investigated the effects of *O. basilicum* in renovascular hypertensive rats by evaluating blood pressure, heart weight/body weight, plasma angiotensin-II and endothelin. The study showed that *O. basilicum* lowered systolic and diastolic blood pressure, reduced cardiac hypertrophy and lower angiotensin and endothelin concentrations. Bora *et al.* (2011) showed that this plant can decrease brain infarct size and lipid peroxidation and suggested that it may be useful clinically to prevent strokes.

Basil as Anti-inflammatory Agent

Linolenic acid in the fixed oil of *O. basilicum* has the capacity to block both the cyclooxygenase and lipoxygenase pathways of arachidonate metabolism and could be responsible for the antiinflammatory activity of *O. basilicum* (Singh, 1998).

Fixed oil of *O. basilicum* was found to possess significant antiinflammatory activity against carrageenan and different other mediator-induced paw edema in rats. Significant inhibitory effect was also observed in castor oil-induced diarrhoea in rats. It also inhibited arachidonic acid- and leukotriene-induced paw edema. The results of antiinflammatory activity of *O. basilicum* support the dual inhibition of arachidonate metabolism as indicated by its activity in inflammation models that are insensitive to selective cyclooxygenase inhibitors. On the basis of these findings, it is possible to conclude that *O. basilicum* may be a useful antiinflammatory agent which block both cyclooxygenase and lipoxygenase pathways of arachidonic acid metabolism (Singh, 1999b).

Fixed oil of *O. basilicum* was found to possess significant antiulcer activity against aspirin, indomethacin, alcohol, histamine, reserpine, serotonin and stress-induced ulceration in experimental animal models. Significant inhibition was also observed in aspirin-induced gastric ulceration and secretion in pylorus ligated rats. The lipoxygenase inhibiting, histamine antagonistic and antisecretory effects of the oil could probably contribute towards antiulcer activity. *O. basilicum* fixed oil may be considered to be a drug of natural origin which possesses both antiinflammatory and anti-ulcer activity (Singh, 1999a).

Basil as Antioxidant (Pushpangadan and George, 2012)

Rosmarinic acid isolated from the leaves of *O. basilicum* was found to be responsible for its antioxidant activity. The nature of antioxidant activity of rosmarinic acid in the liposome system was examined. The results showed that one rosmarinic acid can capture 1.52 radicals, and furthermore, the existence of a synergistic effect between alpha-tocopherol and rosmarinic acid was revealed (Jayasinghe *et al.*, 2003). Acetone and ethanol extracts of *O. basilicum* at various concentrations (50, 100, 250 and 500 in µg/ml) showed antioxidant activities in a concentration dependent manner. However, ethanol extract of *O. basilicum* at the concentration of 500 μg/ml showed 75.87% activity, an antioxidant activity very close to that of 500 μg/ml of α-tocopherol (82.14%), the reference compound. It has been observed that the extract exhibited strong activity with the increase in polar solvent, indicating that polyphenols, flavanone, and flavanoids may play important roles in the activities (Durga *et al.*, 2009). The ethanolic extract of the leaves of *O. basilicum* showed significant antilipid peroxidation effects *in vitro*, besides exhibiting significant activity in super oxide radical and nitric oxide radical scavenging in goat liver (Meera *et al.*, 2009).

Basil as Antimicrobial Agent (Pushpangadan and George, 2012)

Basil showed a strong inhibitory effect against multidrug resistant clinical isolates from the genera *Staphylococcus, Enterococcus* and *Pseudomonas* (Opalchenova and Obreshkova, 2003). Ethanol, methanol, and hexane extracts from *O. basilicum* were investigated for their *in vitro* antimicrobial properties. The result showed that none of the three extracts tested have antifungal activities, but anticandidal and antibacterial effects. Both the hexane and methanol extracts, but not the ethanol extracts, inhibited three isolates out of 23 strains of *Candida albicans* studied. The hexane extract showed a stronger and broader spectrum of antibacterial activity, followed by the methanol and ethanol extracts, which inhibited 10, 9 and 6% of the 146 bacterial strains tested, respectively (Adiguzel *et al.*, 2005). Extracts and purified components of *O. basilicum* were used to identify possible antiviral activities against DNA viruses (herpes viruses (HSV), adenoviruses (ADV) and hepatitis B virus and RNA viruses (coxsackievirus B1 (CVB1) and enterovirus 71 (EV71)). The results showed that crude aqueous and ethanolic extracts of *O. basilicum* and selected purified components, namely apigenin, linalool and ursolic acid, exhibit a broad spectrum of antiviral activity. Of these compounds, ursolic acid showed the strongest activity against HSV-1, ADV-8, CVB1 and EV71, whereas apigenin showed the highest activity against HSV-2, ADV-3, hepatitis B surface antigen and hepatitis B e antigen and linalool showed strongest activity against ADV-II. No activity was noted for carvone, cineole, b-caryophyllene, farnesol, fenchone, geraniol, b-myrcene and a-thujone (Chiang *et al.*, 2005). The essential oil from *O. basilicum* and its purified compounds, especially linalool, have antigiardial activity. Linalool (300 μg/ml), was able to kill 100% parasites after 1 h of incubation, which demonstrates its high antigiardial potential (de Almeida *et al.*, 2007). The antimicrobial activities of chloroform, acetone and two different concentrations of methanol extracts of *O. basilicum* L. were studied. These extracts were tested *in vitro* against 10 bacteria and 4 yeast strains by the disc diffusion method. The results indicated that the methanol extracts of *O. basilicum* exhibited antimicrobial activity against tested microorganisms. While the chloroform and acetone extracts had no effect, the methanol extracts showed inhibition zones against strains of *Pseudomonas aeruginosa, Shigella* sp., *Listeria monocytogenes, Staphylococcus aureus* and two different strains of *Escherichia coli* (Kaya *et al.*, 2008). The volatile oils of *O. basilicum* inhibited the growth of *Klebsiella pneumoniae* at a concentration of 0.51% in the agar; *Streptococcus viridians* and *Staphylococcus albus* at 1.10% and *Pseudomonas aeruginosa* at 10.0%. *Proteus vulgaris* was inhibited at 0.67% by *O. basilicum* in dental isolates (Ahonkhai *et al.*, 2009). The essential oils (10 μL/disc of 1:5, v/v dilution with methanol) and methanol extracts (300 μg/disc) of *O. basilicum* displayed antibacterial activity against *Bacillius cereus, B. subtilis, B. megaterium, Staphylococcus aureus, Listeria monocytogenes, Escherichia coli, Shigella boydii, S. dysenteriae, Vibrio parahaemolyticus, V. mimicus,* and *Salmonella typhi* with their respective zones of inhibition of 11.2-21.1 mm and MIC values of 62.5-500 µg/mL (Hossain *et al.*, 2010). Basil oil had the strongest antimicrobial activity against *Salmonella enteritidis* SE3. Gas chromatography/mass spectrometry analysis revealed that the major constituents of the oil were linalool (64.35%), 1,8-cineole (12.28%), eugenol (3.21%), germacrene D (2.07%), alpha-terpineol (1.64%), and p-cymene (1.03%). When applied in *nham*, a fermented pork sausage, experimentally inoculated with *S. enteritidis* SE3 and stored at 4 °C, basil oil inhibited the bacterium in a dose-dependent fashion. Basil oil at a concentration of 50 ppm reduced the number of bacteria in the food from 5 to 2 log cfu/g after storage for 3 days. An unmeasurable level of the bacterium in the food was observed at days 2 and 3 of storage when 100 and 150 ppm of basil oil was used, respectively (Rattanachaikunsopon and Phumkhachorn, 2010). Sweet basil oil and two of its major constituents (linalool and eugenol), were tested against *Sclerotinia sclerotiorum* (Lib.), *Rhizopus stolonifer* (Ehrenb. exFr.) Vuill and *Mucor* sp. (Fisher) in a closed system. These fungi cause deterioration and heavy decay of peach fruit during marketing, shipping and storage. In the case of basil oil, linalool alone showed a moderate antifungal activity while eugenol showed no activity at all. Mixing the two components in a ratio similar to their concentrations in the original oil was found to enhance the antifungal properties of basil oil indicating a synergistic effect (Edris and Farrag, 2003).

Basil as Anti-nociceptive Agent

Venâncio *et al.*, (2011) reported the antinociceptive effects of essential oil of *O. basilicum* in Swiss mice. The authors suggest that the peripheral and central antinociceptive effects may be associated to inhibition of the biosynthesis of pain mediators such as prostaglandins and prostacyclins.

Basil as Hepatoprotective Agent

Galila *et al.*, (2012) reported hepatoprotective effect of basil on CCl_4-induced liver fibrosis in rats. Basil treatment significantly reduced the liver content of hydroxyproline and significantly increased the activity of hyaluronidase (HAase). The hepatic activity of superoxide dismutase (SOD) was stimulated while the lipid peroxidation was significantly reduced by the effect of basil extract. CCl_4 induced increase in the activities of transaminases [aspartate aminotransferase (AST), alanine aminotransferase (ALT)], and alkaline phosphatase (ALP) were significantly decreased by basil extract. The higher levels of serum urea and creatinine in CCl_4 group were significantly guarded by the protection of basil. Meera *et al.*, (2009) reported that significant hepatoprotective effects were obtained by ethanolic extract of leaves of *O. basilicum* against liver damage induced by H_2O_2 and CCl_4 as evidenced by decreased levels of antioxidant enzymes (enzymatic and non enzymatic).

Basil as Larvicidal Agent (Pushpangadan and George, 2012)

The larvicidal effect of the crude CCl_4, methanol and petroleum ether leaf extracts of *O. basilicum* against *Anopheles stephensi* and *Culex quinquefasciatus* was evaluated. Petroleum ether extract was found to be most effective against the larvae of both mosquitoes, with LC_{50} values of 8.29, 4.57; 87.68, 47.25 ppm and LC_{90} values of 10.06, 6.06;129.32,65.58 against *A. stephensi* and *C. quinquefasciatus* being observed after 24 and 48 h of treatment respectively. These extracts are highly toxic against mosquito larvae from a range of species (Prejwltta *et al.*, 2009). The larval toxicity and smoke repellent potential of *Ocimum basilicum* Linn. at different concentration (2, 4, 6, 8 and 10%) against the different instar (I, II, III and IV) larvae and pupae of *Aedes aegypti* were evaluated. The LC_{50} values of *O. basilicum* for I instar larvae were 3.734, II instar 4.154, III instar 4.664, IV instar 5.124 (Murugan *et al.*, 2007). Laboratory investigation using the plants such as, *Vetiveria zizanioides, O. basilicum* and the microbial pesticide spinosad against the malarial vector *Anopheles stephensi* Liston showed 85% mortality. The observed mortality rate suggests the above extract can be used as biopesticides. The LC_{50} of second, third and fourth instar larvae of *A. stephensi* were 0.276%, 0.285% and 0.305%, respectively (Aarthi and Murugan, 2010). The direct toxicity of the essential oil, *O. basilicum* L. to females of six species of predacious mites of the family phytoseiidae was tested. The phytoseiid mites tested were *Typhlodromus athiasae* Porath and Swirski, *Euseius yousefi* Zaher and El-Borolossy, *Amblyseius zaheri* Yousef and El-Borolossy, *Amblyseius deleoni* (Muma and Denmark), *Amblyseius swirskii* Athias-Henriot and *Amblyseius barkeri* (Hughes). Sweet basil oil was highly toxic to females *E. yousefi* and was relatively intoxic to females A. swirskii. The essential oil has a close toxic effect for predator species, *T. athiasae* and *A. barkeri*. With the exception of *A. zaheri*, females of all predacious mites tested suffered a depression in reproduction and food consumption when treated with sweet basil oil at conc. 2% (Momen and Ame, 2003).

Basil as Flavouring Agent

O. basilicum is commercially cultivated for its green and aromatic leaves which are used dry or fresh as a condiment. Because of a growing preference in the society for naturally derived food flavours and additives, basil is widely used nowadays.

Leaves are used for flavouring purposes (Niir Board, 2005). Sweet basil is used in soups, meat pies, fish, certain cheeses, tomato cocktail, cooked cucumber dishes, cooked peas, squash, and string beans. Chopped basil is sprinkled over lamb chops before cooking. Basil is often used with or as a substitute for oregano in pizza topping, spaghetti sauce, or macaroni and cheese casserole. Basil is an important seasoning in tomato paste products in Italy.

Although not used in large quantities, the oil of sweet basil is employed quite extensively in all kinds of flavours, including those for confectionary, baked goods and in spiced meats, sausages etc. The oil serves also for imparting distinction to flavours in certain dental and oral products.

Sweet basil is used in seasonings of canned spaghetti sauces, meat balls and salads. Sweet basil or its extract is used in combination with other spices, in flavouring confections, baked goods, puddings, condiments, vinegars, ice creams, nonalcoholic beverages, liquors and perfumes (Prakash, 1990).

The essential oil of *Ocimum basilicum* obtained by distillation is used in perfumery, production of aroma and in the food industry as a flavouring agent. It contains cineol, pinene, methyl chavicol, d-camphor and ocimene (Eltohami, 1997).

The major aroma constituents of basil are 3,7- dimethyl-1,6-octadien-3-ol (linalool; 3.94 mg/g), 1-allyl-4methoxy benzene (estragole; 2.03 mg/g), methyl cinnamate (1.28 mg/g), 4-allyl-2-methoxyphenol (eugenol; 0.896 mg/g), and 1,8-cineole (0.288 mg/g) (Lee *et al.,* 2005). Fresh leaves are used in salads. Fresh leaves are also used to flavour vinegar and oils.

Conclusion

O. basilicum has attracted the attention of mankind because of its aroma and medicinal properties. The plant, originally a native of India is now found in all the continents. Recent research finding have confirmed the antibacterial, antifungal, lavicidal, hepatoprotective, anti-diabetic, antiprolifereative, antioxidant, anti-inflammatory, cardioprotective and antinociceptive properties of this plant justifying its use in traditional medicine. Further research on *O basilicum* may lead to the discovery of new bioactive molecules which may lead to the development of novel therapeutic agents.

Acknowledgements

The authors express their sincere thanks to Dr Ashok K Chauhan, Founder President, RBEF and Chairman Amity Group of Institutions and also to Dr Atul Chauhan, President, RBEF for providing necessary facilities and encouragement.

References

Aarthi N and Murugan K 2010. Larvicidal and repellent activity of *Vetiveria zizanioides* L, *Ocimum basilicum* Linn. and the microbial pesticide spinosad against malarial vector, *Anopheles stephensi* Liston (Insecta: Diptera: Culicidae). *Journal of Biopesticides,* 3, 199 - 204.

Adiguzel A, Gulluce M, Sengul M, Ogútcu H, Sahin F and Karaman I 2005. Antimicrobial Effects of *Ocimum basilicum* (Labiatae) Extract. *Turk J Biol.,* 29, 155-160.

Ahonkhai I, Ayinde BA, Edogun O and Uhuwmangho MU 2009. Antimicrobial activities of the volatile oils of *Ocimum basilicum* L. and *Ocimum gratissimum* L. (Lamiaceae) against some aerobic dental isolates. *Pak. J. Pharm. Sci.,* 22, 405-409.

Amrani S, Harnafi H, Bouanani Nel H, Aziz M, Caid HS, Manfredini S, Besco E, Napolitano M and Bravo E 2006. Hypolipidaemic activity of aqueous *Ocimum basilicum* extract in acute hyperlipidaemia induced by triton WR-1339 in rats and its antioxidant property. *Phytother Res.,* 20, 1040-5.

Amrani S, Harnafi H, Gadi D, Mekhfi H, Legssyer A, Aziz M, Martin-Nizard F, and Bosca L 2009. Vasorelaxant and anti-platelet aggregation effects of aqueous *Ocimum basilicum* extract. *J Ethnopharmacol.,* 125(1), 157-62.

Anonymous 1966, Council of Scientific and Industrial Research (CSIR). Wealth of India-Raw materials, Vol.7, Publication Directorate CSIR, New Delhi, 79-89.

Ayurnepal.com available from http://www.ayurnepal.com/en/barbari.html, [Accessed 1 December, 2010]

Baby S and George V 2009. Essential Oils and New Antimicrobial Strategies *In*: Ahmad, I and Aqil, F (Eds). New Strategies Combating Bacterial Infection, WILEY-VCH Verlag GmbH & Co. KGaA, Weinheim, pp. 165- 203.

Benedec D, Pârvu AE, Oniga I, Toiu A, Tiperciuc B. 2007. Effects of *Ocimum basilicum* L. extract on experimental acute inflammation, *Rev Med Chir Soc Med Nat Iasi.*, 111, 1065-1069.

Bonar, A.M. 1985. The Macmillan Treasury of herbs: A complete guide to the cultivation and use of wild and domesticated herbs, Macmillan, New York, NY.

Chiang LC, Ng LT, Cheng PW, Chiang W and Lin CC 2005. Antiviral activities of extracts and selected pure constituents of *Ocimum basilicum*. *Clin Exp Pharmacol Physiol.*, 32, 811-6.

Chopra RN, Nayar SL and Chopra IC 1956. Glossary of Indian Medicinal Plants, CSIR, New Delhi.

Christman S 2010. *Ocimum basilicum*, Floridata, Florida, USA Available from: http://www.floridata.com/ref/o/ocim_bas.cfm, [Accessed 1 December, 2010]

Dastur JF 1970. Medicinal plants of India and Pakistan. D. B. Taraporewala Sons Co. Pvt. Ltd. Bombay.

De Almeida I, Alviano DS, Vieira DP, Alves PB, Blank AF, Lopes AH, Alviano CS, Rosa Mdo S 2007. Antigiardial activity of *Ocimum basilicum* essential oil. *Parasitology Research*, 101, 443-452.

Durga KR, Karthikumar S and Jegatheesan K 2009. Isolation of potential antibacterial and antioxidant compounds from *Acalypha indica* and *Ocimum basilicum*, *J. Medicinal Plants Research*, 3, 703-706.

Edris AE and Farrag ES 2003. Antifungal activity of peppermint and sweet basil essential oils and their major aroma constituents on some plant pathogenic fungi from the vapor phase, Nahrung, 47,117-21.

Eltohami MS 1997. Medicinal and aromatic plants in Sudan, Medicinal, culinary and aromatic plants in the near east. Available from: http://www.fao.org/docrep/x5402e/x5402e16.htm

Farrell K 1985, Spices, Condiments and Seasonings, AVI Publ., Westport, CT.

Galila AY, Nihal ME and Eman FEA 2012. Hepatoprotective effect of basil (*Ocimum basilicum* L.) on CCl4-induced liver fibrosis in rats, *African Journal of Biotechnology*, 11(90), 15702-15711.

Guenther E 1949. The essential oils, Vol 3, Van Norstrand, New York, 395-433, 519-761.

Gulati BC, Duhan SP, Gupta R and Bhattacharya AK 1977, Introduction to French basil (*Ocimum basilicum* L.) in Tarai of Nainital, U.P. (India), Perfumeric and Kosmetics, 58, 165-169.

Harnafia H, Azizb M and Amrania S 2009. Sweet basil (*Ocimum basilicum* L.) improves lipid metabolism in hypercholesterolemic rats, *The European e-Journal of Clinical Nutrition and Metabolism*. 4, e181-186.

Holistic online.com, available from http://www.holistic-online.com/herbal-med/_Herbs/h15.htm, [Accessed 1December, 2010]

Hossain MA, Kabir MJ, Salehuddin SM, Mizanur Rahman SM, Das AK, Singha SK, Alam Md K and Rahman A 2010. Antibacterial properties of essential oils and methanol extracts of sweet basil *Ocimum basilicum* occurring in Bangladesh. *Pharm Biol.*, 48,504-11.

Jayasinghe C, Gotoh N, Aoki T and Wada S 2003. Phenolics composition and antioxidant activity of sweet basil (*Ocimum basilicum* L.). *J Agric Food Chem.*, 51, 4442-9.

Ji-Wen Z, Sheng-Kun L, and Wen-Jun W 2009. The Main Chemical Composition and in vitro Antifungal Activity of the Essential Oils of *Ocimum basilicum* Linn. var. *pilosum* (Willd.) Benth, *Molecules*, 14,273-278.

Kaya I, Yigit N and Benli M 2008. Antimicrobial activity of various extracts of *Ocimum basilicum* L. and observation of the inhibition effect on bacterial cells by use of scanning electron microscopy. *Afr J Tradit Complement Altern Med.*,18, 363-369.

Kirtikar KR and Basu BD 1935. Indian Medicinal Plants. Vol. III 2nd ed. Lalit Mohan Basu, Allahabad. 1959-1968.

Klimankova E, Holadova K, Hajslova J, Cajka T, Poustka J and Koudela M. 2008. Aromaprofiles of five basil (*Ocimum basilicum* L.) cultivars grown under conventional and organic conditions, *Food Chem*, 107, 464-472.

Koba K, Poutouli PW, Christine R, Jean-Pierre C and Komla S 2009. Chemical composition and antimicrobial properties of different basil essential oils chemotypes from Togo, *Bangladesh J Pharmacol*, 4, 1-8.

Lee Seung-Joo, Umano K, Shibamoto T and Lee Kwang-Geun 2005. Identification of volatile components in basil (*Ocimum basilicum* L.) and thyme leaves (*Thymus vulgaris* L.) and their antioxidant properties, *Food Chemistry*, 91,131–137.

Meera R, Devi P, Kameswari B, Madhumitha B and Merlin NJ 2009. Antioxidant and hepatoprotective activities of *Ocimum basilicum* Linn. and *Trigonella foenum-graecum* Linn. Against H_2O_2 and CCl_4 induced hepatotoxicity in goat liver. *Indian J Exp. Biol.*, 47, 584-590.

Momen FM and Ame SAA 2003. Influence of the Sweet Basil *Ocimum basilicum* L. on Some Predacious Mites of the Family Phytoseiidae (Acari: Phytoseiidae). *Acta Phytopathologica et Entomologica Hungarica*, 38, 137-143.

Muralidharan A and Dhananjayan R 2004. Cardiac stimulant activity of *Ocimum basilicum* Linn. Extracts, *Indian J Pharmacol*, 36, 163-166.

Murugan K, Murugan P and Noortheen A 2007. Larvicidal and repellent potential of *Albizzia amara* Boivin and *Ocimum basilicum* Linn. against dengue vector, *Aedes aegypti* (Insecta:Diptera:Culicidae), *Bioresour Technol*, 98, 198-201.

NIIR Board 2005. Compendium of medicinal plants, National Institute of Industrial Research, New Delhi, p.392.

Opalchenova G and Obreshkova D 2003. Comparative studies on the activity of basil-an essential oil from *Ocimum basilicum* L.-against multidrug resistant clinical isolates of the genera *Staphylococcus, Enterococcus* and *Pseudomonas* by using different test methods. *J Microbiol Methods,* 54, 105-10.

Prakash V 1990, Leafy Spices, CRC Press, Boca Raton, Ann Arbor, Boston.

Prejwltta M, Preeti S, Lalit M, Lata B and Srivastava CN 2009, Evaluation of the toxicity of different phytoextracts of *Ocimum basilicum* against *Anopheles stephensi* and *Culex quinquefasciatus. J. Asia-Pacific Entom,* 12,113-115.

Pruthi JS 1976, Spices and Condiments, National Book Trust, New Delhi.

Pushpangadan P and Bradu BL 1995. Basil, (In:) Chadha KL and Gupta R, Advances in Horticulture,Vol.11-Medicinal and Aromatic Plants, Malhotra Publishing House, New Delhi- 110 064.

Pushpangadan P and George V 2012, Basil, (In:) Peter, K. V. (Ed.). Handbook of Herbs and Spices, Vol.1, p.55-72, Woodhead Publishing, Oxford, Cambridge, Philadelphia, New Delhi.

Pushpangadan P, Rajasekharan S and Biju SD 1993. Tulasi, Tropical Botanic Garden and Research Institute, Thiruvananthapuram.

Rattanachaikunsopon P and Phumkhachorn P 2010. Antimicrobial Activity of Basil (*Ocimum basilicum*) Oil against *Salmonella Enteritidis in vitro* and in *Food, Biosci Biotechnol Biochem.,* 74, 1200-4.

Simon JE 1998, Basil. Centre for New Crops and Plant Products, Purdue University. Available from :http://www.hort.purdue.edu/newcrop/CropFactSheets/ basil.html#RTFToC2, [Accessed 1 December, 2010]

Simon JE, Morales MR, Phippen WB, Vieira RF and Hao Z 1999. A source of aroma compounds and a popular culinary and ornamental herb. (In:) (Janick, J. Ed.). Perspectives on new crops and new uses. Alexandria, VA: ASHS Press. 499–505.

Singh S 1998. Comparative evaluation of antiinflammatory potential of fixed oil of different species of *Ocimum* and its possible mechanism of action. *Indian J Exp Biol.,* 36,1028-31.

Singh S 1999a. Evaluation of gastric anti-ulcer activity of fixed oil of *Ocimum basilicum* Linn. and its possible mechanism of action. *Indian J Exp Biol.,* 37, 253-7.

Singh S 1999b. Mechanism of action of antiinflammatory effect of fixed oil of *Ocimum basilicum* Linn. *Indian J Exp Biol.* 37,248-52.

Singh TJ, Gupta PD, Khan SY and Misra KC 1970. Preliminary pharamacological investigation of *Ocimum sanctum* Linn. *Indian J Pharm.,*32, 92-94.

Udayan PS and Balachandran I 2009. Medicinal plants of Arya Vaidya Sala Herb Garden, p. 362. Centre for Medicinal Plants Research, Arya Vaidya Sala, Kottakkal, Kerala.

Venâncio AM, Onofre AS, Lira AF, Alves PB, Blank AF, Antoniolli AR, Marchioro M, Estevam CS and de Araujo BS. 2011, Chemical composition, acute toxicity, and antinociceptive activity of the essential oil of a plant breeding cultivar of basil (*Ocimum basilicum* L.). *Planta Med.*, 77, 825-829.

Waltz L 1996. The Herbal Encyclopedia; http://www.wic.net/waltzark/ herbenc.htm.

Zhang J, Li S and Wu W 2009. The Main Chemical Composition and in vitro Antifungal Activity of the Essential Oils of *Ocimum basilicum* Linn. var. *pilosum* (Willd.) Benth., *Molecules.* 14, 273-278.

5

Bay Laurel

Sanchita, Ashok Sharma

Laurus nobilis, a member of family Lauraceae is native to the shores of the Mediterranean sea. The Mediterranean region consists of forest, woodlands and scrubs vegetation covering the portions of three continents, Europe, Asia and Africa (Sharma *et al.*, 2012). The Mediterranean basin is surrounded by huge laurel forests. It is also known as sweet bay, bay laurel, Grecian laurel, true bay and bay. Theses are evergreen ornamental plants with medicinal properties. These plants have been admired for its beauty and aromatic leaves since Greek and Roman times. The plants are upright, with the dense canopy becoming more rounded at maturity, the lower limbs tend to be retained allowing the tree to remain dense in the lower canopy. Once established, trees may spread to form colonies, the overall texture being medium-coarse to coarse. Although these species are slow to moderately slow growing but not difficult to cultivate. The cold hardiness and intolerance to poorly drained soils are the primary cultural concerns. It grows the best in full sun with afternoon shade, but can tolerate full to partial sun and is tolerant of heat and at least some soil salts. Although plants survive moderate drought, when drought stressed the leaves tend to develop marginal necrosis. These species perform the best on acidic soils and develop chlorosis on the new leaves in high pH sites. The primary landscape asset of these species is its handsome dark glossy evergreen foliage. The flowers are mildly attractive upon close inspection and the foliage is aromatic if crushed. Bay Laurel is a classic staple of the herb or cottage garden, but is also versatile enough to serve as a patio container plant, large evergreen screen or large topiary. It is excellent for Mediterranean or heritage, educational or historical gardens. *Laurus nobilis* is also important as a culinary herb and essential oils are extracted from it for use in perfumery and medicines.

Cytotaxonomic Background

Taxonomical classification

Kingdom:	Plantae
Division:	Magnolids
Class:	Magnoliopsida
Order:	Laurales
Family:	Lauraceae
Genus:	Laurus
Species:	*Laurus nobilis*

The family Lauraceae comprises about 50 genera and 3000 species. Different ploidy levels (2n=36, 42, 48,54, 60, 66, 72) have been reported in *Laurus* (Ehrendorfer *et al.*, 1968) with tetraploidy ($2n = 4x = 48$) being the most frequent karyotype. Thus the chromosome number in most of the laurel is 2n=48. Out of 48, 40 are metacentric and 8 submetacentric chromosomes. The mean chromosome length is 4.01±0.1 µm (Todua; 1984). The plants are unisexual (dioecious) with male and female flowers on different plants. The individual flowers are small with white petals and males with yellow stamens lending an overall creamy white to yellow-white color to the flower clusters. Fruits are roundish black berries that mature in fall. *L. nobilis* shows entomophilous pollination and fleshy fruited seeds are dispersed by birds (Hampe *et al.*, 2003). *L. nobilis* has reported to show the process of vascularization of their carpels. The monocarpellary gynoecium has arisen from multicarpellary codition. The additional number of bundles represents the missing carpels (Sastri; 2006). The stems are medium to stout, stems are bright green and remain so for an extended time, eventually becoming splotched with gray and maturing to a gray-brown color. The buds are divergent; pointed to conical; prominently stalked; initially green, then maturing to a light to medium brown. *L. nobilis* is an upright oval to rounded large shrub or small tree with multiple trunks, reaching 10N to 15N (20N) tall.

Early History

During the middle ages, the branches of laurel tree were traditionally woven into garlands and wreaths to honor an accomplishment or victory. This ceremony is the origin of the term "poet laureate." University undergraduates are known as "bachelors" from the Latin baccalaureus, meaning laurel berry. They were forbidden to marry because it was believed that this would distract them from their studies. By extension of this idea, all unmarried men are referred to as bachelors (Hora; 1981). The laurel crowns adorned ancient Greek and Roman generals and athletes in victory parades. In ancient Rome, it was sacred to the God Apollo and was worn as a 'laurel wreath' by emperors and poets. The Romans believed that a person standing under a laurel tree would be shielded from infection by plague and also from lightning. Laurel wreaths may be worn by healers during healing ceremonies and while treating the sick in order to increase the positive healing energy and protect against negative energy that may be hanging around the sick room. Bay leaf can also be burned in the sick room after the illness has passed to purify it and drive out any residual sickness vibes. Romans called it the plant of good angels. During the middle ages, laurel was believed to provide protection against both lightning and witches (Lust; 1990). From the Mio-Pliocene era only Laurel tree has persisted to the present in Southern Eurasia, despite of other genera (Neolitsea, Lindera, Persea, Cinammomum and others) (Mai; 1989). After considerable range reductions throughout the Neogene, its current distribution is limited to relatively mesic areas in the Mediterranean Basin, the Pontic region (southern Black Sea) and the Macaronesian archipelagos (Santos; 1990). Given its long-standing presence, Laurus represents an excellent model for exploring the evolutionary history of ancient Mediterranean–Macaronesian lineages. Indeed, several authors have emphasized the need of molecular studies involving extant Lauraceae in order to ascertain the biogeographical origin of the Macaronesian laurel forests (Comes; 2004 and Emerson; 2002). Of the four genera of Lauraceae currently inhabiting Macaronesia, namely Apollonias, Ocotea, Persea and Laurus, the latter is the best suited with regard to testing Mediterranean–Macaronesian biogeographical connections, as it is the only one still persisting in the Mediterranean basin (Sánchez; 2009). The bay tree has a long history of folk use in the treatment of many ailments, particularly as an aid to digestion and in the treatment of bronchitis and influenza. It has also been used to treat various types of cancer. Traditionally, the leaf of *L. nobilis* has been used as herbal medicine to treat rheumatism, earaches, indigestion, sprains, and to promote perspiration. Various pharmacological activities such as wound healing activity, neuroprotective activity, antioxidant activity, antiulcerogenic activity, anticonvulsant activity, analgesic and anti-inflammatory, antimutagenic activity, immunostimulant activity, antiviral activity, anticholinergic activity, insect repellent activity, antibacterial activity, antifungal activity and acaricidal activity have been reported (Mansuriya and Patil; 2012).

Recent History

Recent studies have shown a number of new uses of bay laurel as source of flavouring and therapeutic molecules. The essential oils from laurel leaf is generally used in the flavouring industry and also used for the preparation of hair lotion due to its antidandruff activity and for the external treatment of psoriasis (Garbe and Surburg; 1997). The perfumed soaps and candles are manufactured from fruits using its fatty acid content (Hafizoglu and Reunanen; 1993). Leaves, fruits and essential oils of *L. nobilis* have rich medicinal property. The essential oil from laurel also has various antimicrobial activities such as antibacterial and antifungal (Kalemba and Kunicka; 2003). The content of essential oil are shown to be influenced by the area of culture, variety and harvest season (Flaminia *et al.*, 2007). It has been reported that the chemical composition of essential oil of leaves, stem and fruits are different from each other to some extent (Fiorini *et al.* 1997). The content of essntial oil in tissues as well as in plants from different geographical regions varies quantitatively as well as qualitatively (Sangun *et al.*, 2007). The fruits and leaves are not usually administered internally, other than as a stimulant in veterinary practice, but were formerly employed in the treatment of hysteria, amenorrhoea, flatulent colic *etc.* The leaves are also used to treat upper respiratory tract disorders and to ease arthritic aches and pains. The leaves are antiseptic, aromatic, astringent, carminative, diaphoretic, digestive, diuretic, emetic in large doses, emmenagogue, narcotic, parasiticide, stimulant and stomachic. The fruit is antiseptic, aromatic, digestive, narcotic and stimulant. An infusion has been used to improve the appetite and as an emmenagogue. The fruit has also been used in making carminative medicines and was used in the past to promote abortion. A fixed oil from the fruit is used externally to treat sprains, bruises etc, and is sometimes used as ear drops to relieve pain. The essential oil from the leaves has narcotic, antibacterial and fungicidal properties. However, not much work has been done on the genetic improvement of this plant. On the basis of microsatellite based genetic diversity investigation, *L. nobilis* can be separated into two main gene pools, one from western (Tunisia, Algeria and France) and the other from eastern Mediterranean (Turkey). The Algerian, Tunisian and French populations have shown a strong genetic similarity, as compared to the North African laurel populations are recently introduced from north-western Mediterranean stock (Marzouki *et al.*, 2007)). In Turkey, *L. nobilis* is widely distributed, where it is considered the most representative coastal tree. It is a component in coastal maquis, forming dense bushes, and is scattered in the understorey of *Pinus brutia* forests, on rocky slopes and in damp gorges. It is absent only from the mountains, above 800–900 m elevation. Trees with restricted root systems can be used for close planting to walls and homes without fear that the roots will damage foundations. The broad-leaved laurel trees that are good candidates for wall plantings (All about trees, 1982). The wood and bark of laurel have been extracted for different volatile compounds. Two different extraction methods, solvent extraction and simultaneous distillation/extraction (SDE) for wood and bark extraction, respectively. The monoterpenes like α- pinene, β-pinene and 1,8-cineol are the major constituents of bark extract on the contrary of sesquiterpenes in wood part (Kilic and Altuntas; 2006).

Future Prospects and Challenges

Bay laurel has an important place in the world trade because of its diverse uses. The dark green leaves of bay laurel are fragrant and aromatic. After drying, they are broken, cracked or cooked to release their characteristic aroma and flavour. Dried laurel leaves are used as flavouring in soups, fish, meats, stews, puddings, vinegars and beverages. Oil laurel leaves is an essential or volatile oil obtained by steam distillation of bay leaves and oleoresin has replaced the dry leaves in some food preparations. The oil of laurel reaches a content of 1-3 percent on a fresh weight basis. The main constituents of this oil are 1,8-cineole, pinene, sabinene, 1-linalool, eugenol, eugenol acetate, methyleugenol, 1-terpinol acetate, phellandrene, other esters and terpenoids. This oil is generally recognized as being safe for human consumption as a spice, natural flavouring and essential oil extract and is used by the cosmetics industry for creams, perfumes and soaps. Cultivation of bay laurel on commercial level faced a number of challenges which need to be addressed. Major being diseases, insects and pests, development of high yielding varieties and improved harvesting and post harvesting techniques. Different fungal and bacterial diseases such as crown and root rot (Polizzi *et al.*, 2012), leaf spot, wilting disease, powdry mildew (Fox; 2002), southern blight and stem blight are reported in bay laurel. Various pests and insects *Protopulvinaria pyriformis* (Stathas *et al.*, 2009), *Calepitrimerus russoi* (Starzewska; 2008), *Ceroplastes floridensis* (Jendoubi *et. al.*, 2011), *Ceroplastes ceriferus* (Pellizzari *et al.*, 2004), *Icerya purchase* (Watson and Malumphy; 2004), *Ceroplastes japonicas* (Jancar *et al.*, 1999), *Cecidophyopsis russoi* (Vinnik *et al.*, 1998) and *Trioza alacris* (Stanica *et al.*, 1993) are reported to harm *L. nobilis*. The innovative effort should be achieved to overcome these problems. The objectives should be made for the development of disease and pest resistant varieties. Bay leaves are collected from both cultivated and wild plants in many Mediterranean countries. Commercial production centres in the Mediterranean basin include portions of Algeria, France, Greece, Morocco, Turkey and Portugal and Spain (minor producers). Outside the Mediterranean Basin, bay leaves are produced in the Canary Islands, Central America, Mexico and the Southeastern United States (Simon *et al.*, 1984). In general, the variability of dioiceous plants are the main reason for its difficulty in sexual reproduction and seed germination. These problems could be overcome through conventional micropropagation, *in vitro* culture and mass clonal production. The result of application of these techniques in *L. nobilis* depend on the type of cuttings, the date of sampling and culture conditions. The most effective explants are the semiligneous. The presence of auxins in the media significantly increses the rate of rhizogenesis and permits a radial development of the roots (Souayah *et al.*, 2002). Thus the potential morphogenic capacity of *L. nobilis* can be optimized by searching the performing factors in each stage of the breeding technique: by cuttings (the date of sampling, the age of the mother tree, the type of the auxin and the concentrations), by micropropagation (optimizing the different stages) and by using other *in vitro* techniques, like somatic embryogenesis (Rodríguez *et al.*, 1990). Fresh seeds harvested from the tree are best sown as soon as possible. The leaves are also harvested and pruned branches for future use. For post harvesting process, different methods are applied. Drying is one of the methods. Different drying treatments *viz.*, air-drying, oven-drying, freezing and freeze drying are applied on bay leaves for essential oil extraction. Parlak (2012) reported an incremental increase of 5% every year for the demand in world market. The increased demand may cause overexploitation of natural resources. Efforts are required to provide sandard bay laurel product and development of varieties which may fetch higher market values (Parlak and Cuming; 2012).

References

A. Sharma, J. Singh and S. Kumar, 2012. Bay leaves, (In:) K.V. Peter (Ed.) Handbook of herbs and spices, Woodhead Publishing Limited.

F. Ehrendorfer, F. Krendl, E. Habeler and W. Sauer, 1968. Chromosome numbers and evolution in primitive angiosperms, *Taxon*, 17 337–353.

B.T. Todua, 1984. Karyological aspects of the formative process and phylogeny in laurel, Subtropičeskie Kul'tury, 6: 98–101.

A. Hampe, J. Arroyo, P. Jordano and R.J. Petit, 2003. Rangewide phylogeography of a bird-dispersed Eurasian shrub: contrasting Mediterranean and temperate glacial refugia, *Mol Ecol,* 12: 3415-3426.

R.L.N. Sastri, 2006. The vasgularization of the carpel in some ranales, *New Phytologist*, 58: 306-309.

B. Hora, 1981. The Oxford encyclopedia of trees of the world, Oxford University Press, Oxford.

J. Lust, 1990. The Herb Book, 659.

D.H. Mai, 1989. Development and regional differentiation of the European vegetation during the Tertiary, Plant Systematics and Evolution, 162: 79–91.

A. Santos, 1990. Evergreen forests in the Macaronesian region, Council of Europe, Strasbourg, France, 78.

H.P. Comes, 2004. The Mediterranean region – a hotspot for plant biogeographic research, *New Phytologist*, 164: 11–14.

B.C. Emerson, 2002. Evolution on oceanic islands: molecular phylogenetic approaches to understanding pattern and process, *Mol Ecol*, 11: 951-966.

F. Rodríguez-Sánchez, B. Guzmán, A. Valido, P. Vargas and J. Arroyo, 2009. Late Neogene history of the laurel tree (Laurus L., Lauraceae) based on phylogeographical analyses of Mediterranean and Macaronesian populations, *Journal of Biogeography*, 36: 1270–1281.

R. Patrakar, M. Mansuriya and P. Patil, 2012. Phytochemical and pharmacological review on *Laurus Nobilis*, *International Journal of Pharmaceutical and Chemical Sciences* 1: 595-602.

K. Bauer, D. Garbe and H. Surburg, 1997. Common fragrance and flavor materials : preparation, properties and uses, 3rd completely rev. ed., Wiley-VCH, Weinheim, Chichester.

H. Hafizoglu and M. Reunanen, 1993. Studies on the components of *Lauras nobilis* from Turkey with special reference to Laurel berry fat, *European Journal of Lipid Science and Technology*, 95: 304–308.

D. Kalemba and A. Kunicka, 2003. Antibacterial and antifungal properties of essential oils, *Curr Med Chem*, 10: 813-829.

G. Flaminia, M. Tebanoa, P.L. Cionia, L. Ceccarinib, A.S. Riccic and I. Longoc, 2007. Comparison between the conventional method of extraction of essential oil of *Laurus nobilis* L. and a novel method which uses microwaves applied *in situ*, without resorting to an oven, *J Chromatogr A*, 1143: 36–40.

C. Fiorini, Fourasté, I., David, B. and Bessière, J. M., 1997. Composition of the flower, leaf and stem essential oils from *Laurus nobilis*, *Flavour Frag J*, 12.

M.K. Sangun, E. Aydin, M. Timur, H. Karadeniz, M. Caliskan and A. Ozkan, 2007. Comparison of chemical composition of the essential oil of *Laurus nobilis* L. leaves and fruits from different regions of Hatay, Turkey, *Journal of Environmental Biology*/ Academy of Environmental Biology, India, 28: 731-733.

H. Marzouki, N. Nasri, B. Jouaud, C. Bonnet, A. Khaldi, S. Bouzid and B. Fady, 2009. Population genetic structure of *Laurus nobilis* L. inferred from transferred nuclear microsatellites, *Silvane Genetica*, 58: 5-6.

All about trees, Ortho Books, Chevron Chemical Company, San Francisco, 1982.

A. Kilic and E. Altuntas, 2006. Wood and bark volatile compounds of *Laurus nobilis* L., Holz als Roh- und Werkstoff, 64: 317–320.

G. Polizzi, A. Vitale, D. Aiello, V. Guarnaccia, P. Crous and L. Lombard, 2012. First report of *Calonectria ilicicola* causing a new disease on Laurus (*Laurus nobilis*) in Europe, *J Phytopathol*, 160: 41-44.

R.T.V. Fox, 2002. Fungal foes in your garden, *Mycologist*, 16: 132.

G.J. Stathas, P.A. Eliopoulos, G. Japoshvili and D.C. Kontodimas, 2009. Phenological and ecological aspects of *Protopulvinaria pyriformis* (Cockerell) (Hemiptera: Coccidae) in Greece, *J Pest Sci*, 82: 33-39.

J.C. Ostoja-Starzewska, 2008. Calepitrimerus russoi Di Stefano (acari: eriophyidae), found in Britain on imported bay laurel, *Journal of Entomology and Natural History*, 21: 195-200.

H. Jendoubi, P. Suma and A. Russo, 2011. First Records of Ceroplastes Floridensis Comstock (Hemiptera: Coccidae) and *Chrysomphalus Pinnulifer* (Maskell) (Hemiptera: Diaspididae) in Tunisia, *Entomol News*, 122: 247-249.

G. Pellizzari, G. Galbero, N. Mori and C. Antonucci, 2004. Biology of *Ceroplastes ceriferus* (Hemiptera, Coccidae) and trials of control, Informatore Fitopatologico, 54: 39-46.

G.W. Watson and C.P. Malumphy, 2004. *Icerya purchasi* Maskell, cotony cushion scale (Hemiptera: Margarodidae), causing damage to ornamental plants growing outdoors in London, *British Journal of Entomology and Natural History*, 17: 105-109.

M. Jancar, G. Seljak and I. Zezlina, 1999. Distribution of *Ceroplastes japonicus* Green in Slovenia and data of host plants, Slovenskega Posvetovanja o Varstvu Rastlin, 3: 443-449.

E. Vinnik, H. Casteels, F. Goossens and R. Clercq, 1998. Occurrence of eriophyid mites (Acari: Eriophyidae) on *Laurus nobilis* (L.) in Belgium, *Parasitica*, 54: 23-30.

F. Stanica, A. Standardi, D. Hoza and T.A. Tudor, 1993. Studies on micropropagation of laurel (*Laurus nobilis* L.), Horticultura, 35: 83-90.

J.E. Simon, A.F. Chadwick and L.E. Craker, 1984. The scientific literature on selected herbs and medicinal plants of the temperate zone, in: Herbs: an indexed bibliography, 1971-1980, Archon Books, Hamden, Connecticut, pp. 770.

N. Souayah, M.L. Khouja, A. Khaldi, M.N. Rejeb and S. Bouzid, 2002. Breeding improvement of *Laurus nobilis* L. by conventional and *in vitro* propagation techniques, *Journal of Herbs, Spices & Medicinal Plants*, 9: 101-105.

R. Rodríguez, M. Rey, L. Cuozzo and G. Ancora, 1990. *In vitro* propagation of Caper (*Capparis spinosa* L.), *In Vitro Cellular & Developmental Biology*, 26: 531-536.

S. Parlak and D.S. Cuming, 2012. Anatomical examination of root formation on bay laurel (*Laurus nobilis* L.) cuttings, *Journal of Plant Biology Research*, 1: 145-150.

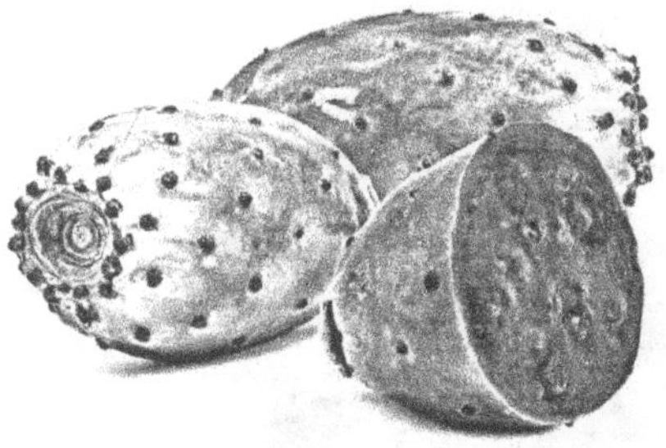

6

Cactus Pears

P K Ghosh, P Bhattacharjee and R S Singhal

Introduction

Cactus pears pear or prickly pear is known for thousands of years to human civilization. It's origin and history are related to ancient Mesomerican civilization, in particular to the Aztec civilization of Mexico (Pimienta, 1990). Cactus pears were first brought to Spain as an exotic flora from the New World by Christopher Columbus (Barbera, 1999). Cactus pear (*Opuntia ficus-indica* (L.) Mill. is a member of the *Cactaceae* family. It has about 400 species and a large number of varieties. It has emerged as a fruit crop in the second half of the last century and is currently extensively cultivated worldwide. Mexico is the largest producer of cactus pears accounting for 79.40% of the total world production. Cactus pears are also cultivated in parts of Africa, Australia, the Mediterranean basin, and in some parts of Asia (Aňorve Morga *et al.*, 2006). A cactus pear plant has been shown in Fig. 1 and the broad taxonomic classification of the plant has been summarized in Table 1.

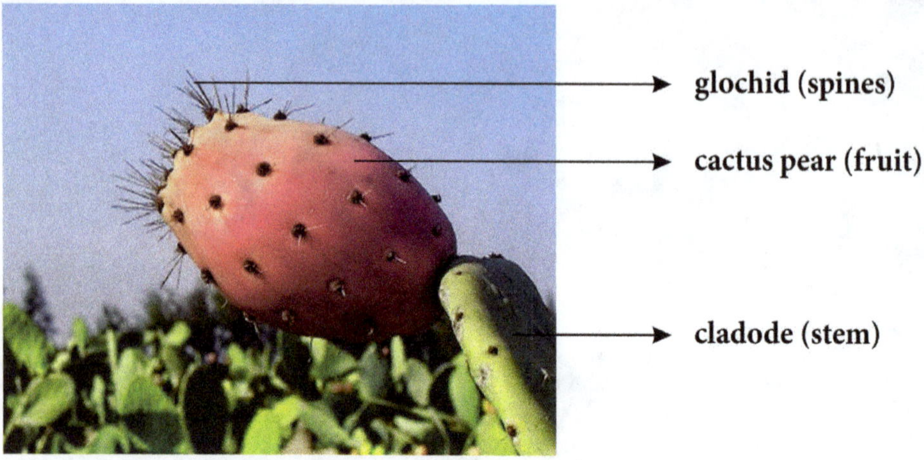

Fig. 1: A cactus pear plant with cladode. (Source: Ghosh *et al.*, 2014)

Table 1: Taxonomic classification of *Opuntia ficus-indica* (L.)

Class	Magnoliopsida
Subclass	Caryophyllidae
Order	Caryophyllidae
Family	Cactaceae
Subfamily	Opuntiodeae
Genus	*Opuntia*
Subgenus	Platyopuntia
Species	*Opuntia ficus-indica* (L.) Mill.

Source: Beccaro *et al.* (2015).

Cactus pears grow as a large bush or a tree-like structure with oblong or ellipitical joints (30-50 cm long), usually with spines; it bears yellow flowers and purple or red fruits (5-10 cm long). Apart from the spiny cactus, another variety, the spineless cactus is said to have been introduced early in this century. This variety is not wild and is mostly grown in gardens in Mexico and in the Mediterranean countries as an edible fruit. This variety is not affected by wild cochineal insects (*Dactylopius* spp.) and therefore cactus pears have been suggested to be suitable for cultivation in famine affected areas as a fodder reserve. However, owing to lack of spines, this variety of cactus needs to be protected by hedges from wild animals and also from fungal attack (The Wealth of India, 1966).

Distribution of Cactus Pears

Cactus pear widely grows in the American continent in southern USA, central America and in the south American countries. Wild and cultivated species of *Opuntia* are also cultivated in Africa, Europe, Australia and India. Fig. 2 shows the worldwide cultivation of *Opuntia* spp. The uncertainty in genus of cactus pears has been associated with polyploidy, high occurrence of interspecific and intergeneric hybridization and high variability of phenotype, depending on environmental conditions. The genus *Opuntia* is divided into four subgenera: *Platyopuntia*, *Cylindropuntia*, *Tephracactus* and *Brasiliopuntia* The subgenus *Platyopuntia* includes 150 -300 described species, including the group *ficus-indicae*, which includes *O. ficus-indica* (L.) Mill, the most widely cultivated cactus pear. Mexico and Italy are the most important cactus pear producers globally (Mondrago´n-Jacobo, 2001).

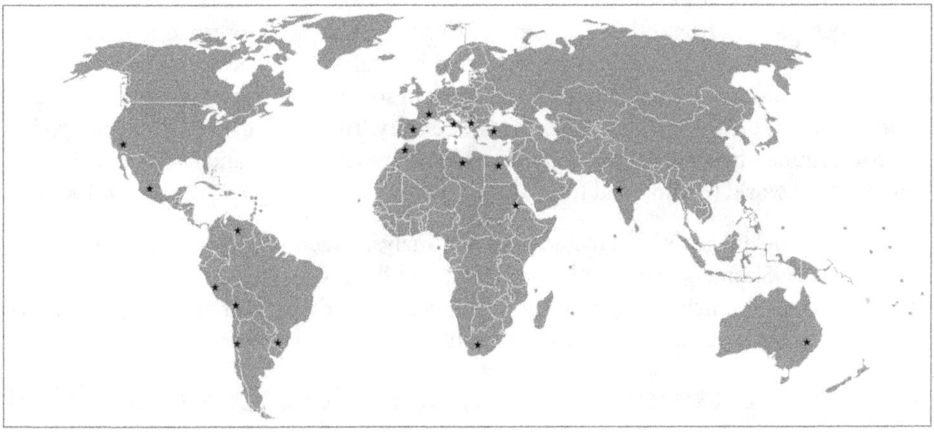

Fig. 2: Worldwide distribution of cultivation of Opuntia spp. [Source: Sáenz (2013)]

Other Names of Cactus Pears

Cactus pears are known by several names across the globe such as *higo chumbo* (chumbo fig) in Spain; *fico d'India* in Italy; *figue de Barbarie* in France; prickly pear in Australia, South Africa and USA; *beles* in Ethiopia; *palma forrageira* in Brazil and *nagphani, anda torra* or *chapathi balli* in India. The phenotype of cactus pears shows high variability according to the climatic conditions (Sáenz, 2013).

Evolutionary Studies on Cactus Pears

The studies on evolutuon of cactus pears has been conducted by distance based algorithm employing Neighbour-joining (NJ) method (Fig. 3). In order to circumvent the limitations of phylogenetic tree, where all branches of dendogram are not fully supported, a network analysis based on NeighborNet (NN) method has been developed by Caruso *et al.* (2010) (Fig. 4). This type of classification is particularly useful for reticulate evolution as is true for *Opuntia*.

From NJ and NN trees (Figs. 3 and 4, respectively), most of the *O. ficus-indica* related species (such as *O. robusta*, *O. elizondoana*, *O. spinulifera*, *O. vulgaris*, *O. quimilo*, *O. oligacantha*, *O. joconostle*, *O. cochenillifera*, and *O. subulata*) are grouped together and are clearly separated from other varieties. Parallel edges in NN especially among the octoploid *Opuntia* spp. indicate the presence of incompatible splits (appearing as boxes) among the cultivated varieties and other cactus pear genotypes of unknown origin. Therefore, a network analysis better explains the relationship between analyzed genotypes, which are a result of complex evolution involving polyploidization, hybridization and recombination (Britton and Rose, 1963).

In NJ and NN analysis, the prickly pears is classified as *O. ficus-indica*, *O. amyclaea*, *O. megacantha*, *O. streptacantha*, *O. fusicaulis*, *O. albicarpa* and *O. leucotricha*. The Mexican varieties display a high level of diversification, whereas the spineless Sicilian varieties have a very narrow genetic base and are closely related to other spineless accessions originating in Israel ('Jerico'), Kenya, South Africa ('Gymno carpo') and the United States ('Texas'). The genetic diversity of cactus pears is particularly evident from NN analysis, where the Sicilian varieties and the above-mentioned spineless accessions are mutually connected and are clearly separated from other accessions. Despite little or no divergence at all eight loci, some of these genotypes could be distinguished by their fruit color or fleshiness and it is likely that the phenotypic variation was the result of somatic mutations of a few clones that occurred in the cultivated regions after the 16th century. This was the time when the *Opuntia* spp. had begun to become naturalized in the Mediterranean region and later in other warm regions of the world (Britton and Rose 1963; Scheinvar 1995; Reyes-Agüero *et al.* 2005)

Other spineless genotypes included in the analysis, such as the Sicilian 'Inerme'; the South African 'Skinners court' and 'Fusicaulis'; and the accession 'Castillo', diverged from the main group of spineless cultivars. These results confirm the hypothesis of Griffith (2004), who suggested that the spineless varieties might have originated from different ancestors.

Botanical Features and Phytochemical Composition of Cactus Pears

Cactus pears are found in various colors such as purple, green, light yellow and light green (Fig. 5). Polyploidy is common in its cells. Within the Opuntioideae family, ploidy levels vary from diploid to octaploid, with the ploid species representing > 64% of the total population (Pinkava *et al.* 1998; Felker *et al.* 2006). The plant can reproduce both sexually and asexually depending on environmental factors (Beccaro *et al.*, 2015).

Climatic conditions required for cultivation of cactus pears

The cactus pears grow in many environmental conditions, ranging from desert areas to areas with low temperatures (Nobel, 1999). The shallow and extensive root system enables the plant to survive low rainfall conditions. Secondary roots formed due to scarce rain, increase the contact surface with the soil, facilitating uptake of water and nutrients. The roots contract radially at the beginning of drought to reduce water loss (Nobel, 1999).

Morphological characteristics and phytochemical composition of cactus pears:

Cladodes: The plant grows up to 3.5-5.0 m high. The stem, known as 'cladodes' or commonly as 'paddles', 'fleshy leaves' or 'racquets' are succulent and the sites of photosynthesis. The cladodes have an elongated or racquet shape that grow 60-70 cm in length depending on the amount of water and nutrients available (Sudzuki *et al.*, 1993). When cladodes are 10-12 cm long, they are tender and eatable. The areoles are found on both sides of the cladodes and are capable of developing into new cladodes, flowers or roots depending on environmental conditions (Sudzoki *et al.*, 1997).

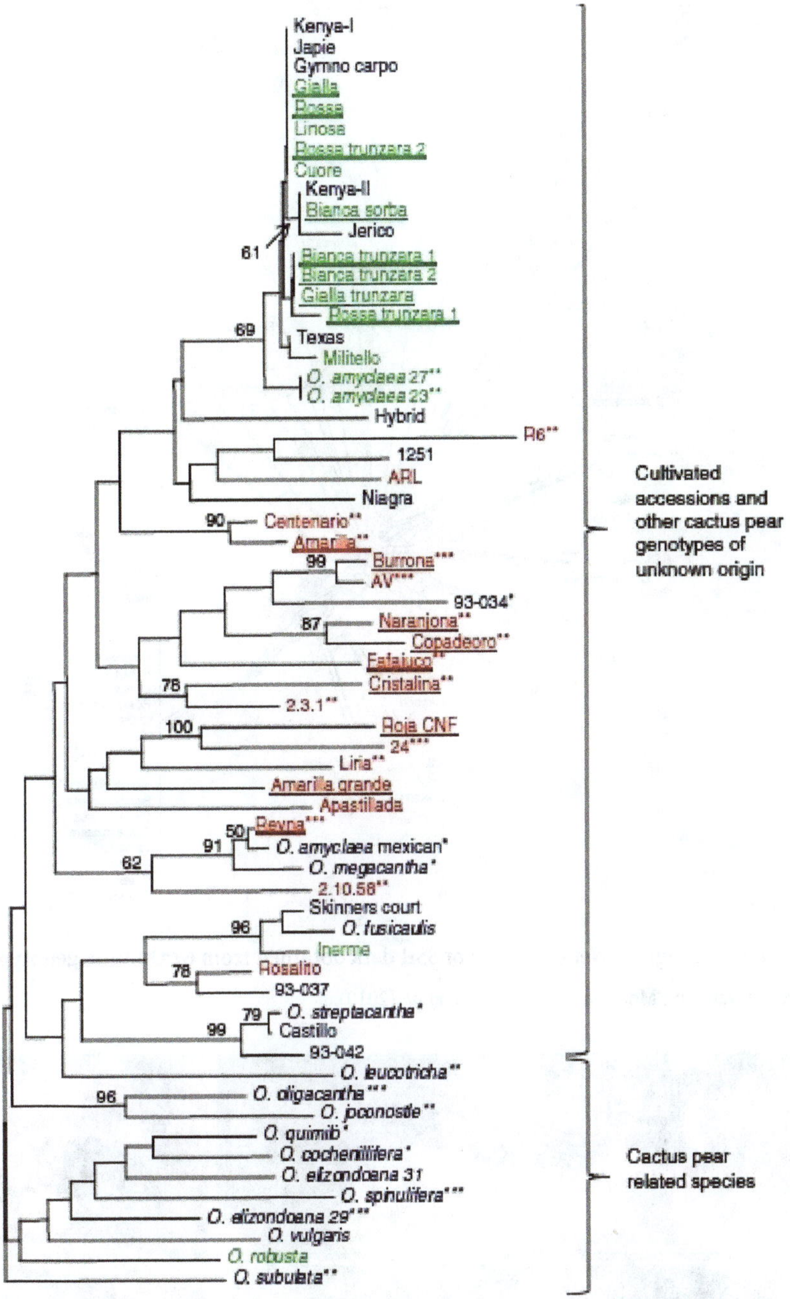

Fig. 3: Neighbor-joining (NJ) dendogram of 62 Opuntia genotypes based on the log-transformed proportion of shared alleles. The numbers at the branch points indicate bootstrap support values [50% (1,000 replicates). Colors indicate different geographic origin (green Italy, red Mexico, blue Israel). Underlined names indicate cultivated varieties. The presence of spines in the cladodes is indicated by asterisks (* few; ** intermediate; * many). Source: Caruso *et al.* (2010).**

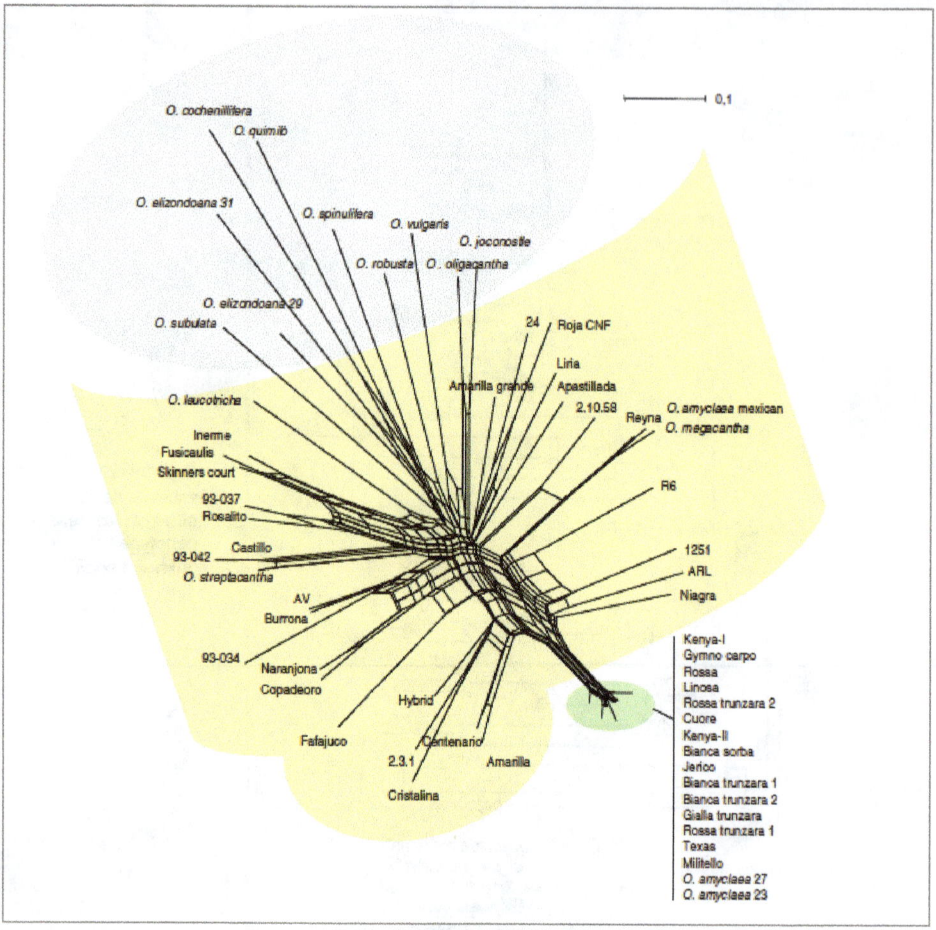

Fig. 4: NeighborNet (NN) tree of SSR data obtained from 62 Opuntia genotypes.

Source: Bryant and Moulton (2004); Caruso *et al.* (2010).

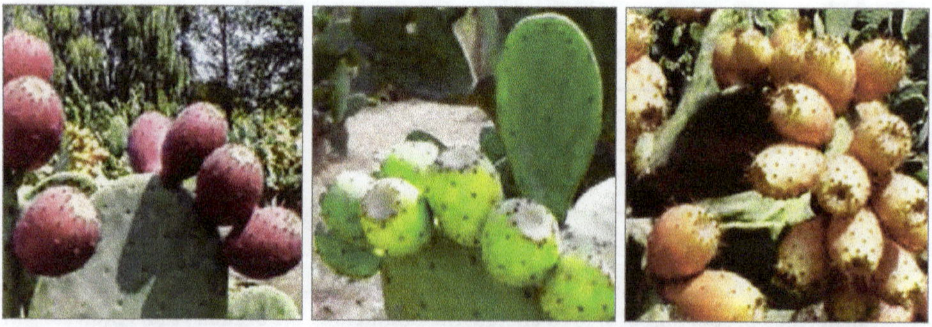

Fig. 5: Cactus pears growing in different colors [Source: Sáenz and Spulveda (2013)].

The cladodes are covered by thick cuticle which is covered by wax or glochids which further reduce water loss. The stem stores abundant water (owing to many parenchymatic cells) which enables the plant to survive droughts. Mucilage is also present in plant tissues which have high water binding capacity (Nobel *et al.*, 1992). The cladodes also have spines, which are modified leaves to reduce transpiration. They have few stomata per unit surface area which remain closed during the day and open at night. During prolonged drought, the stomata remian closed day and night to prevent transpiration. The cladodes undergo crassulacean acid metabolism (CAM) for phytosynthesis. The cladodes are a rich source of amino acids and glutamine (36.12% of all amino acids) is the most abundant amino acid in them (Table 2). Narcissin and iso-quercetin are the major phytochemicals in cladodes (Table 5).

Flowers: The flowers of *Opuntia* spp. have limited bloom duration and appear on one year old cladodes. The color of the flowers varies from yellow to orange, pink, purple, red and white (Anderson, 2001). The development from bud stage to anthesis requires 21-75 days. The flowers have dipurative and diuretic properties and cause relaxation of renal excretory tract (Ammar *et al.*, 2012). The most abundant amino acid in flowers of cactus pears is glutamine at 12.59% of all amino acids. Phytochemical properties and antimicrobial potencies of two common *Opuntia* flowers at different stages of development have been discussed in Tables 3 and 4, respectively. The phenolic content and antioxidant potency are at their highest in the post flowering stage of the flowers (Table 3). Their antimicrobial potency is reportedly the highest against *Staphylococcus aureus* (Table 4). Quercetin 3-*O*-rutinoside and 6 isorhamnetin 3-*O*-robinobioside are the most abundant phytochemicals in flowers of cactus pears as shown in Table 5.

Fruit: The prickly pear fruit resembles pomegranate fruit development in that the fleshy edible parts are derived from seed tissues and therefore, increase in conjunction with seed development. The cactus pear fruit is a false berry consisting of juicy sweet pulp containing many hard coated seeds. The fruit is formed from an inferior ovary of a hermaphroditic, protandrous flower, whose period of differentiation is between 35 to 45 days (Aguilar-Becerril, 1980; Pimiento-Barrios and Engleman, 1985) and 55 days for *O. amyclaea* (Alvarado y Sosa, 1978) and 70 days for *O. ficus-indica* (Gil *et al.*, 1977; Rivera *et al.*, 1981). The time from anthesis to fruit maturity is about 110 days (Alvarado y Sosa, 1978; Lakshminarayana *et al.*, 1979). The fruit may contain 80-300 seeds. The evolution of the fruit follows a double sigmoid growth model and the entire process can take from 80 to 90 days; at high temperatures, there is a quicker ripening while a colder climate slows ripening of the fruits (Nerd and Mizrahi, 1995).

The cactus pear fruit's flavor has been described as faint melon or cucumber-like aroma, mainly constituted by alcohols and esters (Rodríguez *et al.*, 2015). 2-(E/Z)-2,6-nonadien-1-ol and 2-methylbutanoic acid methyl ester is the key aroma among the volatile organic compounds (VOCs) present. The yellow cactus pear fruits have the strongest flavor intensity, followed by red and white cactus pears varieties. VOC extraction is performed by dynamic head space method using forced humidified and charcoal-filtered airstream for headspace generation and by trapping the VOCs in the Tenax column. Analysis of VOCs is carried out by GC-MS (Rodríguez *et al.*, 2015). Analyses revealed that quercetin and isorhamnetin are the most abundant phytochemicals in cactus pear fruit (Table 5).

Pulp: The pulp of Rojo Cenizo and Rojo San Martín varieties of cactus pears of Mexico reportedly has total dietary fiber (145-166 g), vitamin C (3-7 g ascorbic acid), total phenol content [3-6 g gallic acid equivalent (GAE)], betalains (500-3444 mg betanin equivalents), and antioxidant activity (46-67 mM Trolox equivalents (TE)/kg fruit). The peels of Rojo San Martín variety of cactus pear have total phenol content two times greater than the pulp, whereas antioxidant activity of peels of Rojo San Martín and Verde Villanueva varieties are 4-9 times higher than their respective pulps (Jiménez-Aguilar *et al.*, 2015). LC-MS analysis of betanins in cactus pear fruit extracts has shown presence of indicaxanthin, betanin, isobetanin and neobetanin (Fig. 6) (Albano *et al.* 2015). The pulp of cactus pears chiefly consists of phytochemicals such as isorhamnetin and kaempferol (Table 5).

Seed: The seeds constitute about 4-8% by weight of cactus pear pulp. Fruits with hard seeds are less accepted in the market (Inglese *et al.*, 1993). Seed weight or number is correlated with fruit size (Barbera *et al.*, 1992). Development of cactus pear fruit with increased pulp and reduced seeds as well as parthenocarpy in fruits has been successfully carried out by Díaz and Gil (1978). Glutamic acid is the most abundant amino acid (21.68% of all amino acids) in the seeds of cactus pears (Table 2). Sinapoyl-diglucoside is the most abundant phytochemical in seeds of cactus pears (Table 5).

Thorns: The modified leaves i.e. thorns of cactus pears plant play a vital role in regulating transpiration. The presence or absence of thorns are result of stress conditions (Le Houerou, 1996) or are related to juvenility (Nieddu *et al.*, 2006). The naturalized spineless *O. ficus-indica* have converted to spiny forms where the plant have become invasive (Zimmermann, 2010). Owing to uncertainties in taxonomic classification, *O. ficus-indica* is sometimes described as spineless and taxonomically different from other *Opuntia* spp. However this would be an inappropriate basis since the growth habit, presence of spines, the number of spine per areole (bud) and aerole may differ according to growing regions (Rebman and Pinkava, 2001).

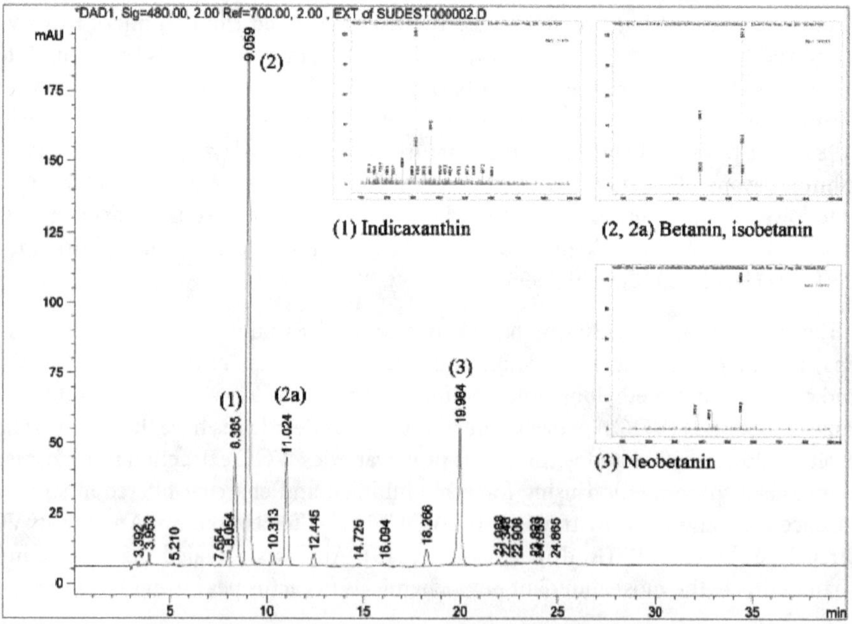

Fig. 6: LC-MS analysis of cactus pear fruit extract (purple variety), MS fragmentation of main peaks is also shown [Source: Albano *et al.* (2015)]

The oil derived from cactus pears is rich in polyunsaturated fatty acids (chiefly, in $C_{18:2}$ *i.e.*, linoleic acid) and also contains monounsaturated and saturated fatty acids. The fatty acid compositions of oil derived from seed, peel, pulp and cladode have been shown in Table 6.

Table 2: L-amino acid content (%) in cladodoes, fruit and seeds of *O. ficus-indica.**

Amino acids	Cladode	Fruit	Seeds
Alanine	1.25	3.17	4.75
Arginine	5.01	1.11	6.63
Aspargine	3.13	1.51	tr
Asparginic acid	4.38	tr	10.42
Glutamic acid	5.43	2.40	21.68
Glutamine	36.12	12.59	tr
Cystine	1.04	0.41	0.37
Histidine	4.18	1.64	3.11
Isoleucine	3.97	1.13	6.20
Leucine	2.71	0.75	9.94
Lysine	5.22	0.63	6.79
Methionine	2.92	2.01	0.70
Phenylalanine	3.55	0.85	5.25
Serine	6.68	6.34	8.46
Threonine	4.18	0.48	1.53
Tyrosine	1.46	0.45	3.09
Trytophan	1.04	0.46	tr
Valine	7.72	1.43	6.02
α-aminobutyric acid	tr	0.04	tr
Proline	tr	46	tr
Taurine	tr	15.79	5.06

Source: Sawaya *et al.* (1983); Feugang *et al.* (2006), El-Mostafa *et al.* (2014); tr- trace amount.

Table 3: Total phenol content and DPPH radical scavenging activity of ethanolic extract of cactus pear flowers during four stages of flowering in cactus pears grown in Tunisia, Africa

Species	Flowering stage	Polyphenol content (mg gallic acid eq./g of extract)	Antioxidant activity IC_{50} of DPPH assay (μg/mL)
Opuntia ficus-indica	A	16.30	205
	B	13.40	240.5
	C	13.30	100
	D	49.00	60

Contd...

Species	Flowering stage	Polyphenol content (mg gallic acid eq./g of extract)	Antioxidant activity IC_{50} of DPPH assay (μg/mL)
Opuntia stricta	A	16.70	147
	B	15.30	62
	C	13.10	110.2
	D	80.90	50.1

A- Vegetative phase, B- initial flowering phase, C- full flowering phase, D- post flowering phase

Source: Ammar *et al.* (2012).

Table 4: Antibacterial and antifungal potencies of *Opuntia* spp. flower extracts in *n*-hexane at different flowering stages by assessment of inhibition diameter.

Microorganism	Opuntia ficus-indica		Opuntia stricta	
	C	D	C	D
Pseudomonas aeruginosa	0	17.6	15.0	14.4
Escherichia coli	0	19.7	16.4	10.5
Staphylococcus aureus	14.8	12.0	11.4	0
Bacillus subtilis	0	0	0	0
Aspergillus niger	0	0	0	11.8
Candida lipolitica	0	0	0	8

C- full flowering phase, D- post flowering phase. Source: Ammar *et al.* (2012).

Table 5: Distribution of major phenols and flavonoids in various parts of *O. ficus-indica*.

Plant Tissue	Main compound identified	Content (mg/100 g)	References
1. Cladode	Narcissin	137.10	Valente *et al.* (2010); Ginestra*et al.* (2009); Guevara-Figueroa *et al.* (2010)
	Coumaric	16.18	
	Iso-quercetin	39.67	
	Ferulic acid	34.77	
2. Flower	Quercetin 3-O-Rutinoside	4900.00	Ammaer *et al.* (2012); Clark *et al.* (1980); Leo *et al.* (2010)
	6 Isorhamnetin 3-O-Robinobioside	4269.00	
	7 Isorhamnetin 3-O-Galactoside	979.00	
	8 Isorhamnetin 3-O-Glucoside	724.00	
3. Fruit	Total phenolic acid	45.70	Kuti (2004); Moussa-Ayoub *et al.* (2011); Jorge et al. (2013)
	Total Flavonoid	6.95	
	Quercetin	4.32	
	Isorhamnetin	2.41	

Contd...

Plant Tissue	Main compound identified	Content (mg/100 g)	References
4. Pulp	Total phenolic acid	218.80	Fernández-López *et al.* (2010); Bensadón *et al.* (2010); Salim *et al.* (2009)
	Isorhamnetin glycosides	50.60	
	Isorhamnetin	4.94	
	Kaempferol	2.70	
5. Seed	Total phenolic acid	89.00	Chougui *et al.* (2013)
	Sinapoyl-diglucoside	23.40	
	Total Tannins	6.60	
	Total Flavonoids	2.60	

Table 6: Comparison of fatty acid composition (g/100 g fatty acids) of *O. ficus-indica*.

Fatty acid ➡ Plant part ⬇	$C_{12:0}$	$C_{14:0}$	$C_{16:0}$	$C_{16:1}$	$C_{18:0}$	$C_{18:1}$	$C_{18:2}$	$C_{18:3}$	$C_{20:0}$	$C_{22:0}$	$C_{22:1}$	$C_{24:0}$
Cactus seed oil*	nd	nd	9.32	1.42	3.11	16.77	70.29	nd	nd	nd	nd	nd
Prickly pear peel†	0.71	1.95	23.1	2.48	2.67	24.1	32.3	9.27	nd	0.50	-	0.41
Fruit pulp oil‡	nd	1.13	34.40	1.62	2.37	10.8	37	12.68	nd	nd	nd	nd
Cladode#	1.33	1.96	13.87	0.24	3.33	11.16	34.87	33.23	nd	nd	nd	nd

*Ramdan and Mörsel (2003a), †Abidi *et al.* (2009), ‡Ramdan and Mörsel (2003b), #Ennouri *et al.* (2005); nd- not detected.

Utility of Cactus Pears

Cactus pears play an important ecological role in combating desertification, in addition to producing nutritious fruits and vegetables. It also provides feed for the livestock, biomass for energy purposes and produces numerous products of industrial utility such as beverages, vegetarian cheese, drugs and cosmetics.

Cactus pears can grow in severely degraded soils, which are inadequate for other crops. *Opuntia* spp. have a great capacity to withstand severe dry conditions. Their root characteristics avoid wind and rain erosion, facilitating their growth in degraded areas also.

Cactus pears have wide range of possible use, such as (i) utility as forage, (ii) as a vegetable (where young cladodes are consumed fresh or cooked), (iii) as a fruit, (the cactus pear fruit consumed in several countries such as in Italy, Tunisia, Morocco, South Africa, Mexico and Chile), (iv) as a source of carminic acid, a natural red dye accepted by health authorities worldwide and (v) for medical applications, where cactus pears have shown promising results in treatment of gastritis, diabetes, hypercholesterolemia and for obesity.

Cactus pears are also a useful resource for many industrial food products such as juices, jams, jellies, sweeteners, alcohols, wines, vinegar, canned fruits, sauces, dietary fiber, pigments and oil obtained from their cladodes, flowers, fruit, peel and seeds (Table 7).

Table 7: Utility of cactus pears in development in food, healthcare and other products

Products		
Cladodes	**Flowers**	**Seeds**
Juices, pickled and brine cured products	The decoction of extracted from flowers has diuretic properties	Seeds are nutritious and could be employed as source of edible oil and fodder for animals after grinding
	Fruit	**Oil from seeds**
Flour, jams and jellies, bakery products, alcohol, sauces and confectionery	Ripe fruits are eaten as such and also employed in production of juices, jams, jellies, sweetmeat and nectars	Mucilage from cladodes, pigments from peel of fruits
Low cost, natural and eco-friendly biosorbent for dyes	Dehydrated fruits and fruit leathers, sweeteners	
Dried cladodes are employed as alternative source of carbon for lactic acid bacteria	Alcohols, wines and vinegars	
Cladodes are employed in manufacture of mortar, plaster and whitewash	Frozen fruit, canned fruit and pulp, Antiseptic, anticancer, hypoglycemic, hypolipidemic effects on humans, employed in phytoremediation, soil fixation, soil mulching and in hedges	
Mucilage or gum with excellent adhesive properties which is almost insoluble in water but swells to a jelly like mass on rehydration		
Cactus pears peels		
Rich source of phytochemicals and used as supplements in food, pharmaceuticals and cosmetics		
Increase milk production in ruminants		

Source: The Wealth of India (1966), Sáenz (2000); Corrales and Flores (2003), Ghosh *et al.* (2014).

Conclusion

Opuntia ficus-indica (L.) Mill. is well suited to grow in arid or semi-arid areas of the world. These plants are drought resistant and are able to reduce transpiration through modified leaves i.e. spines and by waxy coating on the stems i.e. cladodes. The genetic diversity provides the plant opportunity to adapt to the changing environment it grows in. The utility of all the parts of the plant in food, pharmaceutical and cosmetic industries along with low cost of cultivation and maintenance make it a promising future crop for mankind.

References

Abidi, S., Ben Salem, H., Vasta, V. and Priolo, A. 2009. Supplementation with barley or spineless cactus (*Opuntia ficus-indica*) cladodes on digestion, growth and intramuscular fatty acid composition in sheep and goats receiving oaten hay. *Small Rumin. Res.*, 87, 9–16.

Aguilar-Becerril, G. 1980. Effect of various growth regulators on prickly pear (*Opuntia amyclaea*). Professional Thesis, *Chapingo Autonomous University*, Mexico.

Albano C., Negro C., Tommasi N., Gerardi C., Mita G., Miceli A., De bellies L. and Blando, F. 2015. Betalains, phenols and antioxidant capacity in cactus pear [Opuntia ficus-indica (L.) Mill.] fruits from Apulia (south Italy) genotypes. Antioxidants, 4, 269-280.

Alvarado y Sosa, A. L. 1978. Physiology and biochemistry of fruit development of prickly pear (*Opuntia amyclaea* Tenore). M. S. thesis, *Chapingo Postgraduate College*, Chapingo, Mexico.

Ammar, I., Ennouri, M., Khemakhem, B., Yangui, T. and Attia, H. 2012. Variation in chemical composition and biological activities of two species of Opuntia flowers at four stages of flowering. *Ind. Crop. Prod.*, 37, 34-40.

Aňorve Morga J, Aquino EN and Mercado Silva E. 2006. Effect of controlled atmosphere on the preservation of minimally processed cactus pears. *Acta Hort.*, 728, 211–216.

Barbera, G. 1999. History and economic and agro-ecological importance of cactus pear. In: *Agro-ecology, Cultivation and Uses of Cactus Pear*, G. Barbera, P., Inglese and E. Pimienta, (Eds), *FAO Plant Production and Protection Paper*, 132, Rome, pp. 1–12.

Barbera, G., Inglese, P. and La Mantia, T. 1992. Influence of seed content on some characteristics of the fruti of cactus pear (*Opuntia ficus-indica* Mill.). *Act II International Congress of the Tuna and Cochineal*, Santiago, Chile. pp. 8-14.

Beccaro, G. L., Bonvegna, L., Donno, D., Mellano, M. G., Cerutti, A. K., Nieddu, G., I. Chessa, I. and Bounous, G. 2015. *Opuntia* spp. *biodiversity conservation and utilization on the Cape Verde Islands,* Genet Resour. *Crop Evol.*, 62 (1), 21-33.

Bensadón, S., Hervert-Hernández, D., Sáyago-Ayerdi, S. G. and Goñi, I. 2010. By-Products of *Opuntia ficus-indica* as a source of antioxidant dietary fiber. *Plant Food. Hum. Nutr.*, **65,** 210–216.

Britton, N.L. and Rose, J.N. 1963. The Cactaceae, Vol. 1. Dover, New York.

Bryant, D., and Moulton, V. 2004. Neighbor-Net: An agglomerative method for the construction of phylogenetic networks. *Mol. Biol. Evol.*, 21, 255–265.

Caruso M., Currò S., Las Casas G., La Malfa S. and Gentile A. 2010. Microsatellite markers help to assess genetic diversity among *Opuntia-ficus indica* cultivated genotypes and their relation with related species, *Plant Syst. Evol.*, 290, 85-97.

Chougui, N., Tamendjari, A., Hamidj, W., Hallal, S., Barras, A., Richard, T. and Larbat, R. 2013. Oil composition and characterisation of phenolic compounds of *Opuntia ficus-indica* seeds. *Food Chem.*, 139, 796–803.

Clark, W. D., Brown, G. K. and Mays, R. L. 1980. Flower flavonoids of *Opuntia* subgenus Cylindropuntia. *Phytochemistry,* 19, 2042–2043.

Corrales, J. and Flores, C. A. 2003. Current and future tendencies in processing of prickly pear. In: Prickly Pear Cactus: Production, Marketing, Postharvest Handling and Industrialization, V.C.A. Flores (Ed.). Center for Economic, Social and Technological Research of Agro-Industry and World Agriculture, Chapigno Autonomous University, Mexico, pp. 167–215.

De Leo, M., Abreu, M. B. D., Pawlowska, A. M., Cioni, P. L. and Braca, A. 2010. Profiling the chemical content of *Opuntia ficus-indica* flowers by HPLC–PDA-ESI-MS and GC/EIMS analyses. *Phytochem. Lett.*, 3, 48–52.

Díaz, Z., F. and G. Gil S. 1978. Effectiveness of various doses and methods of application of gibberellic acid in the induction of parthenocarpy and growth prickly pear fruit (*Opuntia ficus-indica*, Mill.). *Cienc. Investig. Agrar.*, 5(3), 109-117.

El-Mostafa, K., El-Karrassi, Y., Badreddine, A., Andreoletti, P., Vamecq, J., El Kebbaj, M. S., Latruffe, N., Lizard, G., Nasser, B. and Cherkaoui-Malki, M. (2014). Nopal cactus (Opuntia ficus-indica) as a source of bioactive compounds for nutrition, health and disease. Molecules, 19, 14879–14901.

Ennouri, M., Evelyne, B., Laurence, M. and Hamadi, A. 2005. Fatty acid composition and rheological behavior of prickly pear seed oils. *Food Chem.*, 93, 431–437.

Felker, P., Paterson, A. and Jenderek, M.M. 2006. Forage potential of Opuntia clones maintained by the USDA National Plant Germplasm System (NPGS) collection. *Crop Sci.*, 46, 2161–2168.

Fernández-López, J. A., Almela, L., Obón, J. M. and Castellar, R. 2010. Determination of antioxidant constituents in cactus pear fruits. *Plant Food Hum. Nutr.*, 65, 253–259.

Feugang, J. M., Konarski, P., Zou, D., Stintzing, F. C. and Zou, C. 2006. Nutritional and medicinal use of cactus pear (*Opuntia* spp.) cladodes and fruits. *Front. Biosci.*, 11, 2574–2589.

Ghosh P. K, Bhattacharjee P. and Singhal R. S. 2014, Cactus pears. In: *Future Crops*, Peter K. V. (Ed.), vol. 2, pp. 133-153.

Gil, G. F., Morales, M. and Momberg, A. 1977. Fruit set and fruit development prickly pear (*Opuntia ficus-indica* Mill.) and its relationship with pollination and gibberellic acids and chloroethylphosphonic. *Cienc. Investig. Agrar.*, 4(3), 163-169.

Ginestra, G., Parker, M. L., Bennett, R. N., Robertson, J., Mandalari, G., Narbad, A., Lo Curto, R. B., Bisignano, G., Faulds, C. B. and Waldron, K. W. 2009. Anatomical, chemical, and biochemical characterization of cladodes from prickly pear [*Opuntia ficus-indica* (L.) Mill.]. *J. Agric. Food Chem.*, 57, 10323–10330.

Griffith, M.P. 2004. The origins of an important cactus crop, *Opuntia ficus-indica* (Cactaceae): New molecular evidence. *Am. J. Bot.*, 91, 1915–1921.

Guevara-Figueroa, T., Jiménez-Islas, H., Reyes-Escogido, M. L., Mortensen, A. G., Laursen, B. B., Lin, L. W.; de León-Rodríguez, A., Fomsgaard, I. S. and Barba de la Rosa, A. P. 2010. Proximate composition, phenolic acids, and flavonoids characterization of commercial and wild nopal (*Opuntia* spp.). *J. Food Compos. Anal.*, 23, 525–532.

Inglese, P., Barbera, G. and La Mantia, T. 1993. Research strategies and improvement of cactus pear (*Opuntia ficus-indica*) fruit quality and production. In: P. Felker and J. R. Moss (Eds.), *Proceed. Fourth Annual Texas Prickly Pear Council*, Kingsville, Texas, pp. 24-40.

Jiménez-Aguilar, D. M., López-Martínez, J. M., Hernández-Brenes, C., Gutiérrez-Uribe, J. A. and Welti-Chanes, J. 2015. Dietary fiber, phytochemical composition and antioxidant activity of Mexican commercial varieties of cactus pear. *J. Food Compost. Anal.*, 41, 66-73.

Jorge, A. J., de La Garza, T. H., Alejandro, Z. C., Ruth, B. C. and Noé, A. C. 2013. The optimization of phenolic compounds extraction from cactus pear (*Opuntia ficus-indica*) skin in a reflux system using response surface methodology. *Asian Pac. J. Trop. Biomed.*, 3, 436–442.

Kuti, J. O. 2004. Antioxidant compounds from four Opuntia cactus pear fruit varieties. *Food Chem.*, 85, 527–533.

Lakshminarayana, S., Alvarado y Sosa L. and Barrientos Pérez F. 1979. The development and postharvest physiology of the fruit of prickly pear (*Opuntia amyclaea* Tenore). In: Tropical Foods, G.E. Inglett and G. Charalambous (Eds.), vol. 1, pp. 69-93.

Le Houe´rou, H. N. 1996. The role of cacti (*Opuntia* spp.) in erosion control, land reclamation, rehabilitation and agricultural development in the Mediterranean basin. *J. Arid Environ.*, 33, 135–159.

Mondrago´n-Jacobo, C. 2001. Cactus pear breeding and domestication. *Plant Breed. Rev.*, 20, 135–166.

Nerd A. and Mizrahi Y. 1995. Reproductive biology. In: Agroecology, Cultivation and uses of Cactus Pear, G. Barbera, P. Inglese and B. E. Piemienta (Eds). *FAO Plant Production and Protection Paper*, 132, Rome, Italy.

Nieddu, G and Chessa, I. 1997. Distribution of phenotypic characters within a seedling population from *Opuntia ficus-indica* (cv. "Gialla"). *Acta Hort.*, 438, 37–43.

Nobel, P., Cavelier, J. and Andrade, J. L. 1992. Mucilage in cacti: its apoplastic capacitance, associated solutes and influence on tissue water relations. *J. Exp. Bot.*, 43, 641–648.

Nobel, P. S. 1999. Biología ambiental. (In:) Agro-ecology, Cultivation and Uses of Cactus Pear, G. Barbera, P. Inglese and E. Pimienta (Eds.). *FAO Plant Production and Protection Paper*, 132, Rome, pp. 37–50.

Pimienta, E. 1990. The Prickly Pear. Universidad de Guadalajara, Mexico.

Pimienta-Barrios, E. and Engleman, E.M. 1985. Development and pulp ratio by volume of components in tuna lobule [*Opuntia ficus-indica* (L.) Mill.]. *Agrociencia*, 62, 51-56.

Pinkava, D.J., Rebman, J. and Baker, M. 1998. Chromosome numbers in some cacti of western North America—VII. *Haseltonia*, 6, 32–41

Ramadan, M. F. and Mörsel, J. T. 2003a. Lipid profile of prickly pear pulp fractions. *J. Food Agric. Environ.*, 1, 66–70.

Ramadan, M. F. and Mörsel, J. T. 2003b. Recovered lipids from prickly pear [*Opuntia ficus-indica* (L.) Mill] peel: A good source of polyunsaturated fatty acids, natural antioxidant vitamins and sterols. *Food Chem.*, 83, 447–456.

Rebman J.P. and Pinkava D.J. 2001. Opuntia cacti of North America—an overview. *Fla. Entomol.*, 84, 474–483.

Reyes-Aguero, J. A., Aguirre, J. R. and Herna´ndez, H. M. 2005. Systematic notes and a detailed description of *Opuntia ficus-indica* (L.) Mill. (Cactaceae). *Agrociencia*, 39, 395–408.

Rivera, O., G. Gil, G. Montenegro and G. Avila. 1981. States of differentiation of flower buds of tuna (*Opuntia ficus-indica* Mill.). *Cienc. Investig. Agrar.*, 8 (3), 215-219.

Rodríguez S. A., Díaz S. and Nazareno M. A. (2015). Characterization of volatile organic compounds of *Opuntia* fruit pulp - changes in cactus fruit aroma as a consequence of pulp processing. *Acta Hort.*, DOI: 10.17660/ActaHortic.2015.1067.42.

Sáenz C. 2000. Processing Technologies: An alternative for cactus pear (*Opuntia* spp) fruits and cladodes. *J. Arid Environ.*, 46, 209–225.

Sáenz, C. 2013. Opuntias as a natural resource in: agro-industrial utilization of cactus pear, rural infrastructure and agro-industries division. (In:) Agro industrial utilization of cactus pear. *Food and Agriculture Organization of the United Nations*, Rome 2013.

Sáenz, C. and Sepúlveda, E. 1993. Alternative industrialization of tuna (*Opuntia ficus-indica*). *Alimentos*, 18, 29–32.

Salim, N., Abdelwaheb, C., Rabah, C. and Ahcene, B. 2009. Chemical composition of *Opuntia ficus-indica* (L.) fruit. *Afr. J. Biotechnol.*, 8, 1623–1624.

Sawaya, W. N. and Khan, P. 1982. Chemical characterization of prickly pear seed oil, *Opuntia ficus-indica* . *J. Food Sci.*, 47, 2060–2061.

Scheinvar L. 1995. Taxonomy of utilized opuntias. (In:) Barbera G., Inglese P. and Pimienta-Barrios E. (Eds). Agroecology, Cultivation and Uses of Cactus Pear. FAO plant production and protection paper, 132, FAO, Rome.

Sudzuki, F. 1999. Anatomía y morfología. (In:) G. Barbera, P. Inglese and E. Pimienta, eds. *Agroecología, cultivo y usos del nopal.* FAO Plant Production and Protection, Paper 132. Rome, pp. 29–36.

Sudzuki, F., Muñoz, C. and Berger, H. 1993. Growing prickly pear (cactus pear). Department of Agricultural Reproduction. University of Chile, Santiago.

The Wealth of India 1966. Raw materials, In: *The Wealth of India- a dictionary of Indian Raw Materials and Industrial Products*, vol. VII, Publications and Information Directorate, CSIR, New Delhi, pp. 102-103.

Valente, L. M. M., da Paixão, D., do Nascimento, A. C., dos Santos, P. F. P., Scheinvar, L. A., Moura, M. R. L., Tinoco, L. W., Gomes, L. N. F. and da Silva, J. F. M. 2010. Antiradical activity, nutritional potential and flavonoids of the cladodes of *Opuntia monacantha* (Cactaceae). *Food Chem.*, 123, 1127–1131.

Zimmermann, H. 2010. Managing prickly pear invasions in South Africa. *FAO Cactusnet Newsletter,* 12, 157–166.

7

Cashew

K R M Swamy

Cashew is widely cultivated throughout the tropics for its nuts and is a native of tropical American country : Brazil. It was one of the first fruit trees from the New World to be widely distributed throughout the tropics by the early Portuguese and Spanish adventurers (Purseglove, 1988). At the time of the first Portuguese colonization, the name used by local populations (Tupi Natives of Brazil) for the cashew was "acaju" (nut), which turned into "caju", in Portuguese spelling and "cashew" in English. Most of the names for cashew in Indian languages are also derived from the Portuguese name "caju" (Johnson, 1973). The common Indian names for cashew are, kajubadam (Assamese), hijli badam, kaju (Bengali), hijlibadam, kaajuu, kaju, kajubadam (Hindi), geru, gerumara, godambi (Kannada), kashukavu, kasumav (kasumavu), parangi mavu, paringi maavu (Malayalam), kaju, kajugola (Marathi), Lanka badam, Lanka beej (Oriya), kaju (Punjabi), agnikrita, batada, bhallataka, guchhapuspha, hijli badam, kajutaka, parvati, venamrah, vrkkabijah, vrkkaphalah (Sanskrit), andima, andimankottai, kaḷḷārmā, kolamavu, kolamaavutam, matumancam (nut), mundhiri paruppu, mundiri, munthiri, muntirikkottai munthiriparuppu, saram, tirikai (nut), tirigai, virai-muntirikai (Tamil), grabijamu, (jidi chettu, jidi kaaya, Jidimamidi, Jeedimamidi, Jidi mamidi, munthamamidi (Telugu) and kaajuu (Urdu). The Maconde tribe in Mozambique calls it the Devil's nut. It was offered at wedding banquets as a token of fertility. The cashew has a long history as a useful plant but only in the twentieth century it has become an important tropical tree crop. Small-scale local exploitation of the cashew for its nuts and cashew apples appears to have been the pattern for more than 300 years in Asia and Africa.

Uses of Cashew

In India, use of cashew apples and nuts was adopted by local people, and accounts from Africa are similar; making cashew wine appears to have been a common practice in both Asia and Africa (Johnson, 1973). It was offered at wedding banquets as a token of fertility by the Maconde tribe in Mozambique (Massari, 1994). Cashew tree bears numerous, edible, pear shaped false fruits or pseudo fruits or "accessory fruits" called "cashew apples." A small bean shaped, grey color "true fruit" is firmly adhering to lower end of these apples appearing like a clapper in the bell (Fig. 1). This true fruit is actually a drupe, featuring hard outer shell (cashew nut shell) enclosing a single edible seed or the "cashew kernel" (Fig. 2 & 3). The outer shell is green and leathery and turns an orange red when mature. The inner shell is hard, similar to other nut shells, and contains the edible cashew kernel. The oil enclosed in the nut's shell (cashew nut shell liquid or CNSL) (anacardic acid)) is toxic and can burn the skin. It is used in producing plastics and as a lubricant and insecticide. It is, therefore, the outer shell which is roasted in the processing unit and then, the edible kernel is extracted.

Fig.1: Cashew apple with nut

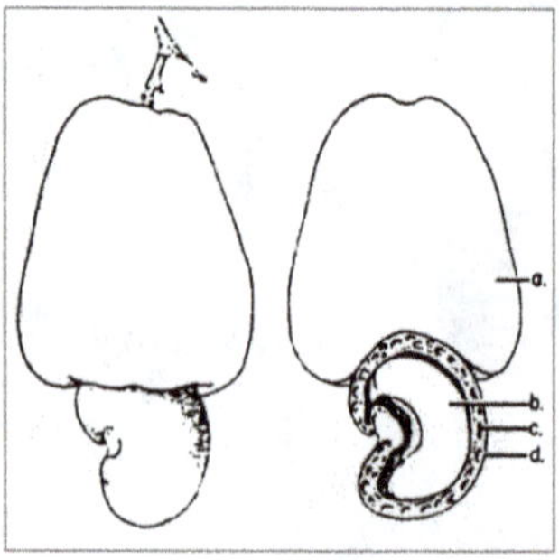

Fig. 2: a) Fleshy hypocarp, b) Seed, c) Secretary cavities of mesocarp, d) Exocarp, c) & d) together constitute the cashew nut shell

Cashew kernels are imported in large quantities by the United States and western European countries. The kernels are salted and eaten as a snack and they are used extensively in the manufacture of candies, cakes, cashew flour and cashew butter. Cashew kernels are highly recommended because, in contrast to the peanut, there is no risk of aflatoxin poisoning (Mitchell and Mori, 1987). The kernels are eaten either fresh or roasted and contain a milky juice which is used in puddings. The kernels are roasted and salted and the dried and broken kernels are sometimes imported to mix with old Madeira as they greatly improve its flavour. In roasting great care must be taken not to let the fumes cover the face or hands *etc.*, as they cause acute inflammation and external poisoning. Ground and mixed with cocoa, the kernels make a good chocolate. Cashew is a versatile nut with its unique combination of unsaturated fatty acids, proteins, carbohydrates, minerals and vitamins. India is the largest producer, processor and exporter of cashew in the world, followed by Brazil and Vietnam. At present cashew is consumed as a snack food in all parts of the world. There is, however, good scope for promoting cashew as a food ingredient as it blends well with every food preparation style. International Nut Council adopted a declaration calling for an international tree nut agreement under the auspices of *United Nations Conference on Trade*

Fig.3: Delicious cashew kernels

and Development (UNCTAD) with the objective of promoting the consumption of edible tree nuts on a worldwide basis (Nayar, 1998).

The cashew nut shell liquid (CNSL), contained in the oil cavities of the fruit wall, is a toxic yet very important by-product of the cashew industry. CNSL consists primarily of cardol, cardanol and anacardic acid. The CNSL is used in the manufacture of plastics, paints, resins and varnishes . In Tanzania, CNSL has been used for making tribal marks and scars on the face and body and in other countries, it is employed in preserving fish nets, protecting books, and in softening chicken eggs. Anacardic acid, the active ingredients in CNSL, is an effective larvicide used in the control of malaria-carrying mosquitoes. Anacardic acid also kills the snail hosts of schistosomes and inhibits the metabolism of several species of bacteria and molds. CNSL is useful as an anthelmintic against ascaridiasis (Mitchell and Mori, 1987). While the kernel of cashew continues to be the main product of commerce, the CNSL is used as an industrial raw material in the manufacture of paints, varnishes, brake linings for automobiles, friction dust etc (Nayar, 1998). The oil must be used with great caution, but has been successfully applied to corns, warts, ringworms, cancerous ulcers and even elephantiasis, and has been used in beauty culture to remove the skin of the face in order to grow a new one.

The hypocarp (apple) is succulent, sweet to very acidic in taste, and very fragrant. It is either eaten raw, compressed and strained to extract juice, fermented into cashew wine, or made into syrup, preserves, chutney, candy, pickles, or used as an ice cream flavour. The hypocarp, often called the cashew apple or pear, is a rich source of vitamin A and vitamin C. In some countries, such as Brazil, the drupe (nut) is less important than the hypocarp (apple). Medical uses of the hypocarp (apple) are varied and include the following: mouth wash and gargle; boiled in sweetened water and used as a treatment for dysentery; treatment for uterine ailments and dropsy; and as a diuretic. Hypocarp tannins are used as an antihypertensive (Mitchell and Mori, 1987). The apple is a red or yellow and has a pleasant subacidic stringent taste, the expressed juice of the fruit makes a good wine, and if distilled, a spirit much better than arrack or rum. The fruit itself is edible, and its juice has been found of service in uterine complaints and dropsy. It is a powerful diuretic. The pseudo fruit of the cashew, the cashew apple, is being utilized to a limited extent in India for conversion into alcoholic drinks and several processed products like cashew apple juice, beverages, jam, jelly, candy and vinegar (Nayar, 1998).

The young leaves are edible and older leaves in combination with other plants are used to treat skin diseases and burns. The bark has a variety of medicinal uses such as the treatment of diarrhea and constipation; a gargle; a treatment for ulcers in the mouth; a febrifuge; a medication to lower blood sugar; and a cure for tooth ache and sore gums. The bark and inflorescences are used in traditional Indian remedies for snake bite. The sap, which oxidizes black upon exposure to the air, is employed as an indelible ink marker for cottons and linens. It is also used as varnish and a flux to solder metals. A gum, which exudes from the bark, is used as a substitute for gum Arabic in book binding. The wood is employed locally for construction, storage cases, boats, wheel hubs, and as a fuel (Mitchell and Mori, 1987). The cashew tree is mainly used as fire-wood but also for certain commercial purposes like manufacture of plywood, particle boards, *etc.* (Nayar, 1998). The black juice of the nut (CNSL) and the milky juice from the tree after incision are made into an indelible marking-ink; the stems of the flowers also give a milky juice which when dried is hard and black and is used as a varnish. A gum is also found in the plant having the same qualities as gum arabic; it is imported from South America under the name of *Cadjii gum*, and used by South American bookbinders, who wash their books with it to keep away moths and ants. The caustic oil found in the layers of the fruit (nut) is sometimes rubbed into the floors of houses in India to keep white ants away.

Nutritional Value of Cashew

The seed (kernel) is highly nutritious (Bakhru, 1988a and 1988b) containing approximately 20 per cent good quality protein, 40 per cent fat (of which about 80% are non-saturated fatty acids) (Agnoloni and Giuliani, 1977) and 26 per cent carbohydrates (IBPGR., 1986).

Nagaraja and Krishnan Nampoothiri (1986) have chemically characterized the cashew apples and kernels from 16 high yielding varieties and have reported a good amount of variability for kernel protein (32.1 – 43.7%), starch (23.1 – 33.1%), total sugar (9.3 – 19.2%), amino acids (34.3 – 50.5%) and phenols (28.2 – 59.3%) and for apple ascorbic acid (144.4 – 269.1 mg. 100 g^{-1} fresh weight of cashew apple), total sugar (5.5 – 8.0%), reducing sugar (5.5 – 7.5%), amino acids (7.5 – 15.1%), phenols (8.2 – 23.0%), and tannins (27.4 – 153.4%). Murthy and Yadava (1972) have studied the oil and carbohydrate content of 24 cashew types and reported considerable variability for oil content of shell (16.5 – 32.9%), oil content of kernel (34.4 – 46.7%), kernel reducing sugar (0.9 – 3.1%), kernel non-reducing sugar (1.3 – 5.7%) and kernel total sugar (2.4 – 8.7%).

The protein content of the kernel varies from 13.3 to 25.03 per cent. The kernels are very nutritious, containing vitamins A, D and K and between 200-210 mg/100 g of vitamin E. Substantial amounts of calcium, phosphorus and iron are also present (Mitchell and Mori, 1987).

Cashew is the most versatile of all nuts and is the most popular nut used by the confectionery industry. In USA alone 87 per cent of the cashewnuts are used in nut salting. Whole kernels and pieces are being used in formulating confections, cakes and cookeries. The nuts are exalbuminous and rich in protein, calcium, phosphorus, unsaturated fats. vitamins (B1, B2, D, E and PP) and low in carbohydrates and saturated fats (Table 1). Hence, they are of high nutritive value. As the nut fats are complete, very active and easily digestible, the nuts can be used by both old and infants alike (Nair *et al.*, 1979).

Table 1: Composition of cashew kernels

Constituents	Percentage*	Constituent Aminoacids	Percentage*
Moisture	5.9	Arginine	10.3
Protein	21.0	Cystine	1.3
Fat	47.0	Histidine	1.8
Carbohydrate	22.0	Lysine	3.3
P	0.45	Methionine	1.3
Ca	0.55	Phenylalanine	4.4
Fe	5.0 mg/100 g	Threonine	2.8
		Tryosine	3.2
		Valine	4.5

The fleshy peduncle of the fruit is called "cashew apple", although most types are rather pear –shaped than apple-shaped. The apple is juicy and sweet when ripe and it varies in size, shape, colour, juice content and taste. Aiyadurai (1966) reported the existence of yellow, red and pink coloured apples. Albuquerque *et al.* (1960) noticed that the yellow apple tended to be heavier, softer and less astringent than red apples. Usually the apple is 10 times heavier than the nut. The apple is a rich source of vitamin C (250 mg/100 g fresh weight) which is 5 times higher than the vitamin C content of an orange (IBPGR., 1986). Since cashew apples perish within a few days after harvest they are only sold in local markets or processed as juice. In Brazil, Mozambique and Indonesia cashew apple is also important: it is eaten fresh or mixed in fruit salads and a drink is prepared from the juice. Cashew wine (slightly fermented juice) is enjoyed at harvest time and can be distilled to produce strong alcoholic drinks (Van Eijnatten, 1991). The apple is very juicy and the expressed juice has a brix of 12-14° containing 10.15 – 12.5% sugars (mostly reducing) and about 0.35% acid (as malic). It is known for its rich vitamin C content, upto five times that of citrus fruit. The apple is eaten as such by sucking the juice and discarding the residual fibrous mass. Sugar or salt is added sometimes for reducing the astringency (Nair *et al.*, 1979) (Table 2).

Table 2: Chemical composition of cashew apple

Constituents	Percentage
Moisture	87.8
Proteins	0.2
Fat	0.1
Carbohydrate	11.6
Calcium	0.01
P	0.01
Fe	0.2 mg/100 g
Vitamin C	261.5 mg/100 g
Minerals	0.2
β-carotene	0.09

Health Benefits of Cashew

Cashew is a versatile tree nut. It is, in fact, a precious gift of nature to mankind. The cashew kernel is a unique combination of fats, proteins, carbohydrates, minerals and vitamins. Cashew contains 47 per cent fat, but 82 per cent of this fat is unsaturated fatty acids. The unsaturated fat content of cashew not only eliminates the possibility of the increase of cholesterol, but also balances or reduces the cholesterol level in the blood. Cashew also contains 21 per cent proteins and 22 per cent carbohydrates and the right combination of amino acids, minerals and vitamins and therefore nutritionally, it stands on a par with milk, eggs and meat. As cashew has a very low content of carbohydrates, almost as low as 1 per cent soluble sugar, the consumer of cashew is privileged to get a sweet taste without having to worry about excess calories. Cashewnuts do not lead to obesity and help to control diabetes. In short, it is a good appetizer, an excellent nerve tonic, a stimulant and a body builder. Cashew is indigenous to Brazil, but India is the country that nourished this crop and made it a commodity of international trade and acclaim. Even today, India is the largest producer, processor, exporter and second largest consumer of cashew kernels in the world (Nayar, 1998).India with its extensive orchards and modern processing machinery, is the chief center of cashew production. Although *A. occidentale* is cultivated primarily for its kernel and fleshy hypocarp, it is occasionally used for reforestation in tropical America and in Dahomey (Mitchell and Mori, 1987). Research carried out at the University of Bologna has in fact indicated the presence in cashew kernel of numerous vitamins including vitamin E, considered by many to be aphrodisiac (Massari, 1994).

Cytotaxonomic Background

Basic Botany of Cashew

Cashew belongs to the family Anacardiaceae, the genus *Anacardium* and species *occidentale*. A taxonomic treatment of *Anacardium* (Anacardiaceae; Anacardieae), a Latin American genus of trees, shrubs and geoxylic subshrubs, is provided by Mitchell and Mori (1987). Anacardiaceae Lindl., the cashew family, is an economically important, primarily pantropically distributed family of 82 genera and over 700 species. This family is well known for its cultivated edible fruits and seeds (mangos, pistachios, and cashews), dermatitis causing taxa (*e.g., Comocladia, Metopium, Semecarpus, Toxicodendron, etc.*), and lacquer plants (*Toxicodendron* and *Gluta* spp.). Two genera, *Anacardium* L. and *Semecarpus* L. f., have an enlarged edible hypocarp subtending the drupe. One species of *Anacardium, A. microsepalum* Loesener, lacks the hypocarp and grows in the flooded forests of the Amazon where it may be fish dispersed (Mitchell and Mori, 1987; Pell, 2004). Anacardiaceae is a moderately large family consisting of Ca 74 genera and 600 species. It is subdivided into five tribes, namely Anacardieae , Spondiadeae, Semecarpeae, Rhoeae, and Dobineae. The tribe Anacardieae consists of 8 genera, namely, *Androtium, Buchanania, Bouea, Gluta, Swintonia, Mangifera, Fegimanra* and *Anacardium* (Mitchell and Mori, 1987).

Anacardium is one of the most economically important genera in the Anacardiaceae. This is due to *Anacardium occidentale* (the cashew of commerce), which yields: roasted cashew nuts (seeds), which are a major third world export to industrialized nations; cashew apples (hypocarps), which are consumed locally or used to make a widely marketed juice in South America, especially Brazil; and cashew nut shell liquid, which has medical and industrial applications. Some of the other species have economic potential but they are currently under-utilized. *A. excelsum* is used for construction and as a shade tree for coffee and cocoa plantations. *A. giganteum* is a locally important timber, and its hypocarps are relished by local people. The spectacular white leaves associated with the inflorescences of *A. spruceanum* make it a tree with excellent ornamental potential. *A. humile*, a subshurb closely related to *A. occidentale*, possess edible hypocarps and seeds. Selective breeding for higher quality hypocarps and seeds, and hybridizations with *A. occidentale*, could yield subshrubs with fruits that could be mechanically harvested. The economic potential of the other two subshrubs, *A. nanum* and *A. corymbosum* also should be investigated. (Mitchell and Mori, 1987).

Species of *Anacardium*

According to Bailey (1958) *Anacardium* is a small genus of eight species indigenous to South America. However, Agnoloni and Giuliani (1977) and Johnson (1973) have recognised eleven and sixteen species, respectively. Valeriano (1972) named five different species, namely, *A. occidentale* L., *A. pumilum* St Hilaire, *A. giganteum* Hanca, *A. rhinocarpus* and *A. spruceanum* Benth. He also suggests recognition of only two species namely *A. nanum* and *A. giganteum* which can further be sub-divided based on the colour (yellow or red), and shape (round, pear-shaped or elongated) of the pseudo-fruit. He also considers the division into dwarf and giant species to be the only way to classify cashew in a rational and practical way. His arguments are based on the characteristics of pseudo fruits. However the description provided by Peixoto (1960) separates recognition of more than two species. It appears from the published accounts that *A. occidentale* L. is the only species which has been introduced outside the New World. Within Central and South America as many as 20 species of *Anacardium* are known to exist (Nair *et al.* 1979). *Anacardium* L. is a small genus of trees, shrubs and sub-shrubs indigenous to the tropics. Some authors report 10 species and others until 20 species. It is native to Latin America, having a primary center of diversity in Amazon and a secondary center in the Planalto of Brazil. Only one species is an incipient domesticate commonly known as cashew, (*Anacardium occidentale* L.) (CGIAR, 2015). *Anacardium* species names and synonyms approved by most authorities are as under (MMPND, 2001):

- *Anacardium corymbosum* Barbosa Rodriguez
- *Anacardium excelsum* (Bertero & Balbis ex Kunth) Skeels
- *Anacardium giganteum* Hancock ex Engler
- *Anacardium humile* St. Hilaire
- *Anacardium latifolium* Lam. -> *Semecarpus anacardium* L.f.
- *Anacardium longifolium* Lam. -> *Semecarpus cuneiformis* Blanco
- *Anacardium macrocarpa* Engler
- *Anacardium microcarpum* Ducke -> *Anacardium occidentale* L.
- *Anacardium nanum* St. Hilaire
- *Anacardium negrense* Pires & Fróes
- *Anacardium occidentale* L.
- *Anacardium officinarum* Gaertn. -> Semecarpus anacardium L.f.
- *Anacardium orientale* auct. ex Steud. -> Semecarpus anacardium L.f.
- *Anacardium pumilum* St. Hilaire
- *Anacardium rhinocarpus* (Bertero & Balb. ex Kunth) DC.
- *Anacardium spruceanum* Bentham ex Engler

However, Mitchell and Mori (1987) recognised ten species in the genus *Anacardium*, one of which, *A. fruticosum*, is described as new. The species of Anacardium and there synonyms are presented in Table 3. The genus has a primary centre of diversity in Amazonia and a secondary centre in the Planalto of Brazil. All known species of the *Anacardium* genus can be found in the American continent, only four of them (*A. coracoli, A. encardium, A. excelsum, A. rhinocarpus*) do not exist in Brazil. There, the high number of wild species suggests that the North-East is the site of origin for *Anacardium* genus and namely for *Ancardium occidentale* L. In fact, here different forms of cashew can be found with a high variability for local populations, namely along the coast and dune areas. Now-a-days most species belonging to the *Anacardium* genus are found everywhere in Brazil (NOMISMA, 1994). Ascenso (1986) reported that cashew (*A. occidentale* L.) is the only species in the genus which attained economic importance. The *Anacardium* genus appeared to have originated in the Amazon region of Brazil and hence speciation followed different geographic patterns. Although of tremendous economic importance, no work has been done with the intraspecific classification of *A. occidentale*. A few cultivars have been named and the only major distinction established is that between yellow and red hypocarp cultivars. Some trees have been selected for nut quality, whereas others have been selected for hypocarp size, colour, succulence and flavour (Mitchell and Mori, 1987).

Table 3: Species of Anacardium and their synonyms

Sl. No.	Species	Synonym (s)
1.	*A. occidentale* Linnaeus	*Cassuvium pomiferum* Lamarck *Acajuba occidentale* (L) Gartner *Anacardium occidentale* var. *americanum* de Candole *A. occidentale* var. *indicum* de Candole *A. mediterraneum* Vellozo *A. curatellifolium* A. St. Hilaire *Cassuvium reniforme* Blanco *A. occidentale* var. *longifolium* Presl *A. occidentale* var. *gardneri* Engler *A. subcardatum* Presl *A. microcarpum* Ducke *A. rondonianum* Machado *A. amilcarianum* Machado *A. kuhlmannianum* Machado *A. othonianum* Rizzini
2.	*A. giganteum* Hancock ex Engler	*A. giganteum*. Loudon ex Steudel.
3.	*A. humile* St. Hilaire	*Monodynamus humilus* Pohl *A. subterraneum* Liais *A. pumilum* St. Hilaire ex Engler *A. pumilum* var. *petiolata* Engler *A. humile* var *subacutum* Engler *A. humile* Martius
4.	*A. microsepalum* Loesener	*A. negrense* Pires & Froes
5.	*A. excelsum* (Bertero & Balbis ex Kunth Skeels)	*Rhinocarpus excelsa* Bertero & Balbis ex Kunth *Anacardium rhinocarpus* de Candole *Anacardia rhinocarpa* St. Lager *Anacardium caracoli* Mutis
6.	*A. parvifolium* Ducke	*A. tenuifolium* Ducke
7.	*A. corymbosum* Barbosa Rodrignes	--
8.	*A. spruceanum* Bentham ex Engler.	*Anacardium brasiliense* Barbosa Rodrigues
9.	*A.nanum* St.Hilaire	*A. pumila* Walpers
10.	*A. fruticosum* Mitchell & Mori	--

(Mitchell and Mori, 1987)

Brief Description of Species of Anacardium

A brief description of ten species of *Anacadium* genus recognised by Mitchell and Mori (1987) are detailed below:

1) *Anacardium occidentale* Linnaeus (Cashew; caju)

A. occidentale is probably an indigenous element of the savannas of Colombia, Venezuela, and the Guianas. It is clearly a native and occasionally a dominant feature, of the *cerrados* (savanna-like vegetation) of Central and Amazonian Brazil. The *cerrado* populations of *A. occidentale* differ from the *restinga* populations by having undulate, thickly coriaceous leaves with short, stout petioles. The hypocarps (cashew apples) of *cerrdo* trees are usually smaller and sometimes have a more acidic flavour than those of the *restinga*. The natural distribution of *A. occidentale* extends from northern South America, the West Indies, or South America west of the Andes. *A. occidentale* originally evolved in the *cerrados* of Central Brazil and later colonized the more recent *restingas* of the coat. Central Brazil is a center of diversity for *Anacardium* where the distribution of *A. occidentale* overlaps the ranges for *A. humile, A. nanum*, and *A. corymbosum. A. humile*, the closest relative of the cultivated cashew, is closer morphologically to the *cerrado* ecotype than it is to the *restinga* and cultivated populations of *A. occidentale. A. occidentale* grows in the coastal *restinga* of eastern Brazil, the thorny *caatinga* of north eastern Brazil, and the *cerrados* of Central and Amazonian Brazil. It is also present in the savannas of the Guianas, Venezuela and Colombia. In India, East Africa, and Malaysia, where *A. occidentale* is an introduced cultivar, it readily invades and becomes established in native vegetation. Ecologically the cashew is very adaptable as it can tolerate extended periods of drought and poor soils (pH 4.5 – 6.5). The most important limiting factors in its growth are water-logged or calcareous soils and frosts. Cold has the most severe effect on young trees and this probably limits the species to less than 1500 m above seas level. In the Greater Antilles, *A. occidentale* flowers from December to August with peak flowering from February to Aril. A secondary flowering peak occurs in August. Fruits usually are produced from February to August. In the Lesser Antilles, flowering occurs from November to August with peak flowering from February to May. The Central American populations generally flower from January to April and fruiting from December to July. In Colombia and Venezuela flowering and fruiting are either distributed throughout the year or varies regionally, and, in the Guianas, flowering and fruiting occur throughout the year with peaks in April to July and from September to November. The eastern Amazonian populations of *A. occidentale* flower from December to February and from May to September. In western Amazonia, flowering occurs throughout the year with peaks from February to June and from August to October. In northeastern Brazil flowering occurs throughout the year with peak from June to December. In southeastern Brazil, flowering is throughout the year with peaks in January and February and September and October. Fruiting is primarily from November to April. In Central Brazil, flowering mostly occurs from June to October with the majority of the fruits produced in October. In general, throughout its range *A. occidentale* flowers most profusely during the dry season. In many areas, two fruits crops are produced yearly. Flowering is controlled by several environmental cues,. An increase in sunshine and moisture stress concomitant with a decrease in relative humidity following the end of the rainy season induces bud break. Then a new flush of leaves is produced which is directly followed by flowering. Low temperatures, however, delay flowering . Staminate flowers open before bisexual ones. Most cashew flowers open between 0600 and 1800 with peak anthesis occurring between 1100 to 1230. The stigmas are receptive as soon as the flowers open and the anthers dehisce one to five hours later. *A. occidentale* is highly out-crossed. Experiments have shown that maximum fruits set (80%) is obtained by crossing emasculated flowers with pollen from another plants, whereas self pollinated flowers gave a much lower fruit set (40%). The flowers are pollinated by bees, wasps, ants, flies, and possibly humming birds. In cashew plantations, most of the pollinations is accomplished by honey bees. Natural populations of *A. occidentale* are pollinated by bees and butterflies. The primary dispersal agents of the fruits of *A. occidentale* are probably frugivorous bats. Bats fly into a tree or shrub, seize the fruit, eat the fleshy hypocarp, and then discard the poisonous drupe. Bat dispersal also has been reported for *A. excelsum* and bats are probably the primary dispersal agents for most species of the genus. However, water has been suggested as a secondary means of dispersal. Acaju, derived from the Tupi Indians of Brazil, became caju in Portuguese and variants of caju in tropical Asian aand African countries. The English name cashew is clearly derieved from the Portuguese caju. In Brazil the cashew nut is called castanha and the hypocarp is frequently referred to as caju manso. In India cashew is also known as Agnikrita, guchapushpa, hajli badam, jidi-mamadi, kaju, kaju kalinga, kajutaka, kere-mara, parangimaru, prithagabija, xophahara and vrittapatra. *A. occidentale* is a polymorphic speices. The *restinga* ecotype of *A. occidentale* of eastern coastal Brazil is easily differentiated from *A. humile*. However, the cerrado ecotype of *A. occidentale* sometimes overlaps in leaf morphology with *A. humile*. The principal difference between *A. occidentale* and *A. humile* is that the former is always a tree and the latter is a subshrub with a massive underground root system and rigidly ascending branches. The *cerrado* ecotype of *A. occidentale* frequently has broadly obovate leaves with an obtuse or slightly auriculate base, whereas the leaves of *A. humile* are generally oblanceolate with attenuate bases. Moreover, the majority of flowers of *A. occidentale* have 9-10 stamens while the majority of flowers of *A. humile* have 7-8 stamens. The types of *A. curatellaefolium* St. Hil., *A. rondonianum* Machado, *A. kuhlamannianum* Machado, and *A. othonianum* Rizzini exemplify the *cerado* ecotype of *A. occidentale*. Hypocarp pyriform, much larger in cultivated forms than in wild populations, 5-20 x 2-8 cm, yellow, orange, or red. Drupe subreniform, 2-3.5 x 1-2 cm, gray or brown at maturity. The mature hypocarp is often glabrous. Drupe symmetry bilateral, fruit wall (pericarp), 3-4 mm thick, pericarp consists of exocarp, mesocarp and endoscarp. The exocarp uniseriate, the cell walls thick, lignified, the cuticle thin. Mesocarp is the thickest part of pericarp. Secretary cavities dominant feature of mature pericarp which contain yellow, caustic liquid (cashew nut shell liquid) consisting of phenols such as anacardiol and cardol. Endocarp mostly of mechanical tissue. Seeds edible, 2-4.5 x 0.9-3.7 cm. Endosperm scanty or poorly developed. Testa with thick layer of crushed, lignified parenchyma. Cotyledons with very thin cuticle.

Some authors say that *A. microcarpum* is synonym of *A. occidentale. A. microcarpum* is a wild cashew relative featuring 3-4.5 cm fruits that are smaller than the more common cashew apple. As with other cashew species, the fruit is the hard nut, with the bulbous pseudofruit (the apple) ripening to red and having an acidic sweet flavor. A small tree, usually growing to 3-7.5 cm. Fruits form in large clusters, with well-sized panicles. Overall appearance is much like the cashew apple tree. Fruits and nuts are used much like the regular cashew. Fresh pseudofruits are edible and the roasted nuts are edible and quite tasty. The pseudofruits are harvested and sold in local markets. It is native to scrublands and non flood plain zones of the lower Amazon region of Brazil.

2) *Anacardium giganteum* Hancock ex Engler (caju)

Trees, to 40 m x 300 cm. Trunk cylindrical, bark very thick, gray, moderately coarse with vertical fissures, the inner bark pinkish-brown. Hypocarp pyriform, 1.3 x 1-5 cm, red. Drupe subreniform, black, 27 x 18 mm. The distribution is from the Pacific coast of Colombia and Loreto, Peru south to northern Mato Grosso and east to Surinam and Maranhao, Brazil. A large tree growing in moist, *terra firme* forests. It flowers from November to January and in June and August. In the state of Para, Brazil, it normally flowers at the beginning of the wet season (December – January) and the fruits mature in the middle of the wet season (March). Individual trees will frequently flower every other year. The flowers of *A. giganteum* change colour after pollination from yellow to white to dark red. The single fertile stamen is only 0.5 mm long in unpollinated bisexual flowers, but it increases to 4.5 – 5 mm long and dehisces after pollination (*i.e.*, in flowers with red corollas). This suggests that self pollination is inhibited by protogyny. The hypocarp is eaten by spider and capuchin monkeys. The seed is sometimes dispersed endozoochorously by tortoises. The wood is easily worked and finishes smoothly. The ripe hypocarp is edible and a fine red wine can be prepared from its juice. However, the quality of the ripe hypocarp varies from very sweet and tasty to extremely acidic and usually they are too sour for eating. *A. giganteum* is abundant in the forests of Parque Nacional do Tumucumaque and Amapa, Brazil where it is an important item in the diet of the native Indians. They consume the hypocarps mixed with the flour of mandioca (*Manihot esculenta* Crantz). The seeds are toxic when raw but edible when roasted, and are said to be as delicious as those of the commercial cashew.

3) *Anacardium humile* St. Hilaire (caju anao)

Subshrub 30 to 150 cm tall, with large underground trunk and rigidly ascending branches. Hypocarp obconical to pyriform, 1-3 x 1-2 cm, red or yellow when ripe. Drupe subreniform, 1.3 – 2.3 x 1.-1.7 cm, green gray, or dark brown at maturity. The distribution extends from Santa Cruz, Bolivia south to eastern Paraguay and in Brazil from southeastern Rondonia and northern Goias to Parana. A very common subshrub of savanna like vegetation (*campo* and *cerrado*) between 100 and 1200 m alt. It often grows in very dense patches usually 40 to 75 cm in height. The patches are the result of an extensive lateral branch system. *A. humile* has a very long and thick tap root that penetrates deeply though the very hard soil of the *cerrado* to reach the low water table, especially during the dry season. Radiating from the central subterranean axis are large plagiotropic branches from which are produced orthotropic shoots that emerge above the ground. The above ground parts of *A. humile* consist of short, rigidly ascending branches, tight clusters of leaves and terminal or axillay inflorescences. The prostrate form of *A. humile* is well adapted to the frequent and severe fires, seasonally dry environment, poor soils, and the low water table of the *cerrado*. The majority of the underground parts of *A. humile* represent a subterranean trunk of stem, not root. Flowering occurs primarily between July and October , and peak fruiting takes place in October and November at the beginning of the wet season. *A. humile* is pollinated by bees and butterflies. The hypocarps are eaten by parakeets, bats, and terrestrial mammals. The hypocarp is eaten raw by natives. It is also made into preserves, candies, juice, liquors and wine in the same ways as that of *A. occidentale*. The roasted seed ('nut') is edible.

4) *Anacardium microsepalum* Loesener

Trees, to 20 m x 50 cm. Bark smooth with scattered lenticels, the inner bark reddish-brown, forming a resinous exudate when cut. A lenticel is a porous tissue consisting of cells with large intercellular spaces in the periderm of the secondarily thickened organs and the bark of woody stems. *Hypocarp absent.* Drupe reniform, 20-30 x 19-26 mm, glabrous, green at maturity, Pedicel not accrescent. The distribution extends in Central Amazonian Brazil from the upper Rio Negro south to Rondonia and east to Borba and Manaus. A medium-sized to large tree in seasonally inundated forests. Its fruits may be dispersed by water. The seeds of *A. microsepalum* are eaten by fish. The flowers appear from December to June and the fruits are present from December to June. The roasted cashew seeds ('nuts') are eaten in Northwestern Amazonia. *A. microsepalum* is the only species of the genus without a fleshy hypocarp (Fig.4).

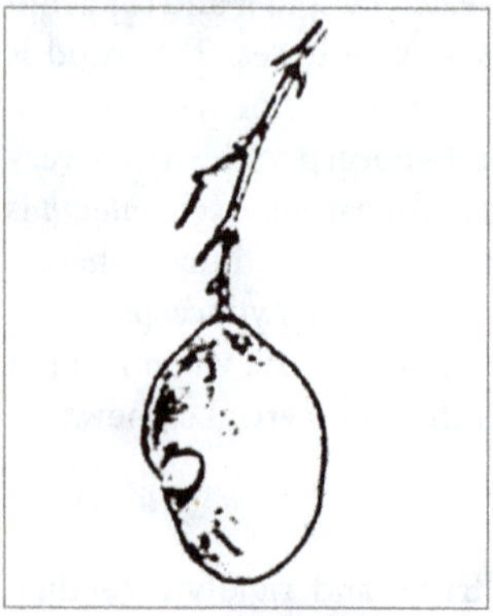

Fig. 4: *Anacardium microsepalum* (Note absence of hypocarp)

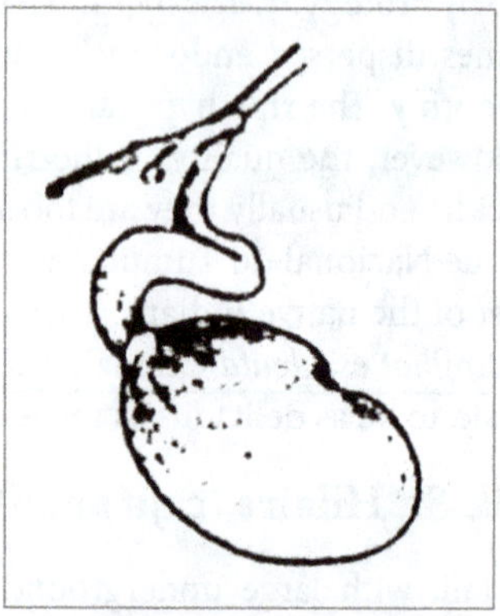

Fig. 5a: *Anacardium excelsum* (Note slender, sigmoid hypocarp)

Fig. 5b: *Anacardium excelsum* (Note slender, sigmoid hypocarp)

5) *Anacardium excelsum* (Bertero & Balbis ex Kunth) Skeels (Wild cashew)

Tree, to 50 m x 30 cm, with cylindrical trunk, *slightly swollen at base. Hypocarp slender, sigmoid,* 2-4 x 1.5 – 2 cm, green. Drupe reniform, 23-34 x 14-20 mm, glabrous, green at maturity (Figs. 5 & 6). The distribution is from southern Honduras south to Los Rios, Ecuador and east to Aragua, Venenzuela. This is one of the largest trees in tropical America, *A. excelsum* grows in moist upland forests and in gallery forests adjacent to dry forests generally below 1000 m alt. It is frequently a dominant species in primary and secondary forests. *A. excelsum* usually drops its leaves in the early dry season. The flower appear from January to April (May) shortly after the new leaves are flushed and the fruits mature from March to May. On Barro Colorado Island peak flowering is in late February and the seeds generally mature in April. *A. excelsum* is pollinated primarily by settling moths. The hybpcarps are eaten by coatis, white-faced and howler monkeys; the bat

Fig. 6: *Anacardium excelsum* (Note slightly swollen trunk at base)

Micronycteris hirsute and other bats. The hypocarps dropped to the ground by howler monkeys are eaten by basilisk lizards (*Basilisicus basiliscus*). Neither the monkeys nor the lizards eat the drupes and further it washypothesized that the "toxic fruits and edible pedicels may be an evolutionary response of *Anacardium* to insure the dispersal of fruits without injury to the fruit". However, it is reported that the monkeys and lizards never carry the fruit far from the parent tree and suggested that birds and bats may be the primary dispersal agents. The hypocarp of *Anacardium* is carried by bats to feeding roosts, where the untouched seeds are dropped after the frugivorous bat eats the receptacle. Exudates from the trunk of *A. exelsum* are an important items in the diet of the Panamanian tamarin (*Saguinus oedipus geofrooyi*). The macerated bark is used as barbasco (fish poison) in Panama and the wood is employed for bridge boards, fence posts, concrete forms, rough construction, general carpentry, boxes and crates. However, the wood is susceptible to dry rot and termite attack and is hard to work. *A. excelsum* has potential as a plantation crop. The seeds (nuts) are toxic when raw but edible when roasted and the hypocarps are edible. *A. excelsum* is cultivated in Cuba and Ecuador.

6) *Anacardium parvifolium* Ducke

Large trees, 22 to 40 m x 100 cm, Bark smooth. Hypocarp pyriform. 2 x 1.4 cm, red. Drupe reniform, 13 x 10 mm. The distribution extends in Amazonian Brazil from extreme western Amazonas east to Maranhao and Amapa. A large tree of moist primary forests growing in flood plains and upland sites. It flowers from May to November and fruits have been collected in June.

7) *Anacardium corymbosum* Barbosa Rodrigues

Subshrub 50 to 150 cm tall, with large underground trunk and rigidly ascending branches. Hypocarp subreniform, 1.5 – 2 x 1 – 1.7 cm, dark brown at maturity. The distribution is endemic to south central Mato Grosso, Brazil. A subshrub of savannas (*campo* and *cerrado*) between 100 and 800 m alt. Flowering occurs from June through October and fruiting commences in October. The hypocarps, according to local informants, are eaten by birds and mammals. The hypocarp is eaten raw by local people in Brazil.

8) *Anacardiuim spruceanum* Bentham ex Engler (Forest cashew; caju assu)

Tree to 35 m x 100 cm, with cylindrical trunk slightly swollen at base. Bark smooth. Hypocarp obconical or pyriform, 100 x 6-15 mm, very juicy, white, red, or yellow, with strong resinous smell. Drupe reniform, 14-15 x 13-20 mm, black at maturity. The distribution is from Bolivar, Venezuela south to Rondonia, Brazil and Pando, Bolivia and east through the Guianas to eastern Para, Brazil. A large tree in moist primary and old secondary forests growing in flood plains and on upland sites. *A. spruceanum* is a relatively uncommon tree. Out of 1000 trees sampled in Camaipi, Amapa, only five individuals of *A. spruceanum* were encountered, and in Saul, French Guiana not one tree was found in a survey of 800 tress. Scattered large individuals of *A. spruceanum* were found outside of the transect in the Saul study. *A. spruceanum* flowers from April to September and from November to January with peak flowering in July and August. Mature fruits appear at the beginning of the wet season. This species is probably dispersed by bats. The green and white foliage of the outer branches associated with the inflorescences give the tree a magnificent appearance when in flower and therefore *A. spruceanum* has been recommended as an ornamental for tropical climates. Populations of *A. spruceanum* with white hypocarps appear to be concentrated in French Guiana and Amapa, Brazil, whereas specimens with red hypocarps have been colleted from the area around Manaus, Brazil. *A. brasiliense* is treated here as a synonym of *A. spruceanum* because of Barbosa Rodrigues reference to the leaves as being of a different colour than in other species of the genus. Certainly, he was referring to the showy white leaves and foliaceous bracts subtending the rachises of the inflorescence of *A. spruceanum*, a feature unique to this species.

9) *Anacardium nanum* St. Hilaire

Subsrub, 30 to 150 cm tall, with large underground trunk, 35 to 65 cm diam. The distribution extends in Brazil from central Goias and the Distrito Federal south to central and western Minas Gerais. A subshurb of savannas (*campos* and *cerrados*) between 700 and 900 m alt. It flowers from May to August and is pollinated by bees and butterflies.

10) *Anacardium fruticosum* Mitchell and Mori

Low spreading tree, 2-3 m tall. The distribution is known only from the upper Mazaruni river basin, Guyana. A shrub to low spreading tree in savannas from 460 to 1250 m alt. It flowers from June to October. The fruits have not yet been collected. This species is very similar to *A. parvifolium*, differing primarily in being a shrub or low spreading tree 2-3 m tall growing in savannas and bearing relatively large, coriaceous leaves. *A. parvifolium*, on the other hand, is a tall rain forest tree with chartaceous, smaller leaves.

Cytogenetic Studies

Cytology of *Anacardium occidentale* L. has not been studied in detail. The chromosome number is reported only for *A. occidentale*. This morphologically polymorphic species also exhibits choromosome polymorphism (Mitchell and Mori, 1987). Chromosome numbers reported in the literature range from 2n=24 (Khosla *et al.*, 1973; Goldblatt, 1984), 2n=30 (Machado, 1944), 2n=40, (Simmonds, 1954; Goldblatt, 1984) to 2n=42 (Darlington and Janaki Ammal, 1945; Khosla *et al.*, 1973; Goldblatt, 1984; Purseglove, 1988). Such chromosome polymorphism is well known in many domesticated trees (Khosla *et al.*, 1973).

By comparison with other tropical industrial crops as oil palm, coffee, cacao and tea, very little cashew-improvement research has been done, owing to lack of adequate knowledge of cytology and genetics of the crop. The importance of cytological information to crop improvement cannot be overemphasized. Cytological studies have a lot to help in resolving the origin and evolution of plant species. Since the basis of improvement is based on variation, and variation has both genetic and non-genetic components, comparative work should provide useful data for solving the problems of low fertility, incompatibility *etc*, which are key components in tree crop breeding. These two variations bring about changes in the chromosome either structurally or morphologically are bound to affect the DNA components and therefore have genetic consequence. It has also been known that chromosomal or cytological studies help in determining the path of evolution of new species. Cytogenetics has been employed in agriculture for the development of improved cultivars especially in identifying the cause(s) of infertility in organism.

Cytological and breeding investigations in cashew are few compared with other crops. Hutchinson and Dalziel (1954) reported diploid chromosome number of 2n = 42 in cashew. Aliyu, and Awopetu (2007) quoted Deckers *et al.*(2001) that Cashew is a dicotyledonous evergreen tree with a morphologically polymorphous species that has been reported with polymorphic chromosome number 2n = 24, 30, 40, 42. Two major populations of cashew (Brazilian and Indian) were studied by Aliyu, and Awopetu (2007). The study revealed that Indian accessions are potentially prolific and consistent in fruiting, but with characteristic small-medium sized nuts and low-premium kernels. Meanwhile, Brazilian accessions produce large nuts with potentially good quality kernels, but very low and inconsistent in fruit yield. Based on the report of this diversity work, a recurrent selection breeding strategy has been developed that will involve the use of hybridization of identifiable promising genotypes as parents. The wide variability in germplasm collections of cashew offers opportunities for the exploitation of useful genes for improvement of the crop. However, to achieve the desired improvement through this breeding strategy, knowledge of basic cytology (number, structure and behaviour of chromosomes and pollen grain fertility) of these selections is very essential. The mitotic chromosomes of the studied accessions are presented in Fig.7-10 with their corresponding schematic drawings. Diploid chromosome number of 2n = 42 was recorded among the two selected cashew populations. The karyotypic observations on the chromosomes were recorded among the Indian and Brazilian populations comprising mostly of medium and large jumbo sized nuts. The total length of the homologous chromosomes recorded for the Indian cashew population was found to be 51.10 μm, and chromosomes were designated 1 - 21, according to decreasing lengths. Karyotyping was based on the absolute length of the chromosome and ratio of chromosome short arms to long arms (centromeric location). The chromosome compliment gave a karyotypic formulae of 6Asm + 1Am + 4Bsm + 5Bm + 5Cm, while A represent chromosome >3.00 μm, B = 1.50 - 2.99 μm and C < 1.49 μm. Meanwhile, the chromosome lengths ranged between 1.00 and 4.20 μm for the shortest and the longest respectively. Based on the morphology of the chromosomes, the complement comprises of 6 long submetacentric, 1 long metacentric, 4 intermediate submetacentric, 5 intermediate metacentric and 5 small metacentric chromosomes with regular mitotic division (Fig. 8). The mitotic metaphase chromosome of Brazilian cashew population is presented in Fig. 9. The total length of the homologous chromosomes recorded for the Brazilian cashew population was found to be 56.00 μm. Individual chromosome length ranged between 1.00 and 4.50 μm. The chromosome karyotype was very similar to that of Indian population comprising, 6Asm + 1Am + 1Ast +9Bm+ 2Bsm + 2Cm. It however shows that the complement includes, 6 long submetacentric, 1 long metacentric, 1 long subtelocentric 9 intermediate metacentric, 2 intermediate submetacentric and 2 small metacentric chromosomes with regular mitotic division (Fig. 10). Apart from the slight variation in the chromosome length observed between the two populations, the chromosome stainability and behaviour during different stages of cell division was quite similar and regular divisions were recorded in their anaphase phases (Fig. 9 and 10). Regular meiotic divisions of 21 bivalents (Fig. 11) were commonly recorded among the two studied population with common occurrence tetrads in their pollen mother cells (Fig. 12). This observation is an indication of potential high fertility of the pollen grain of cashew acession in these two populations. Few triads signifying iregularities during meiosis (microsporogenesis) was obtained in the pollen mother cells of

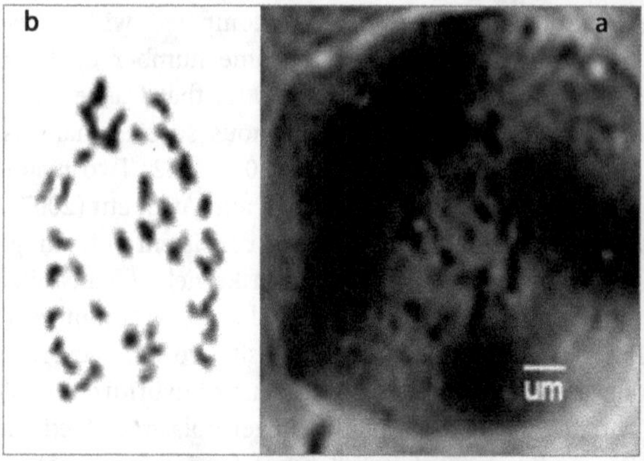

Fig. 7 (a): Mitotic chromosomes of Indian cashew accession. (b): The corresponding schematic drawings. Magnification: x400.

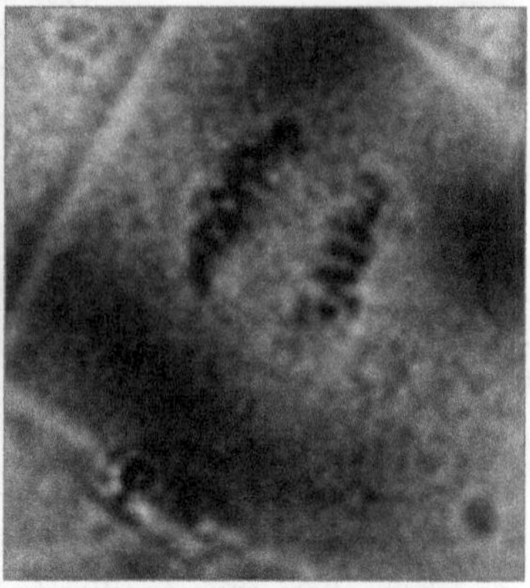

Fig. 8: Late anaphase of Indian cashew accession showing regular mitotic chromosome division. Magnification: x400.

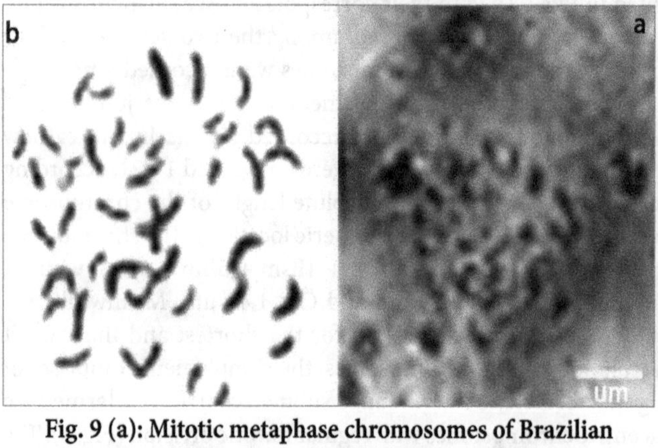

Fig. 9 (a): Mitotic metaphase chromosomes of Brazilian cashew accession. (b): The corresponding schematic drawings. Magnification: x400.

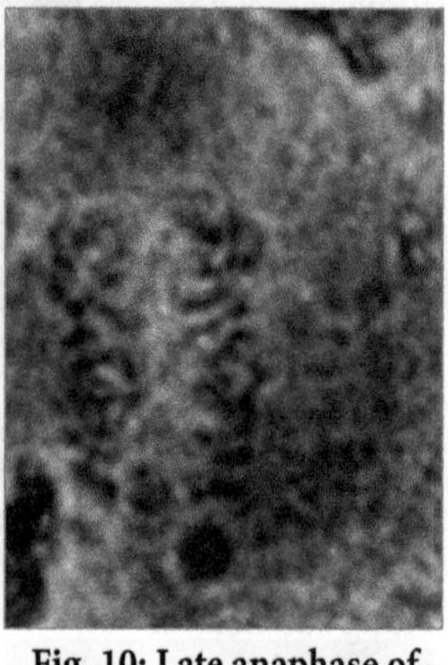

Fig. 10: Late anaphase of Brazilian cashew accession showing regular mitotic chromosome division. Magnification: x400.

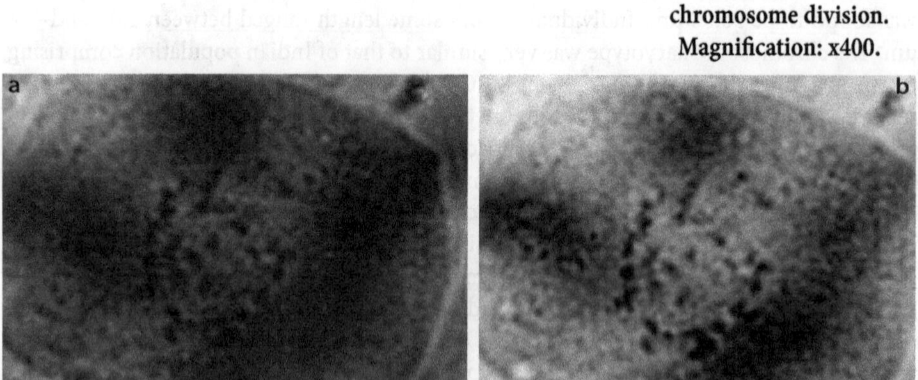

Fig. 11: Mieotic cell division in Indian (a) and Brazilian (b) cashew accessions showing 21 bivalents

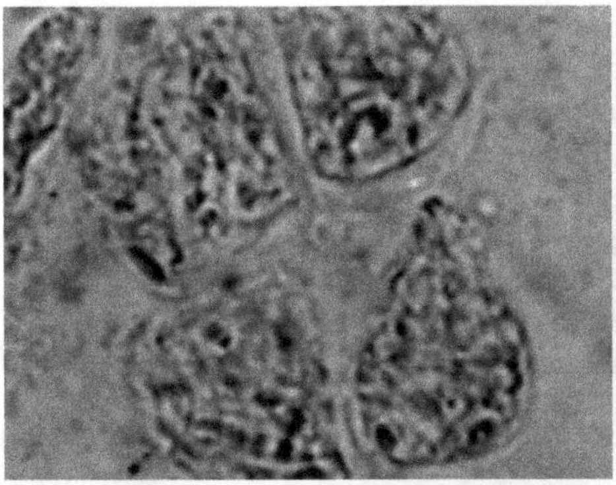

Fig. 12: Tetrads in pollen mother cell showing regular meiotic division. Magnification: x400.

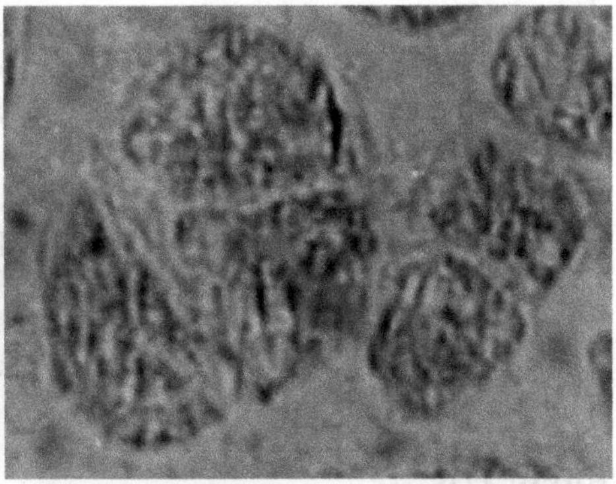

Fig. 13: Triads showing irregularities in microsporogenesis in Brazilian accessions. Magnification: x400.

three Brazilian accessions (Figure 13). The occurrence of triads in the pollen mother cells of these three accessions would probably have implication on the fertility of their pollens. The chromosome lengths recorded in haploid set in this study tends to group into 3 sets (A B and C of karyotypic formulae), probably corroborating the 7x basic chromosome number. Mabberley (1997) reported basic chromosome number of between 7 and 16 in *Anacardiaceae.* The relative similarity to previous findings probably suggests that the crop species is relatively stable, with very little changes in the chromosomes. The similarity in chromosome number of the accessions in the two evaluated cashew populations also suggest that the materials are morphologically and genomically close, and probably have a common progenitor. The chromosome morphology of these two populations was very similar with respect to the range and gradation of chromosome length and the position of centromere. The chromosomes are mostly metacentric and submetacentric with regular mitotic cell division. Although the chromosomes are relatively small, slight differences in their size and morphology were detected. Most of the accessions analysed for the two populations have symmetric karyotypes, composed mainly of metacentric pairs with several submetacentrics. The presence of a pair of subtelocentric chromosome in the Brazilian population would seem to be exceptional in the cashew cultivars. This fact probably suggests intraspecies variation, accompanied by few changes in the karyotype constitiution of the species. Degree of genomic closeness observed suggests that gene exchange among the accessions could be possible. Regularity of the chromosome behaviour during meiotic division and free flow of gene exchange observed in the fruit set study attest to the commonness of the progenitor of these accessions. Close similarity in the chromosome number, behaviour and structure of these accessions corroborate high degree of similarity in the tree morphological appearance. Regularity of chromosome behavior during meiotic cell division probably attests to the high pollen grain fertility recorded among the selections. It is concluded that despite the increased cultivation of cashew as a commodity crop in sub-Saharan Africa, Asia and South America there are few chromosome studies on it. The present study investigates number, structure and behavior of chromosome in cashew populations growing in Nigeria. Cytological examination of these populations revealed a diploid and haploid chromosomes of $2n = 42$ and $n = 21$ respectively. The karyotypes were mostly symmetric, composed mainly of metacentric pairs and several submetacentrics. Similarity in the morphology, number and behavior of the chromosomes in the accessions from different populations or origin attests to the degree of genetic closeness of the selections. This probably indicates high potential for use as parents in the breeding and improvement of cashew with very limited cross-incompatibility barriers (free gene exchange). Polymorphism in chromosome number was not recorded among these cashew selections. The diploid chromosome number of 2n = 42 and haploid of n = 21 obtained in this study agrees with Hutchinson and Dalziel (1954), but disagree with the polymorphism of 2n = 24, 30, 40, and 42 reported by Deckers *et al.* (2001). Archack *et al.* (2003b) quoted Purseglove (1988) that cashew has a chromosome number of 2n = 42; however, the ploidy is unclear. The accessions are probably polyploids with basic chromosome number x = 7.

Early History

Origin, Distribution and Domestication

The genus *Anacardium* is a native to latin America and has a primary centre of diversity in Amazonia and secondary one in the Planalto of Brazil. Behrens (1998) described the crop as a tropical tree species that originated from South America. Natural occurrences of cashew have been reported for Mexico to Peru and in the West Indies. The crop was introduced into India, the East Indies and Africa by the Portuguese explorer in the 16th century. Thereafter, exploitation of cashew for its fruits (nut and apple) among local people appears to have been the pattern for more than 400 years in Asia and Africa (Mitchell and Mori, 1987). *Anacardium* includes ten species which are naturally distributed from Honduras south to Parana, Brazil and eastern Paraguay. The genus is found west of the Andes in South America only in Venezuela, Colombia, and Ecuador. *Anacardium occidentale* is cultivated or adventive throughout the New and Old World tropics. The genus has two centres of diversity –Central Amazonia and the Planalto of Brazil. This is illustrated by the occurrence of four species in the vicinity of Manaus and by three species occupying the same habitat in the Federal District, Brazil. The following five distribution patterns of *Anacardium* were reported by Mitchell and Mori (1987):

1. *A. excelsum* is isolated taxonomically and geographically from its congeners (species within the same genus) by the Andes. The uplift of the Andes was probably the driving force in the early differentiation of *A. excelsum* from the rest of the genus.
2. *A. giganteum* and *A. spruceanum* have Amazonian-Guyanan distributions.
3. *A. occidentale*, which is the most widespread species in the genus, has disjunct populations in the Planalto of Brazil, the *restingas* of eastern Brazil, the savannas of the Amazon basin, and the Llanos grasslands of Colombia and Venezuela. It should be kept in mind, however, that the natural distribution of this species is obscured by its widespread cultivation in both the Old and New World.
4. Three closely related species, *A humile*, *A. nanum* and *A. corymbosum*, are restricted to the Planalto of central Brazil.
5. Two species of *Anacardium* are narrow endemics. *A. corymbosum*, which is restricted to south-central Mato Grosso of Brazil is an allospecies (geographically separated species that constitute a superspecies) of *A. nanum* and *A. fruticosum* (a new species) is endemic to the upper Mazaruni River basin in Guyana. It is closely related to the Amazonian *A. parvifolium*.

A. occidentale is native to tropical America where its natural distribution is unclear because of its long and intimate association with man. The problem of its origin and distribution has been investigated by Johnson (1973) who suggested that it originated in the *restinga* found in sandy soil along the Eastern Brazil of Northeastern Brazil. *Restinga* is a spit and a distinct type of coastal tropical and subtropical moist broadleaf forest found in Eastern Brazil. *Restingas* form on sandy, acidic, and nutrient-poor soils, and are characterized by medium sized trees and shrubs adapted to the drier and nutrient-poor conditions. Johnson is probably correct in assuming that the cultivated form of *A. occidentale* came from Eastern Brazil, because cashew trees cultivated in the Old and New Worlds are identical in appearance to native trees found in *restinga* vegetation. In particular, cultivated and wild populations of cashew from Eastern Brazil share chartaceous (resembling paper) leaf blades and long petioles. *A. occidentale* is probably an indigenous element of the savannas of Colombia, Venezuela and the Guyanas. A *savanna* or savannah is a grassland ecosystem characterised by the trees being sufficiently widely spaced so that the canopy does not close. It is clearly a native, and occasionally a dominant feature of the *cerrados* (savanna-like vegetation) of central and Amazonian Brazil. The *cerrado* populations of *A. occidentale* differ from the restinga populations by having undulate, thickly coriaceous leaves with short, stout petioles. The hypocarps (cashew apples) of *cerrado* trees are usually smaller and sometimes have a more acidic flavour than those of the *restinga*. The natural distribution of *A. occidentale* extends from northern South America south to Sao Paulo, Brazil. It is probably not native to Central America, the West Indies, or South America west of the Andes. It is believed that *A. occidentale* originally evolved in the *cerrados* of Central Brazil and later colonized the more recent *restingas* of the coast. Central Brazil is a center of diversity for *Anacardium* where the distribution of *A. occidentale* overlaps the ranges of *A. humile*, *A. nanum* and *A. corymbosum*. *A. humile*, the closest relative of the cultivated cashew, is closer morphologically to the *cerrado* ecotype than it is to the *restinga* and cultivated populations of *A. occidentale* (Mitchell and Mori, 1987).

The earliest reports of cashew are from Brazil coming from French, Portuguese and Dutch observers (Johnson, 1973). The French naturalist and monk, *Andre Thevet* was the first to describe, in 1557, a wild plant extremely common in Brazil, the cashew tree and its fruits. He recounted that cashew apples and their juice were consumed and that the nuts were roasted in fires and the kernels eaten. Andre Thevet provided the first drawing of the cashew showing the local people harvesting fruits and squeezing juice from the cashew apples into a large jar (Johnson, 1973; NOMISMA, 1994) (Fig. 14). There are indications that the local Tupi Indians had used cashew fruits for centuries. They probably played a major role in the species dispersion in their temporary migrations towards the coast of north-eastern Brazil, where a considerable intraspecific variation has been recorded (Ascenso, 1986). The entire cashew fruit, nut and peduncle, will float when mature. This could account, in Brazil, for coastward dispersal of the species by rivers draining north and east. Fruit bats may also have been involved in seed movement. Within the Amazon forests fruit bats are the most important agents of seed dispersal of tree species (Johnson, 1973). From its origin in Northeastern Brazil, cashew spread into South and Central America (Van Eijnatten, 1991).

The Eastern portion of the Amazon river figures prominently in distributions of many plants and animals, many of which are found either exclusively to the North or South of the river. However, in the case of *Anacardium*, all Amazonian species are found on both

Fig.14: Engraving of Tupí Indians harvesting cashew fruits in André Thevet's Les Singularitez de la France Antarctique; (Paris, 1557)

sides of the Amazon river. The reason for this is probably the ease with which bats, large birds, and water (in the case of *A. microsepalum*) carry fruits across water barriers (Mitchell and Mori, 1987). *A. occidentale* is cultivated and adventive throughout the Old and New World Tropics where the geographical limits of its cultivation are latitudes 27°N and 28°S, respectively (Nambiar, 1977).

The presence of cashew in other continents is to be attributed to man's intervention (Johnson 1973). The Portuguese discovered cashew in Brazil and spread first to Mozambique (Africa) and later into India between sixteenth and seventeenth centuries (De Castro, 1994). According to Agnoloni and Giuliani (1977), it arrived in Africa during the second half of the sixteenth century, first on the east coast and then on the west and lastly in the islands. In Africa, although it can be guessed that the cashew was introduced at an early period by the Portuguese, there are no records which provide specific dates. Dispersal of the cashew in east Africa may in part be due to the elephant, whose fondness for fruits is well known (Johnson, 1973). Attracted by the colour of the false fruit (apple), they swallowed this together with the nut which was too hard to be digested. This was then expelled with their droppings, a natural manure, and trodden far enough into the ground by the animals following along behind to root and grow into a seedling first and then a tree. This is how the cashew was spread along the East coast of Africa facing the Indian Ocean (Massari, 1994). The spreading of the cashew within the South American continent was gradual and spontaneous (NOMISMA, 1994). It is believed that the Portuguese brought the cashew to India, between 1563 and 1578. It was first described in gardens in Cochin on the Malabar coast. Following its introduction into southwestern India, the cashew probably diffused throughout the Indian subcontinent to some degree by means of birds, bats, but most importantly, human elements. Cochin served as a dispersal point for the cashew in India, and perhaps for south-east Asia as well (Johnson, 1973). According to Johnson (1973) the reason for the introduction is not documented, although the popular explanation is that it was for the purpose of checking soil erosion in the coastal areas of India. This interpretation, however, smacks of a twentieth century concept being applied to a 16th century event. Portuguese learned of the reported medicinal properties of the cashew and also that the juice of the cashew apple could be fermented into a good wine. It seems plausible, therefore, that they visualized the cashew as a crop of potential value to India. After India, it was introduced into South-Eastern Asia (NOMISMA, 1994). Dispersal in Southeast Asia appears to have been aided by monkeys. Whether the cashew reached the Philippines via India is uncertain. It may have come directly from the New World on the Manila Galleons (Johnson, 1973). The cashew later spread to Australia and some parts of the North American continent, such as Florida. Finally, its present diffusion can be geographically located between 31° North latitude and 31° South latitude, both as a wild species and under cultivation (NOMISMA, 1994). At present cashew is cultivated in many tropical countries, mainly in coastal areas (Ascenso, 1986; Van Eijnatten, 1991). In the 19th century, proper plantations were established and the tree then spread to a number of other countries in Africa, Asia and Latin America (Massari, 1994). Though cashew is indigenous to Brazil, India is the country that nourished this crop and made it a commodity of international trade and acclaim. Even today, India is the largest producer, processor, exporter and second largest consumer of cashew kernels in the world (Nayar, 1998).

Recent History

It was not until the early years of the 20th century that international trade in cashew kernels began with the first exports from India. A very slow beginning, but recent decades have seen the cashew become an important commercial tree crop (Johnson, 1973). Cashew production takes place mainly in the Central and South American zone, Asia and Oceanic zone and African zone. The Asiatic zone includes India as the major producer besides China, Indonesia, Malaysia, Philippines, Thailand, Vietnam, Sri Lanka and Myanmar. In the African Zone, Mozambique, Tanzania and Kenya are the major producers, besides, minor countries such as Benin, Guinea Bissau, Ivory Coast, Madagascar, Nigeria, Ghana, Senegal and Togo. In the Latin American Zone, the primary producers of cashew comprises of Brazil., Columbia, Costa Rica, Honduras, Salvador, Gautemala, Panama and Venezuela. Traditionally cashew has been cultivated on commercial scale in Brazil, India, Tanzania, Mozambique, Kenya and Madagascar, while in the recent years plantations are also raised in South East Asian countries like Vietnam, Myanmar, Thailand, on commercial scale. In India cashew is grown mainly in Maharashtra, Goa, Karnataka and Kerala along the west coast and Tamil Nadu, Andhra Pradesh, Odisha and West Bengal along the east coast. To a limited extent it is grown in Manipur, Meghalaya, Tripura, Andaman and Nicobar Islands and Chhattisgarh.

India has been exporting the cashew kernels since 1950s. Over the years, both the export earnings as well as quantity of kernels has been increasing. Thus, India has been importing raw nuts from African countries to meet the demand of cashew processing industries. Export earnings has been on the increase since 1955. The demand for raw nuts has increased worldwide, particularly during the past two decades, and the demand for cashew nuts far exceeds the current production. Foreign exchange earned by the country through export of cashew kernels during 2011-12 touched Rs. 4,390 crore as per statistics provided by the Cashew Export Promotion Council of India (CEPCI). Industry sources termed it a 'big jump' with respect to the Rs 2,819 crore earned from kernel exports during 2010-11 (*Anon.*, 2012).

Collection of Genetic Resources

Eventhough no reliable records of the introductions are available, it is presumed that the initial introductions in the Malabar Coast of Kerala were from only few trees and due to the hardy nature of the crop it has spread to all the coastal regions of India naturally. All these introductions were belonging to *Anacardium occidentale*. The initial empahsis was only on the establishment of plantations of seedling origin. As cashew is primarly a cross pollinated crop, it is highly heterozygous. Considerable segregation has resulted in large variation in the populations (Bhaskara Rao and Bhat, 1996). The early attempts for germplasm collections were made during the early 1950s with the sanctioning of Ad-hoc Schemes in the then composite States of Madras, Travancore, Cochin and Bombay. The research stations started under these Ad-hoc Schemes in Kerala (Kottarakkara), Karnataka (Ullal), Tamil Nadu (Vridhachalam), Anhdra Pradesh (Bapatla) and Maharashtra (Vengurla) took up the programme of collection of locally available elite plants for evaluation and further selection. These were the first attempts in collection of cashew germplasm in India. Many of the other research centres which were established subsequently have collected the seeds of germplasm from the centres, namely, Bapatla, Kottarakkara, Ullal, Vridhachalam and Vengurla. While making the initial collections of germplasm these centres have confined their survey mainly to the respective States and hence they represent the local germplasm available in the States. Since the inception of All India Coordinated Spices and Cashew Improvement Project in 1971, Central Plantation Crops Research Institute (CPCRI) Regional Station, Vittal, also took up the programme of cashew germplasm collection which mainly consisted of the seedling progenies of the collections which are available at Bapatla, Vridhachalam, Vengurla, Anakkayam and a few collections made locally from Karnataka (Swamy, 2011).

Subsequent to the establishment of National Research Centre for Cashew (NRCC) at Puttur (presently, ICAR-Directorate of Cashew Research-DCR), the germplasm collection through seeds has been discontinued and only the vegetative material (scion sticks) are being collected, the clones (softwood grafts) are prepared and conserved in the National Cashew Field Gene Bank (NCFGB). A coordinated approach was brought in the cashew gemplasm collection by organising joint survey teams consisting of scientists of NRCC/ ICAR-DCR and the All India Coordinated Research Project on Cashew (AICRP on Cashew) centres of the respective States. Since 1986, cashew germplasm collection surveys are being under taken by NRCC/ ICAR-DCR in collaboration with AICRP on cashew centres (9) and the cashew growing States, namely, Karnataka, Kerala, Maharashtra, Goa, Tamil Nadu, Andhra Pradesh, Odisha and West Bengal have been surveyed for germplasm collection (Swamy, 2011).

Conservation of Genetic Resources

In situ conservation of cashew germplasm is done only in the Amazon forests of Brazil (original home of cashew). However, *ex situ* conservation is generally being followed in cashew. Like in all other tree crops, the cashew germplasm is maintained as an active collection in the field gene bank which locks up the land for a long time. Further, field maintenance also has the inherent danger of losing the valuable collections during the natural calamities (as it often happens in the East Coast, especially at Bapatla centre due to cyclones). In order to overcome this danger as well as to effectively utilize the land for other field experiments, suitable *in vitro* preservation / cryo-preservation methods with the aid of biotechnology are to be developed. An effort in this direction is being made by establishing Biotechnology laboratory at DCR Puttur (Swamy, 2011).

Presently a total of 1225 Cashew germplasm accessions have been collected and conserved in National Cashew Field Gene Bank (NCFGB) at DCR Puttur (527) and in Regional Cashew Field Gene Banks (RCFGBs) at various centres of AICRP on Cashew, namely, Bapatla (132), Bhubaneswar (98), Jhargram (119), Vridhachalam (208), Madakkathara (130), Pilicode (43), Vengurla (302), Chintamani (128), Jagdalpur (65) (Bhat *et al.*, 2010; (Swamy, 2011).

Evaluation and Documentation of Genetic Resources

Germplasm evaluation, in the broad sense and in the context of genetic resources, is the description of the material in a collection. It covers the whole range of activities starting from the receipt of new samples by the curator and growing these, characterisation and preliminary evaluation and also further detailed evaluation and documentation. The evaluation of gemrplasm is an important part of a variety development programme designed to utilize germplasm resources. As new accessions become available, it is important to identify and characterize the new materials so that researchers may incorporate lines with desirable characters into utilization programmes. The first steps of this evaluation are (i) to compile the passport data and (ii) to characterize the accessions for easily recognised traits (minimum descriptors) (Swamy, 2011).

Characterization and Cataloguing of Germplasm

For systematic characterisation of cashew gemrplasm each accession was grown in the field gene bank (@6 grafts/accession; spacing 6 m x 6 m) at RCFGB, Puttur, Karnataka. Agronomic recommended practices were adopted. Observations were recorded on 3 selected plants in each accession after 10th year of planting and after obtaining 6 annual harvests for 68 descriptors. For cashew germplasm, IPBGR (presently IPGRI) "Cashew Descriptors" (IBPGR, 1986) available were followed. Cashew germplasm collection in the NCFGB at NRCC/ DCR, Puttur has 527 Accessions. A total of 285 accessions have been characterized as per IPGRI descriptors. Three germplasm catalogues for 255 accessions have been brought out (Swamy *et al,* 1997, 1998 and 2000; Bhat *et al.*, 2010). These are the first efforts made in characterisation of clonal accessions of cashew in the world. By adopting K-clustering algorithm using centroid distance method under Indostat Statistical package the 255 accessions were grouped into 22 clusters based on 19 descriptors. From each group a 10 per cent sample size was randomly selected as core entries.

Genetic Diversity in the Germplasm

Wide variations among the accessions due to the cross pollinated nature of the crop was reported for the 292 germplasm accessions of seedling origin, which were evaluated at CPCRI Regional Station, Vittal/ Shantigodu. A considerable variability was observed for some of the economic characters such as flowering season (October – January), flowering duration (40 - 127 days), harvesting duration (30 - 105 days), number of nuts/panicle (1 – 8), weight/apple (30 - 150 g), weight/nut (2.4 - 18.0 g), apple to nut ratio (4:1 - 12:1), shelling percentage (19.0 - 35.0), weight/kernel (0.5 - 4.5 g), kernel count/lb (100 – 900), shell thickness (1.5 - 5.0 mm) and mean yield/plant (0.50 - 11.75 kg) (Bhaskara Rao and Swamy, 1994). Nagaraja *et al.* (2007) have charecterizes 33 varieties and 79 germplasm accessions of cashew for biochemical composition of apple juice (tannin, flavonoids, sugars, ascorbic acid and organoleptic acceptability) and pomace (protein, carbohydrate, sugar, tannin and crude fibre). Based on the variability noted, quality indices have been developed and the varieties and accessions with desirable qualities identified. Among the varieties and accessions analysed, the variety Jhargram-1 and the accession NRC-190 were the best for cashew apple juice, while the varieties NRC Selection-1, BLA 139-1 and Bhubaneshwar-1 and the accessions NRC-160 and NRC-247 were found to have the best cashew apple pomace characteristics. Anik *et al.* (2002) have estimated the genetic diversity among 90 germplasm accessions of cashew using random amplified polymorphic DNA (RAPD) markers. A dendrogram was constructed using Ward's squared euclidean distance method which confirmed that the diversity of Indian cashew collection can be considered to be "moderate" to "high". A core collection has been identified based on the study which represents the same diversity as the entire population. This could be the first step towards more efficient management of cashew germplasm in India.

Molecular Markers / DNA Fingerprinting

Molecular markers like DNA markers [randomly amplified polymorphic DNA (RAPD), inter simple sequence repeats (ISSR), restriction fragment length polymorphism (RFLP), amplified fragment length polymorphism AFLPs)] and biochemical markers (isozyme protein) can be used for characterization of germplasm and somaclonal variants. DNA fingerprinting of varieties using RAPD markers is being done at NRC for DNA fingerprinting, New Delhi and Department of Horticulture, UAS, Bangalore. DNA finger printing of cashew germplasm is being carried out in collaboration with Division of Horticulture, UAS, Bangalore under DST funded project. Leaf samples of 153 accessions (NRC 1 – 153) of cashew have been supplied from NCGB. RAPD markers generated by 7 selected operon primers (10 base-long) which produced 123 consistent, unambiguous, repeatable bands ranging from 400 bp to 3 Kbp were used to assess the diversity among 90 accessions. About 40 per cent of the samples were repeated to confirm the results. The analysis of the results (squared euclidean distance, Ward's clustering) revealed that the diversity in India cashew germplasm is moderate but not narrow as reported. Attempts have been made to fingerprint cashew genetic markers (Archak *et al*, 2003a, 2003b; Thimmappaiah *et al.*, 2009a, 2009b).

Archak *et al.* (2003a) have developed the molecular profiles of 35 Indian accessions of cashew (24 varieties and 11 hybrids) with increased yield and excellent nut characters) using a combination of five RAPD and four ISSR primers pre-selected for maximum discrimination and repeatability. A total of 94 markers were generated which discriminated all the varieties with a probability of identical match by chance of $2.8 \times 10{-}11$. There was no correlation between the relationships based on molecular data and the pedigree of the varieties. Narrow range of average similarity values among major cashew breeding centres with only 3.6% of molecular variance partitioned between them was attributed to the exchange of genetic material in developing varieties. Difference in the average similarity coefficients between selections and hybrids was low indicating the need and scope for identification of more parental lines in enhancing the effectiveness of hybridisation programme. Archak *et al.* (2003b) have also analysed nineteen cashew accessions (8 varieties and 11 accessions) with 50 random primers, 12 ISSR primers and 6 AFLP primer pairs to compare the efficiency and utility of these techniques for detecting variation in cashew germplasm. Each marker system could discriminate between all of the accessions, albeit with varied efficiency of polymorphism detection. AFLP exhibited maximum discrimination efficiency with a genotype index of 1. The utility of each molecular marker technique, expressed as marker index, was estimated as a function of average band informativeness and effective multiplex ratio. Marker index was calculated to be more than 10 times higher in AFLP than in RAPD and ISSR. Similarity matrices were determined based on the data generated by molecular and morphometric analyses, and compared for congruency. AFLP displayed no correspondence with RAPD and ISSR. Correlation between ISSR and RAPD similarity matrices was low but significant ($r = 0.63$; $p < 0.005$). The similarity matrix based on morphometric markers exhibited no correlation with any of the molecular markers. AFLP, with its superior marker utility, was concluded to be the marker of choice for cashew genetic analysis. At DCR 239 accessions have been fingerprinted by Thimmappaiah *et al* (2009a). Low-level diversity has been observed in 40 elite varieties using RAPD, ISSR and SSR markers. Moderate diversity has been reported in cashew population using protein isoenzyme electrophoretic analysis (Thimmappaiah *et al.*, 2009a, and 2009b).

Assigning Indigenous Collection Numbers (IC No.) to Germplasm

In order to safeguard our national interest in the field of plant genetic resources, national identity numbers/ indigenous collection numbers (IC Nos) are being assigned to crop gemrplasm by the National Bureau of Plant Genetic Resources (NBPGR), New Delhi. In order to obtain IC numbers for the cashew germplasm holding in the country, the passport data for 1149 cashew accessions maintained at NCFGB and RCFGBs in the country have been compiled and submitted to NBPGR, New Delhi. IC Numbers have been assigned by NBPGR for 433 clonal accessions of cashew that are being conserved in NCFGB at Puttur and for the 716 accessions of cashew accessions which are being maintained at RCFGBs, namely, Chintamani (53), Vengurla (142), Pilicode (64), Madakkathara (73), Vridhachalam (250), Bapatla (80), Bhubaneswar (5) and Jhargram (49) (Swamy *et al.*, 2002; Swamy, 2011).

Registration of Unique Germplasm

Nine novel unique germplasm accessions of cashew maintained at DCR Puttur, namely NRC-59, 111, 116, 120, 121, 140, 142, 152, and 201, have been registered with the National Bureau of Plant Genetic Resources (NBPGR), New Delhi. These unique accessions have been assigned the INGR No. 03080 to 03088 by NBPGR (NBPGR, 2006; Swamy, 2011).

Utilization of Genetic Resources

By utilizing the existing cashew germplasm in the country, the following 40 high-yielding varieties of cashew have been developed and released by the Directorate of Cashew Research, Puttur, and various Agricultural Universities (Abdul Salam and Bhaskara Rao, 2001; Bhat *et al.* (2010): DCR Puttur (NRCC Selection-1, NRCC Selection-2 and Bhaskara); Bapatla (BPP-1 to BPP-6 and BPP-8); Vridhachalam (VRI-1, VRI-2, VRI-3 and VRI-5); Bhubaneswar (Bhubaneswar-1); Jhargram (Jhargram-1); Vengurla (Vengurla-1 to Vengurla-7); Goa (Goa-1 and Goa-2); Madakkathara [Anakkayam-1, Madak-1 (BLA-39-4), Madak-2 (NDR-2-1), K-22-1, Kanaka, Dhana, Priyanka and Amrutha]; Ullal (Ullal-1, Ullal-2, Ullal-3, Ullal-4, UN-50); Chintamani (Chintamani-1 and Chintamani-2). Of these, 13 are hybrids and 27 are selections.

In addition to these varieties/ selections, 13 hybrids have been developed by utilizing gemrplasm accessions as parents in the hybridisation programme. Till recently the emphasis has been on selecting or producing a hybrid having high yield only and size of the nut was not given much importance till 1980s. Currently efforts are made at all the cashew research stations in the country to identify few accessions with bold nuts for improving the nut size in the prolific bearing varieties. Crop improvement programme through hybridization is receiving greater attention in almost all the cashew research centres of India. Crop improvement programme in Australia also is centered around development of hybrids wherein thousands of hybrids are produced using parents of wide genetic diversity obtained from different countries especially from India and Brazil (Chacko, *et al.*, 1990 and Chacko, 1993). A total of 65 accessions in NCFGB have also been utilized as parents (female 55 and male 10) in the hybridisation programme under the ad-hoc research scheme entitled "Network programme on hybridisation in cashew", which was in operation at NRCC/ DCR and AICRP on Cashew Centres (Bhubaneswar, Vengurla, Madakkathara and Vridhachalam) (Swamy, 2011).

Cultivars Developed in Brazil

Brazil clearly needs a more productive germplasm as yields per unit are now very low. "Cashew Comum" and "Cashew Anao Precoce" are the two populations or "types" responsible for these results: the first is characterized by big trees (10-20 m), fruit and cashew apple size variability and production differences from one tree to another. It is the commonest type, in English speaking countries it is called "*Brazilian common*". The second is smaller and its foliage less, it is an early plant with a smaller fruit and cashew apple. In English speaking countries it is called "*Brazilian dwarf*". Both varieties have undergone genetic treatment to fulfil a number of requirements: to increase unit production, with a higher number of fruits per inflorescences, to increase inflorescence per plant and average fruit weights; to obtain resistance to some plant disease such as anthracnose, mainly responsible for production fall in Brazil; to improve fruit quality namely by decreasing tannin contents; to obtain a long flowering and thus prolong the harvesting season, so as to have a well defined harvesting period (NOMISMA, 1994). Recently research by the EMBRAPA-CNPCa carried out in the Ceara State has made it possible to detect, isolate and multiply 4"anao precoce" clones, with excellent production and quality features.

The four clones of precocious dwarf cashew tree presently available for planting were developed through individual phenotypic selection from germplasm existing in the Experimental Farm of Pacajus, Ceara, which is native of the municipality of Maranguape, Ceara, by utilization of the polycross method for increasing variability and clonal selection in order to identify the best clones. These clones are characterized by low stature, permitting the mannual harvesting of almost all fruits, and making possible the utilization of the peduncle which is not possible in cashew orchards of the common type; precocious development, hastening the first profitable production from the 4th to the 2nd year; seasonal precociousness, initiating fructification at least 30 days earlier than that of the common type; longer productive cycle, by around 60 days, than the common cashew, thus increasing utilization of the peduncle. Low stature, precociousness and productive potential permit utilization of management techniques impossible in traditional orchards, such as pruning and control of pests and diseases, resulting in greater productivity, reducing the time necessary for recuperation of the capital invested from the 17th year to the 7th. This tree modernizes cashew culture: productivity increases from 250 kg/ha in traditional plantings to 1300 kg/ha and may reach yields above 2000 kg/ha (NOMISMA, 1994) (Table 4) (Fig.15).

Table 4. Salient features of four cashew clones.

Clones	*Colour of cashew apple*	*Nut Average weight (g)*	*Average Prdouction (kg/ha/yr)*				
			1st year	*2nd year*	*3rd year*	*4th year*	*5th year*
CPP06	Yellow	6.5	14	82	755	783	905
CPP09	Yellowish	8.5	51	184	367	367	712
CCP76	Reddish	9.0	31	163	306	307	571
CCP1001	Red	5.5	65	367	557	1187	1493

CCP= Caju Clone de Pacajus

Source: EMBRAPA-CNCPa, technical communication 1991

Fig.15: Brazilian Dwarf cashew (Cajueiro Anao Precoce)

Perspective

One of the immediate priorities of the Indian cashew germplasm programme is to enhance the genetic variability in the collections, as all the plantations were raised from the limited initial introductions and genetic base available in the country is very narrow. Therefore, introductions from Central America and Brazil where considerable variability including the dwarf types is reported and is also the original home of cashew is essential. It is suggested that these introductions may be made from Brazil which is the original home of cashew. Introduction of *A. gigantium* from Surinam which was reported to have the biggest apple (200 g) will be advantageous, especially in states like Goa where the cashew apple utilization contributes substantially to the economy of the state. Introduction of dwarf clones reported from Brazil (Ascenso, 1986) will be useful in developing high density orchards which will enhance productivity considerably. One of the main problems in cashew is that all the existing germplasm is susceptible to stem and root borer which kills the trees. There is a need to screen the allied species (for their suitability as root stocks) which have relatively hard wood and also posses smooth bark. It is suggested that introduction from Brazil of other species like *A. rhinocarpus* and *A. spruceanum* which are reported to possess hard wood will be useful for testing their suitability as root stocks.

References

Abdul Salam, M. and Bhaskara Rao, E.V.V. (Eds.). 2001. *Cashew varietal wealth of India.* Directorate of Cashew and Cocoa Development, Govt. of India, Ministry of Agriculture, Cochin. 101 pp.

Agnoloni, M. and Giuliani, F. 1977. *Cashew cultivation.* Library of Tropical Agriculture, Ministry of Foreign Affairs. Instituto Agronomico Per L' oltremare, Florence. 168 pp.

Aiydurai, S.G. 1966. *A review of research on Spices and Cashewnut in India*. Indian Council of Agriculotural Research, New Delhi. pp.228.

Albuquerque, S.D.S. , Hassan, M.V. and Shetty, K.R. 1960. *Mysore Agric. J.*, 35:2-8.

Aliyu, O. M. and Awopetu, J. A. 2007. Chromosome studies in Cashew (*Anacardium occidentale* L.). *African Journal of Biotechnology*, 6(2): 131-136.

Anik, L.D., Bhaskara Rao, E.V.V., Swamy, K.R.M., Bhat, M.G., Theertha Prasad, D. and Suresh N.Sondur. 2002. Using RAPDs to assess the diversity in Indian cashew (*Anacardium occidentale* L.) germplasm. *J.Hort.Sci.& Biotechnol.*, 77 (1): 41-47.

Anonymous. 2012. Rise in earnings from cashew kernel exports. *The Hindu* (English Daily), Kollam. September 14, 2012.

Archak, S., Gaikwad, A.B., Gautam, D., Rao, E.V.V.B., Swamy, K.R.M. and Karihaloo, J.L. 2003a. DNA fingerprinting of Indian cashew (*Anacardium occidentale* L.) varieties using RAPD and ISSR techniques. *Euphytica*, 230: 397–404.

Archak, S., Gaikwad, A.B., Gautam, D., Rao, E.V.V.B. , Swamy, K.R.M. and Karihaloo, J.L. 2003b. Comparative assessment of DNA fingerprinting techniques (RAPD, ISSR and AFLP) for genetic analysis of cashew (*Anacardium occidentale* L.) accessions of India. *Genome*, 46: 362-369.

Ascenso, J.C. 1986. Potential of the cashew crop-1. *Agric. Intl.*, 38(11):324-327.

Bailey, L. H. 1958. *The Standard Encyclopedia of Horticulture* Vol.1. (17th edition). Mac Millan, New York. 1200 pp.

Bakhru, H.K. 1988a. *Indian Cashew J.*, 18(1):7-8.

Bakhru, H.K. 1988b. *Indian Cashew J.*, 18(3):8.

Behrens, R. 1998. About the spacing of cashewnut tree. Proceeding of the International Cashew and Coconut conference, 17-21 February 1997, Dar es Salam, Tanzania. BioHybrids International Ltd., Reading, Uk. pp: 48-52.

Bhaskara Rao, E. V. V. and Swamy, K. R. M. 1994. Genetic resources of cashew. In: K L Chadha and P Rethinam (eds.). *Advances in Horticulure Vol.9 - Plantation and Spice Crops* - Part-1. Malhotra Publishing House, New Delhi. pp. 79-97.

Bhat, M.G., Nagaraja, K.V. and Rupa, T.R. 2010. Cashew research in India. *J. Hortl. Sci.*, 5(1): 1-16.

CGIAR. 2015. Anacardium species. *ciatweb.ciat.cgiar.org/ipgri/fruits_from.../species%20 Anacardium.htm*

Chacko, E.K. 1993. *Genetic improvement of cashew through hybridisation and evaluation of hybrid progenies* - A final report prepared for the Rural Industries R & D Corporation (Mimeographed), CSIRO Division of Horticulture, PMB 44, Winnellie, NT 0821, Australia.

Chacko, E. K., Baker, I. and Downton, J. 1990. Towards a sustainable cashew industry for Australia. *J. Australian Inst. Agric. Sci.* 3 (5): 39 -43.

Darlington, C.D. and Janaki Ammal, E. K. 1945. *Chromosome Atlas of cultivated plants.* Allen and Unwin., London. 397pp.

De Castro, P. 1994. Summary of the study. In: .A. M. Delogu and G. Haeuster (Eds.). *The world cashew economy* NOMISMA, L'Inchiostroblu, Bologna, Italy. pp. 11-12.

Goldblatt, P. (**Ed.**). 1984. *Index to plant chromosome numbers 1979 – 1981.* Missouri Bot. Garden.

Hutchinson, J.J. and Dalziel, M. 1954. Flora of West Africa. Crown Agent for Overseas Government and Administration, Millbank, London, S.W. I. Vol. 1: 428-429.

IBPGR. 1986. *Cashew Descriptors.* International Board for Plant Genetic Resources (presently, International Plant Genetic Resource Institute), Rome, Italy. 33pp.

Johnson, D. 1973. The botany, origin, and spread of the cashew *Anacardium occidentale* L. *J. Plantn. Crops,* 1: 1-7.

Khosla, P. K., Sareen, T. S. and Mehra, P. N. 1973. Cytological studies on Himalayan *Anacardiaceae. Nucleus,* 16:205-209.

Machado, O. 1944. Estudos novos sobre uma planta velha cajueiro (*Anacardium occidentale* L.). *Rodriguesia,* 8: 19-48.

Massari, F. (1994). (**In**): The World Cashew Economy (Delogu, A.M. and Haeuster, G. eds.). NOMISMA. Marino Cantelli Printing House, L'Inchiostroblu, Bologna, Italy. pp. 3-4.

Mitchell, J.D. and Mori, S.A. 1987. The cashew and its relatives (*Anacardium*: Anacardiaceae). *Memoirs of The New York Botanical Garden,* 42 : 1-76.

MMPND. 2001. Sorting Anacardium names. Multilingual Multiscript Plant Name Database. http://www.plantnames.unimelb.edu.au/Sorting/Anacardium.html

Murthy,K.N. and Yadava, B.B.R. 1972. Note on the oil and carbohydrate contents of varieties of cashew nut (*Anacardium occidentale*). *Indian J. Agric. Sci.,* 42(10):960-961.

Nagaraja, K.V. and Krishnan Nampoothiri, V.M. 1986. Chemical characterization of high-yielding varieties of cashew (*Anacardium occidentale*). Qual. Plant *Plant Foods Humn. Nutr.,* 36:201-206.

Nagaraja, K.V., Bhuvaneshwari, S. and Swamy, K.R.M. 2007. Biochemical characterization of cashew (*Anacardium occidentale* L.) apple juice and pomace in India. *Plant Genetic Resources Newsletter,* No. 149: 9-13.

Nair, M.K., Bhaskara Rao, E.V.V., Nambiar, K.K.N. and Nambiar, M.C. (Eds). 1979. *Cashew* (*Anacardium occidentale* L.). Monograph on Plantation Crops-1. , CPCRI, Kasaragod. pp. 169.

Nambiar, M.C. 1977. (**In**): *Ecophysiology of Tropical Crops. Symposium* (Alvim, T. and Kozlowski, T.T. Eds). Academic Press, Inc., New York. pp. 461 – 478.

Nayar, K.G. 1998. (**In**): *Proceedings of International Cashew and Coconut Conference. Trees for Life - the Key to Development* (Topper, C.T., Caligari, P.D.S., Kulaya, A.K., Shomari, S.H., Kasuga, L.J., Masawe, P.A.L. and Mpunami, A.A. **Eds**). Bio Hybrids International Ltd., Reading, U.K. .pp .195 - 199.

NBPGR. 2006. Plant Germplasm Registration (1996 to 2004). *Information Bull.,* 01: 16-17. NBPGR, New Delhi.

NOMISMA. 1994. *The World Cashew Economy* (Eds. Delogu, A.M. and Haeuster, G.). Marino Cantelli Printing House, L'Inchiostroblu, Bologna, Italy. 218 pp.

Peixoto, A. 1960. *Caju.* Servico de Informacao Agricola, Ministerio de Agricultura, Rio de Janerio (Brazil).

Pell, Susan Katherine. 2004. Molecular systematic of the cashew family (Anacardiaceae). *PhD Thesis, Louisiana State University,* Louisiana, USA . pp. 193.

Purseglove, J.W. (edr.). 1988. *Tropical crops - Dicotyledons.* English Language Book Society, Longman, London. pp. 18-32.

Simmonds, N. W. 1954. Chromosome behaviour in some tropical plants. *Heredity.* 8:139.

Swamy, K.R.M. 2011. Genetic wealth of cashew in India. (**In**): Souvenir, DCR, Puttur.

Swamy, K. R. M., Bhaskara Rao, E. V. V. and Bhat, M. G. 1997. *Catalogue of minimum descriptors of cashew (Anacardium occidentale* L.*) germplasm accessions -I.* National Research Centre for Cashew, Puttur, Karnataka. (June 1997). 41 pp.

Swamy, K. R. M., Bhaskara Rao, E .V .V. and Bhat, M. G. 1998. *Catalogue of minimum descriptors of cahsew (Anacardium occidentale* L.) *germplasm accessions-II.* National Research Centre for Cashew, Puttur, Karnataka (October 1998). 54 pp.

Swamy, K. R. M., Bhaskara Rao, E. V. V. and Bhat, M. G. 2000. *Catalogue of minimum descriptors of cashew (Anacardium occidentale* L.) *germplasm accessions -III.* National Research Centre for Cashew, Puttur, Karnataka (Oct 2000). 54 pp.

Swamy K.R.M., Bhaskara Rao, E.V.V. and Bhat, M.G. 2002. *Status of cashew germplasm collection in India. NRCC Tech Bull.* No.7, February 2002 .pp. 48.

Thimmappaiah, Santhosh, W.G., Shobha, D. and Melwyn, G.S. 2009a. Assessment of genetic diversity in cashew germplasm using RAPD and ISSR markers. *Sci.Hort.,* 120:411-417.

Thimmappaiah., Melwyn, G.S., Shobha, D. and Shirly, Raichal Anil. 2009b. Assessment of cashew species for molecular diversity. *J. Plant. Crops,* 37:146-151.

Valeriano, C. 1972. O cajueiro. *Boletim do Instituto Biologico de Bahia (Brazil)* 11(1):19-58.

Van Eijnatten, C. L. M. 1991. *Anacardium occidentale* L. In: E. W. M. Verheij and R. E. Coronel (eds.). *Plant resources of South-East Asia-No.2. Edible fruits and nuts.* Pudoc-DLO, Wageningen, The Netherlands. pp. 446.

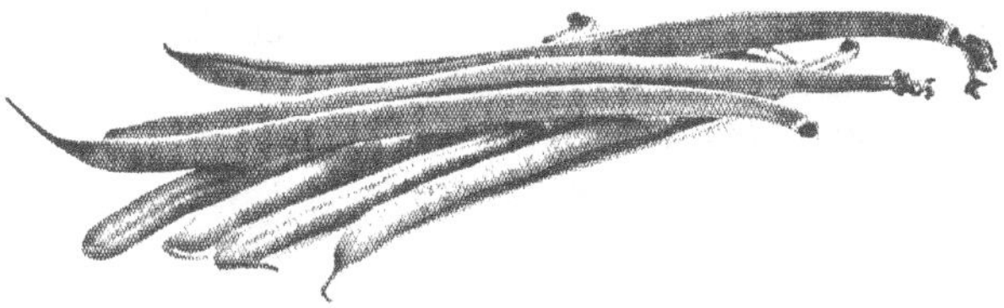

8

French bean

Avijit Kr Dutta

French bean (*Phaseolus vulgaris* L.), an important affiliate of the family Fabaceae (Leguminosae)[the third largest family of flowering plants after the Orchidaceae and Asteraceae, with 800 genera and 20,000 species (Lewis *et al.*, 2005)],is predominantly self-pollinated due to its papilionaceous, cleistogamous flower structure. It is the most important legume worldwide for human consumption (Singh *et al.*, 2007). French bean is an important cool season legume vegetable grown for its tender pods, shelled green beans and dry beans (*Rajmah beans*). According to Kara *et al.* (2009), it is consumed by people from all income levels and serves as a primary source of dietary protein for people in the lower income group (Wortmann *et al.*, 2004). Among domesticated plant species, *Phaseolus vulgaris* is the most important source of protein for direct human consumption (Singh, 2001; Broughton *et al.*, 2003). Beans provide essential proteins (20 to 25%) in the human diet, complementing other food sources like maize and rice (Broughton *et al.*, 2003). The major protein of *P. vulgaris* is phaseolin that helps to regulate blood sugar levels. Therefore, diabetics and people who suffer from hypoglycaemia can benefit from eating green beans. Eating beans may reduce the risk for developing certain types of cancers due to their contribution of bioactive compounds to the diet, including flavonoids, tannins, phenolic compounds and other antioxidants (Amarowicz and Pegg, 2008). The high fibre content of common bean aids digestion and a diet rich in beans can lower cholesterol levels, especially the triglycerides. French beans are also useful in controlling certain cardiac problems as well. *Phaseolus vulgaris* emanates from two centres of domestication of the species, Mesoamerica and the Andes (Gepts and Bliss, 1988; Zeven, 1997). The cultivation of French beans was started 7000 years ago, by the Indian tribes settled in Tehuacan Valley of Mexico and in Callejon de Huaylas, Peru. When Christopher Columbus returned from his second voyage to the New World in the year 1493, he brought French beans with him in the Mediterranean region. French beans were considered to be rare to find and expensive but soon became one of the commonly used beans in the 19th century. In France,French beans were introduced in the year 1597 by the conquistadors, which were then spread to other parts of the world by Spanish and Portuguese traders.

Domestication and Origin

Members of the Fabaceae family were domesticated as grain legumes in conjunction with the domestication of grasses for cereals (de Candolle, 1884; Vavilov, 1951; Smartt, 1990; Zohary and Hopf, 2000; Abbo *et al.*, 2012). Linnaeus assigned the origin of common bean to India; de Candolle (1882) himself expressed doubts that common bean had been domesticated in the Americas. It was only during and after the World War-II that it became generally admitted that common bean originated in the Americas. The conclusion was based on the discovery of wild common bean in Argentina (Burkart and Brücher, 1953) and Guatemala (McBryde, 1947) and archaeological remains in the Americas (Kaplan and Kaplan, 1988).There are two subspecies of *Phaseolus sylvestris* in Mexico and Central America; and *aborigineus* in South America. These were domesticated in Mexico and South America, respectively, giving rise to the Mexican and South American gene pools of domesticated common beans. *Phaseolus vulgaris*-L.wasoriginated in South Mexico and Central American Centres (Vavilov, 1951).The genus *Phaseolus*L. is a member of the economically important sub-tribe Phaseolinae (tribe Phaseoleae) and includes more than 80 wild species extensively distributed throughout the Americas (except for Alaska, west and north Canada, western USA, Chile and southern Argentina), from south eastern Canada, southeastern and southwestern USA, Mexico, Central America, West Indies and to mostly eastern South America. Of these species, just over 60 are distributed mostly in the Mexican uplands (Delgado-Salinas *et al.*, 2006). However, according to the most recent monograph of the genus (Freytag and Debouck, 2002 as quoted in Smýkal *et al.*, 2015), there are 76 species of *Phaseolus*, all distributed in the New World with a center of diversity in Mexico. Botanical, archaeological, biochemical and molecular evidence indicate that two major events occurred in the domestication process of the species, one in Mesoamerica and another in the Andes, which made up the two main centres of the origin of this species. Several species of *Phaseolus* were domesticated prehistorically in North and South America. The domestication of *P. vulgaris* might have occurred in Brazil and Northern Argentina from a wild form of which *P. aborigineus* is a modern survivor (Evans, 1976).Over a period of at least 7,000 – 8,000 years, French bean had evolved from a wild growing vine *viz., Phaseolus aborigineus* distributed in the highlands of middle America and Andes (Brucher, 1988). Crossing experiments conducted by Yarnell (1965) had confirmed *P. aborigineus* (found in North West Argentina) as the progenitor of *P. vulgaris.* Sturtevant (1919) assembled impressive evidence from the literature prior to 1887 in support of the widespread use of *Phaseolus vulgaris* by the American Indian at the time of early explorations and colonization.The common bean originated in the new world, principally Central and South America (Kaplan, 1981). The domesticated species used in Mesoamerica include common bean (*P. vulgaris*), lima bean (*P. lunatus*), tepary bean

(*P. acutifolius*) and scarlet runner bean (*P. coccineus*), the first being the most economically important.Common beans (*P. vulgaris*) were by far the most abundant domesticated legume in the Ocampo caves (Kaplan and MacNeish, 1960) [Ocampo Caves is the collective name for three rock shelters located near the town of Ocampo in Tamaulipas State, Mexico, called Romero's Cave, Valenzuela's Cave and Ojo de Agua Cave]. Wild common beans have a very wide geographical distribution, ranging from northern Mexico to northern Argentina (Gepts, 1996). Molecular studies have revealed that the common bean was domesticated independently both in the Andes and in Mesoamerica (Chacón *et al.*, 2005; Gepts, 1996), and recent research has further shown that the latter likely originated in the Río Lerma - Río Grande de Santiago Basin in west-central Mexico (Kwak *et al.*, 2009).The primary centers of origin for the *Phaseolus* cultigens are in the New World; however, for each crop there has been a different spread outside the original range, with greater spread for common bean than for any of the other species. The expansion and the pathways of distribution of the common bean out of the American domestication centers were very complex and they involved several introductions from the Americas that were combined with exchanges between continents and within continents among several countries (Belluci*et al.*, 2014). Outside of America, the common bean populations were more 'free' to pass through new evolutionary pathways that were not possible in the American center of origin, due to the spatial isolation between these two gene pools. As a consequence, several continents and countries have been proposed as secondary centers of diversification for common bean (*P. vulgaris*). Ithas secondary centers of diversity in central-eastern and southern Africa (Gepts and Bliss, 1988), Brazil (Burle*et al.*, 2010), China (Zhang *et al.*, 2008), Europe (Angioi*et al.*, 2010; Santalla*et al.*, 2002), the Caribbean and India (Asfaw*et al.*, 2009; Blair *et al.*, 2010; Sharma *et al.*, 2013), while tepary bean has spread very little outside its original range in northern Mexico, spreading only to parts of Central America (Blair *et al.*, 2012).

Evolutionary History

The current organization of genetic diversity in the cultivated gene pool of common bean is the result of evolution under both natural conditions(*i.e.*, prior to domestication) and cultivation. Before domestication, wild *P. vulgaris* had already diverged into two major gene pools, each with its characteristic geographic distribution,in Mesoamerica [the southern part of Central America, together with Colombia and Venezuela are not included in the traditional definition of Mesoamerica, which encompasses roughly the southern half of Mexico and the northern half of CentralAmerica. Nevertheless, analyses with bio-chemical and molecular markers show that wild bean populations in these areas are closely related tothose ofthe actual Mesoamerican area (Gepts, 1998)] and the Andes (southern Peru, Bolivia and Argentina)[Fig.-1]. The presence of geographically isolated gene pools in *Phaseolus vulgaris* that originated from at least two independent domestication events and the overlapping distribution with other domesticated and wild species that have different mating systems and are at various degrees of reproductive isolation make *P. vulgaris* and the genus *Phaseolus* a unique model for studies of plant evolution (Papa *et al.*, 2006). During the process of domestication in common bean, several morphological changes have occurred. The cultivated French bean is an erect growing plant with determinate branching whereas the wild type is indeterminate and profusely branched. The cultivated types have smaller numbers of nodes on the main axis while the wild forms have more nodes. The inter node length is relatively shorter in the cultivated types. Although, French bean cultivars are products from multiple domestications in the American Continent (Van Schoonhoven and Voysest, 1991). Archaeological, morphological, biochemical and molecular evidence has suggested two well-defined Andean and Mesoamerican centres of common bean origins, which were confirmed through morphological markers (Gepts *et al.*, 1986; Singh *et al.*, 1991; Chacón *et al.*, 2005; Delacruz *et al.*, 2005), isozyme markers (Koenig and Gepts, 1989; Santalla *et al.*, 2004; Chacón *et al.*, 2005), phaseolin types by eletrophoretic profiles (Gepts *et al.*, 1986; Pereira and Souza, 1992; Maciel *et al.*, 1999; Solano, 2005), and molecular markers of RFLP (Chacón *et al.*, 2005), AFLP (Maciel *et al.*, 2003), RAPD (Beebe *et al.*, 2000) and ISSR (Delacruz *et al.*, 2005). The primary centre of the Central American type is characterized by cultivars with predominantly the 'S' phaseolin type and smaller seeds (<25 g/100 seeds). The other primary centre, the Andean, is characterized by cultivars with the 'T' phaseolin type and larger seeds (>40 g/100 seeds) [Singh *et al.*, 1991]. The expression of the phaseolin phenotype is not influenced by environment (Brown *et al.*, 1981) or by human selection or any other type of selection (Gepts *et al.*, 1986).

Definitely in Mesoamerica the distribution of *P. vulgaris* overlaps with that of *P. coccineus* and *P. polyanthus*. Molecular studies have shown that *Phaseolus polyanthus*, which was formally, included in *P. coccineus*, is intermediate in its morphological features between these other two species (Hernandez-Xolocotzi *et al.*, 1959) and a hybrid origin has indeed been suggested (Pifiero and Eguiarte, 1988; Kloz, 1971; Llaca *et al.*, 1994). At the molecular level, *P. polyanthus* is closer to *P. coccineus* by nuclear DNA comparison (Pifiero and Eguiarte, 1988; Delgado-Salinas *et al.*, 1999) but more similar to *P. vulgaris* chloroplast DNA comparison (Llaca *et al.*, 1994). Thus,*Phaseolus polyanthus* probably originated from a cross that involved *P. vulgaris* as the maternal parent, with successive backcrosses to *P. coccineus*as the paternal donor (Schmit *et al.*,1993; Llaca *et al.*, 1994). This interpretation is reliable with studies showing that in artificial crosses between *P. coccineus* and *P. vulgaris*, fertile F_1 progeny can be produced, particularly when *P. vulgaris* is the maternal parent (Singh, 2001; Broughton *et al.*, 2003). This suggests that introgression between *P. coccineus* and *P. vulgaris* occurred in the evolutionary history of both species in Mesoamerica.A nucleus of diversity is located in Ecuador and northern Peru (Coulibaly, 1999; Kami *et al.*, 1995), from which wild beans dispersed both northwards and southwards to form two geographically distinct gene pools in Mesoamerica and the Southern Andes (Gepts, 1998). In turn, post-domestication divergence gave rise to three domesticated races in each of these two gene-pools (Singh *et al.*, 1991) [Fig.-2].

The five domesticated species of bean belong to two distinct lineages. The *Phaseolus vulgaris* group includes: *P. vulgaris, P. coccineus, P. polyanthus* and *P. acutifolius*. The fifth cultivated species, *Phaseolus lunatus,* is part of a separate, very well-defined clade which includes the South American and oceanic island diversification of *Phaseolus viz. P. augusti* Harms, *P. bolivianus* Piper, *P. lignosus* Britton, *P. mollis* Hook., *P. pachyrrhizoides* Harms, *P. rosei* Piper and *P. viridis* Piper (Caicedo *et al.*, 1999; Delgado- Salinas *et al.*, 1999). A group containing *Phaseolus viridis* (from Oaxaca, Mexico) and *P. lignosus* (from Bermuda) is sister to the rest of this primarily South American group. The Andean *Phaseoluspachyrrhizoides*, *P. augusti* and *P. bolivianus* form a monophyletic group which is sister to a lineage containing both Mesoamerican and Andean accessions of *P. lunatus*, as well as one accession of *P. mollis*, a species endemic to the Galapagos Islands (a volcanic archipelago in the Pacific Ocean), a province of Ecuador. *Phaseolus mollis* and a Peruvian accession of *P. lunatus* are resolved as a sister group, suggesting a mainland origin for the Galapagos species. *Phaseolus rosei* collected at the type locality (Chimborazo, Ecuador) falls within the Andean group of wild Lima beans. *Phaseolus rosei* could thus be an Andean wild form of Lima bean (which would make its correct name *P. lunatus*) [Toro *et al.*, 1993; Caicedo*et al.*, 1999].

Fig. 1: Distribution of wild *Phaseolus vulgaris* in Latin America (adapted from Gepts, 1998).

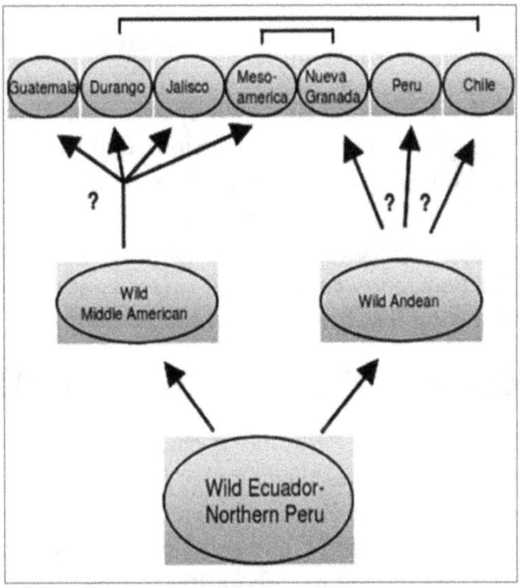

Fig. 2: Domestication of the Andean- and middle American- gene pools lead to four races in the Middle Americas and three races amongst the Andean gene-pool [adapted fromBeebe *et al.*, 2000,as cited in Broughton *et al.*, 2003].

Taxonomy and Cytogenetics

The taxonomic hierarchy of French bean is Kingdom: Plantae(plants); Subkingdom: Tracheobionta (vascularplants); Superdivision:Spermatophyta (seed plants); Division: Magnoliophyta (flowering plants); Class: Magnoliopsida (dicotyledons); Subclass: Rosidae; Order: Fabales; Family:Fabaceae (pea family); Genus: *Phaseolus* L. (bean) and Species:*Phaseolus vulgaris* L.(kidney bean/common bean/dwarf bean/haricot bean/snap bean/string bean/garden bean). The genus *Phaseolus* (was adopted and made official by Linnaeus in 1753)includes more than 80 wild species of these species; just over 60 are distributed mostly in the Mexican uplands (Delgado-Salinas *et al.*, 2006). Five of these were domesticated and brought in cultivation, the 'tepary' or 'escumite' bean (*P. acutifolius*-A. Gray): amazingly adapted to arid locations; the 'ayocote' or 'scarlet runner bean' (*P. coccineus* L.): splendid climber with beautiful blossoms, well adapted to temperate cold environments; the 'pallar' or 'lima bean' (*P. lunatus* L.): a versatile vine occurring in seasonally dry or humid forests; the 'gordo' or 'year bean' (*P. dumosus*-Macfadyen):well adapted to temperate-humid sites, locally important to different human groups; however, still neglected, and the world-wide cultivated 'frijol', 'poroto' or common bean (*P. vulgaris*-L.) [Delgado-Salinas, 2014].The first reports on chromosome numbers in *Phaseolus* go back to 1925, when Karpetschenko obtained $2n = 22$ for *P. acutifolius* A. Gray, *P. coccineus* L., *P. lunatus* L. and *P. vulgaris* L. The earliest records pertaining to cytological studies in *Phaseolus* reveal that the somatic chromosome number (2n= 2x= 22) is 22 and it was confirmed by Weinstein (1926).Some authors (Sarbhoy, 1977; Sarbhoy, 1980; Sinha and Roy, 1979a; Sinha and Roy, 1979b) have pointed out that the main factors involved in the karyotypic evolution of the genus are pericentric and paracentric inversions, translocations and the loss or gain of chromatin. Biswas and Dana (1975) observed that the chromosomes of the tetraploid *Phaseolus* species and *P. calaratus* (presently *Vigna umbellata*) differ through translocations, inversions, duplication and deficiencies. Further, cytological studies (Sarbhoy, 1977; Sarbhoy, 1978) concluded that polyploidy has not played a significant part in the evolution of *Phaseolus*.Pushpa*et al.* (1979) through karyomorphology [combine study of the karyotype and chromosome morphology of organisms] of three related species, *viz., Phaseolus aborigineus, P. vulgaris* and *P. angularis* (now *Vignaangularis*) confirmed that all having the chromosome number 2n = 22.

Cross-ability and Interspecific Hybridization

The genus *Phaseolus* is characterized by two main poles (Maréchal*et al.*, 1978): the first one includes the complex *P. vulgaris* - *P. coccineus*, [three species *viz.P. coccineus, P. polyanthus* and *P. vulgaris* belong to the same evolutionary branch (Schmit *et al.*, 1993; Schmit *et al.*, 1995)] and at the opposite end is *P. lunatus* (Fofana *et al.*, 1999; Fofana *et al.*, 2001; Maquet *et al.*, 1999). Interspecific hybridization in *Phaseolus* has been extensively reviewed (Smartt, 1979; Smartt, 1990; Hucl and Scoles, 1985; Debouck, 1991). Interspecies hybridization has allowed to specify the proximity of a large number of species in this genus. Hybridization of *P. vulgaris* with *P. acutifolius* was first investigated by Honma (1956) in order to incorporate common bean blight resistance into *P. vulgaris*.Harlan and de Wet (1971) with modifications (Smartt, 1990, as cited in Maquet *et al.*, 1997) distinguish three main centres of origin: primary gene pool - the equivalent to the concept of a species, whose members are free to cross-fertilize and consists of cultivated and wild forms of the species; secondary gene pool - the equivalent of all the species that can give viable or partially fertile hybrids with the primary one; and tertiary gene pool- that includes types among which artificial hybridization is possible, but the resulting hybrids are sterile, abnormal or lethal. From several studies it was revealed that *Phaseolus vulgaris* is the primary gene pool; *P. coccineus, P. costaricensis,* and *P. polyanthus* (syn.*P. dumosus*) are secondary gene pools, whereas, *P. acutifolius, P. lunatus,* and other *Phaseolus* species are considered as the tertiary ones (Debouck and Smartt, 1995; Debouck, 1999; Debouck, 2000). Al-Yasiri and Coyne (1966) crossed seven species of *Phaseolus,* namely *P. calcaratus, P. mungo* (now *Vigna mungo), P. angularis, P. lunatus, P. coccineus, P. acutifolius,* and *P. vulgaris* in all possible combinations and found that *P. vulgaris* x *P. coccineus* was the only compatible cross (where hybrid seeds were obtained). *P. polyanthus* crosses more easily with *P. coccineus* and related forms than with *P. vulgaris*, particularly if the latter is the pollen donor (Baudoin *et al.*, 2001). *P. polyanthus* belongs to the *P. vulgaris* clade, but its nuclear genome has been introgressed with *P. coccineus* genes and this limits its use in interspecific hybridizations. Especially when *P. vulgaris* is used as the female parent, crosses between *P. vulgaris* and *P. costaricensis* are simple to perform without embryo rescue, but it is not clear whether *P. coccineus*genes have contaminated the nuclear genome of *P. costaricensis* (Debouck, 1999). According to Tomlekova (2012), *P. coccineus* is difficult to hybridize with other *Phaseolus* species but hybridizes successfully with *P. vulgaris*. Similarly, *P. lunatus* is difficult to hybridize with the common bean. *P. acutifolius*, cultivated mainly for mature grain, is valuable for its resistance to bacterial blight and drought tolerance. It hybridizes with great difficulty with common bean (Poryazov *et al.*, 1988, as cited in Kumanov *et al.*, 1988). While the successful hybridization have been reported through *in vitro* methods of embryo-cultures (Nikolova *et al.*, 1986). The major reproductive barrier to interspecific hybridization amongst the genus *Phaseolus* occurs post-fertilization, especially during early embryo development (Baudoin *et al.*,1995). Infertility in *P. polyanthus* x *P. vulgaris* crosses results from early nutritional barriers that are related to a deficient endosperm tissue development while in reciprocal crosses, endothelium proliferation and to some extent, hypertrophy of the vascular elements are causes of early embryo abortion (Lecomte *et al.*,1998; Geerts *et al.*,1999; Geerts, 2001; Geerts *et al.*, 2002). Although several hybrids between *P. vulgaris* and species belonging to its tertiary gene pool can only be obtained by embryo rescue, most infertility results from male sterility which is caused by incomplete chromosomal pairing in Metaphase-I (Baudoin *et al.*,1995).

References

Abbo, S.; Lev-Yadun, S. and Gopher A. 2012. Plant Domestication and Crop Evolution in the Near East: On Events and Processes. *Critical Reviews in Plant Sciences*, 31: 241–257.

Al-Yasiri, S. A. and Coyne, D. P. 1966. Interspecific hybridization in the genus *Phaseolus. Crop Science*, 6:59-60.

Amarowicz, R. and Pegg, R. B. 2008. Legumes as a source of natural antioxidants.*European Journal of Lipid Science and Technology*,110:865-878.

Angioi, S. A.; Rau, D.; Attene, G.; Nanni, L.; Bellucci, E.; Logozzo, G.; Negri, V.; Spagnoletti-Zeuli, P. L. and Papa, R. 2010. Beans in Europe: origin and structure of the European landraces of *Phaseolus vulgaris* L. *Theoretical and Applied Genetics*, 121:829-843.

Asfaw, A.; Blair, M. W. and Almekinders, C. 2009. Genetic diversity and population structure of common bean (*Phaseolus vulgaris* L.) landraces from the East African Highlands. *Theoretical and Applied Genetics*, 120: 1–12.

Baudoin, J. P.; Camarena, F. and Lobo, M. 1995. Amélioration de quatreespèces de légumineusesalimentairestropicales*Phaseolus vulgaris, P. coccineus, P. polyanthus* et*P. lunatus*. Sélection intra- etinterspécifique. In: "Quelavenir pour l'amélioration des plantes?".QuatrièmesJournéesscientifiques du réseauBiotechnologievégétale de l'UREF, Namur, 18-21 Octobre 1993. Ed. par J. Dubois, Y. Demarly, AUPELF-UREF. John LibbeyEurotext, Paris, pp. 31-49.

Baudoin, J. P.; Camarena, F.; Lobo, M. and Mergeai, G. 2001.Breeding *Phaseolus*for intercrop combinations in Andean highlands.*In*: Broadening the genetic basis of the crop. Cooper, H. D.; Spillane, C. and Hodgkin, T. (Eds.). CAB International (Wallingford, U.K.), pp. 373-384.

Beebe, S.; Skroch, P. W.; Duque, M. C.; Pedraza, F. and Nienhuis, J. 2000.Structure of genetic diversity among common bean landraces of Central American origin based on correspondence analysis of RAPD. *Crop Science*, 40: 262-272.

Belluci, E.; Nanni, L.; Biagetti, E.; Bitocchi, E.; Giardini, A.; Rau, D.; Rodriguez, M.; Attene, G. and Papa, R. 2014. Common bean origin, evolution and spread from America. *Legume Perspectives*, 2: 12-16.

Biswas, M. R. and Dana, S. 1975. Interchange in a triploid species hybrid of *Phaseolus. Indian Agriculturist*, 19: 273–274.

Blair, M. W.; Gonzales, L.F.; Kimani, P. and Butare, L. 2010.Inter-genepool introgression, genetic diversity and nutritional quality of common bean (*Phaseolus vulgaris* L.) landraces from Central Africa.*Theoretical and Applied Genetics*, 121: 237–248.

Blair, M. W.; Pantoja, W. and Munoz, L. C. 2012. First use of microsatellite markers in a large collection of cultivated and wild accessions of tepary bean (*Phaseolus acutifolius* A. Gray).*Theoretical and Applied Genetics,* 125: 1137–1147.

Broughton, W. J.; Harnandez, G.; Blair, M.; Beebe, S.; Geptsand, P. and Vanderleyden, J. 2003.Beans (*Phaseolus* spp.): Model food legumes. *Plant and Soil*, 252: 55-128.

Brown, J. W. S.; Ma, Y.; Bliss, F. A. and Hall, T. C. 1981.Genetic variation in the subunits of globulin-1 storage protein of French bean.*Theoretical and Applied Genetics*, 59: 83-88.

Brucher, H. 1988.The wild ancestor of *Phaseolus vulgaris* in South America.*In*: Genetic resources of *Phaseolus* beans: their maintenance, domestications, evolution and utilization, Gepts, P. (Ed.). Kulwer, Netherland, pp. 185 – 214.

Burkart, A. and Brücher, H. 1953. *Phaseolus aborigineus* Burkart, die mutmasslicheandine Stammform der Kulturbohne. *Züchter* [in German], 23: 65-72.

Burle, M. L.; Fonseca, J. R.; Kami, J. A. and Gepts, P. 2010.Microsatellite diversity and genetic structure among common bean (*Phaseolus vulgaris* L.) landraces in Brazil, a secondary center of diversity. *Theoretical and Applied Genetics*, 121:801-813.

Caicedo, A. L.; Gaitan, E.; Duque, M. C.; Toro Chica, O., Debouck, D.G. andTohme, J. 1999. AFLP fingerprinting of *Phaseolus lunatus*L. and related wild species from South America. *Crop Science*, 39: 1497–1507.

Chacón, M. I.;Pickersgill, S. B. and Debouck, D. G. 2005. Domestication Patterns in the Common Bean (*Phaseolus vulgaris* L.) and the Origin of the Mesoamerican and Andean Cultivated Races. *Theoretical and Applied Genetics*, 110(3):432-444.

Coulibaly, S. 1999. *PCR*-derived analysis of genetic diversity and relationships within the *Phaseolus vulgaris* L. complex and in *Vigna unguiculata*L. Unpublished Ph.D. Thesis, University of California, Davis.

de Candolle, A. 1882. L'origine des plantescultivées. English translation: The origin of cultivated plants. Appleton, New York, USA.

de Candolle, A. 1884. *Origin of cultivated plants*. Kesinger Publishing LCC, Whitefish, MT, 2006.

Debouck, D. 1991. Systematics and morphology. (In:) A. van Schoonhoven and O. Voysest (Eds.) Common beans: research for crop improvement. C.A.B. International, Wallingford, UK and CIAT, Cali, Colombia. pp. 55-118.

Debouck, D. G. and Smartt, J. 1995. Beans, *Phaseolus* spp. (Leguminosae –Papilionoideae). *In*: Smartt, J. and Simmonds, N.W. (Eds.) *Evolution of Crop Plants, 2nd Edn.*, Longman, London, pp. 287–294.

Debouck, D.G. 1999. Diversity in *Phaseolus*species in relation to the common bean.*In*: Singh, S.P. (Ed.) *Common Bean Improvement in the Twenty-first Century*. Kluwer Academic Publishers, Dordrecht, The Netherlands, pp. 25–52.

Debouck, D.G. 2000. Biodiversity, ecology and genetic resources of *Phaseolus* beans – seven answered and unanswered questions. (In:) *Wild Legumes*, Oono, K. (Ed.). National Institute of Biological Resources, Tsukuba, Japan, pp. 95–123.

Delacruz, E. P.; Gepts, P.; Marín, P. C. G. and Villareal, D. Z. 2005.Spatial distribution of genetic diversity in wild populations of *Phaseolus vulgaris* L. from Guanajuato and Michoacán, México. *Genetic Resources and Crop Evolution*, 52: 589-599.

Delgado-Salinas, A. 2014. Biodiversity and systematics of *Phaseolus* L. (Leguminosae). *Legume Perspectives*, 2: 5-7.

Delgado-Salinas, A.; Bibler, R. and Lavin, M. 2006. Phylogeny of the genus *Phaseolus* (Leguminosae): a recent diversification of an ancient landscape. *Systematic Botany*, 31:779-791.

Delgado-Salinas, A.; Turley, T.; Richman, A. and Lavin, M. 1999. Phylogenetic analysis of the cultivated and wild species of *Phaseolus* (Fabaceae).*Systematic Botany*, 24: 438-460.

Evans, A. M. 1976. Beans: *Phaseolus* spp. *In*: Evolution of Crop Plants, Simmonds, N. W. (Ed.), Longman, London, pp. 168-172.

Fofana, B.; Baudoin, J. P.; Vekemans, X.; Debouck, D. G. and du Jardin, P. 1999. Molecular evidence for an Andean origin and a secondary gene pool for the Lima bean (*Phaseolus lunatus*) using chloroplast DNA.*Theoretical and Applied Genetics*, 98: 202–212.

Fofana, B.; du Jardin, P. and Baudoin, J. P. 2001. Genetic diversity in the Lima bean (*PhaseoluslunatusL.*) as revealed by chloroplast DNA (cpDNA) variations. *Genetic Resources* and *Crop Evolution*, 48: 437–445.

Freytag, G. F. and Debouck, D. G. 2002. Review of Taxonomy, distribution, and ecology of the genus *Phaseolus* (Leguminosae Papilionoideae) in North America, Mexico, and Central America. *Botanical Miscellany*, 23:1-30.

Geerts, P. 2001. Study of embryo development in *Phaseolus* in order to obtain interspecific hybrids. Ph. D. thesis. Gembloux Agricultural University (Belgium), pp. 183.

Geerts, P.; Mergai, G. and Baudoin, J. P. 1999. Rescue of early heart-shaped embryos and plant regeneration of *Phaseolus polyanthus* Greenm. and *P. vulgaris* L. *Biotechnologie, Agronomie, Société et Environnement* (*Biotechnology, Agronomy, Society* and *Environment*), 3(3):141-148.

Geerts, P.; Toussaint, A.; Mergeai, G. and Baudoin, J. P. 2002.Study of the early abortion in reciprocal crosses between *Phaseolus vulgaris* and *Phaseolus polyanthus* Greenn. *Biotechnologie, Agronomie, Sociétéet Environnement* (*Biotechnology, Agronomy, Society* and *Environment*), 6: 109-119.

Gepts, P. 1996. Origin and Evolution of Cultivated *Phaseolus* Species. (In): *Advances in Legume Systematics 8: Legumes of Economic Importance*, Pickersgill, B. and Lock, J. M. (Eds.) Royal Botanic Gardens, Kew. pp. 65-74.

Gepts, P. 1998. Origin and Evolution of Common Bean: Past Events and Recent Trends. *HortScience*, 33(7): 1124-1130.

Gepts, P. and Bliss, F. A. 1988. Dissemination pathways of common bean (*Phaseolus vulgaris*, Fabaceae) deduced from phaseolin electrophoretic variability. II. Europe and Africa. *Economic Botany*, 42: 86-104.

Gepts, P.; Osborn, T. C.; Rashka, K. and Bliss, F. A. 1986. Phaseolin-protein variability in wild forms and landraces of the common beans (*Phaseolus vulgaris*): Evidence for multiple centers of domestication. *Economic Botany*, 40: 451-468.

Harlan, J. R. and de Wet, J. M. J. 1971. Toward a rational classification of cultivated plants, *Taxon*, 20: 509-517.

Hernandez-Xolocotzi, E.; Miranda Colin, S. and Prwyer, C. 1959. El origen de *Phaseoluscoccineus*L. darwinianus Hernandez-X & Miranda C. subspecies nova. *Revista de la Sociedad Mexicana de Historia Natural*, 20: 99-121.

Honma, S. 1956. A bean interspecific hybrid. *Journal of Heredity*, 47: 217-220.

Hucl, P. and Scoles, G. J. 1985. Interspecific hybridization in the common bean: a review. *HortScience*, 20:352-357.

Kami, J.; Becerra Velásquez, B.; Debouck, D. G. and Gepts, P. 1995. Identification of presumed ancestral DNA sequences of phaseolin in *Phaseolus vulgaris*. *Proceedings of the National Academy of Sciences of the United States of America*, 92: 1101–1104.

Kaplan, L. 1981. What is the origin of the common bean? *Economic Botany*, 35: 240–254.

Kaplan, L. and Kaplan, L. N. 1988. *Phaseolus* in archaeology.*In*: Genetic Resources of *Phaseolus* Beans: their Maintenance, Domestication, Evolution and Utilization, Gepts, P. (Ed.). *Current Plant Science and Biotechnology in Agriculture*, Vol. 5.Kluwer Academic Publishers, Dordrecht. pp. 125-142.

Kaplan, L. and MacNeish, R. S. 1960. Prehistoric Bean Remains from the Caves in the Ocampo Region of Tamaulipas, Mexico.*Botanical Museum Leaflets*, (Harvard University, Cambridge, Massachusetts), 19 (2): 33-56.

Kara, R.; Dalton, T. J. and Featherstone, A. M. 2009. A Nonparametric Efficiency Analysis of Bean Producers from North and South Kivu. Prepared for the Southern Agricultural Economics Association Annual Meeting, Atlanta, Georgia.

Kloz, J. 1971. Serology of the Leguminosae.*In:*Harborne, J. P.; Boulter, D. and Turner, B. L. (Eds.), Chemotaxonomy of the Legunimosae, Academic Press, London, UK. pp. 309-365.

Koenig, R. and Gepts, P. 1989. Segregation and linkage of genes for seed proteins, isozymes, and morphological traits in common bean (*Phaseolus vulgaris*). *Journal of Heredity*, 80: 455-459.

Kumanov, B.; Poryazov, I.; Uzunova, E.; Popov, D.; Kostov, D. and Velev, B. 1988. *Legumes*, Zemizdat, Sofia, pp. 157.

Kwak, M.; Kami, J. A. and Gepts, P. 2009. The Putative Mesoamerican Domestication Center of *Phaseolus vulgaris* is Located in the Lerma-Santiago Basin of Mexico. *Crop Science*, 49:554-563.

Lecomte, B.; Longly, B.; Crabbe, J. and Baudoin, J. P. 1998. Etude comparative du développement de l'ovule chez deuxespèces de *Phaseolus*: *P. polyanthus* et*P. vulgaris*. Développement de l'ovuledans le genre *Phaseolus.Biotechnologie, Agronomie, Sociétéet Environnement* (*Biotechnology, Agronomy, Society* and *Environment*), 2(1): 77-84.

Lewis, G.; Schrire, B.; Mackinder, B. and Lock, M. 2005. *Legumes of the World*. Royal Botanic Gardens, Kew, The United Kingdom.

Llaca, V.; Delgado, S. A. and Gepts, P. (1994). Chloroplast DNA as an evolutionary marker in the *Phaseolus vulgaris* complex.*Theoretical and Applied Genetics*, 88: 646-652.

Maciel, F. L.; Echeverrigaray, S.; Gerald, L. T. S. and Grazziotin, F. G. 2003. Genetic relationships and diversity among Brazilian cultivars and landraces of common beans (*Phaseolus vulgaris* L.) revealed by AFLP markers. *Genetic Resources and Crop Evolution*, 50: 887-893.

Maciel, F. L.; Gerald, L. T. S. and Echeverrigaray, S. 1999. Variation of phaseolin andother soluble proteins among cultivars and landraces of common beans of south-Brazil. *Journal of Genetics and Breeding*, 53: 149-154.

Maquet, A.; Zoro, Bi I.; Delvaux, M.; Wathelet, B. and Baudoin, J. P. 1997. Genetic structure of a Lima Bean base collection using allozyme markers, *Theoretical and Applied Genetics*, 95: 980-991.

Maquet, A.; Vekemans, X. and Baudoin, J. P. 1999. Phylogenetic study on wild allies of Lima bean and implication on its origin. *Plant* Systematics and *Evolution*, 218: 43–54.

Maréchal, R.; Mascherpa, J. M. and Stainier, F. 1978. *Etude taxonomique d'un groupecomplexed'espèces des genres Phaseolus et Vigna (Papilionaceae) sur la base de donnéesmorphologiques et polliniques, traitéesparl'analyseinformatique*, C.J.B.G., Genève, *Boissicra*, 28: 1-273.

McBryde, F.W. 1947. Cultural and historical geography of southwest Guatemala. *Smithsonian Institution Publications*, 4:1-184.

Nikolova, V.; Poryazov, I. and Rodeva, V. 1986. Hybridization between *P. acutifolius*Gray and *P. vulgaris* L. *Annual Reports of the Bean Improvement Cooperative*, 29: 124-125.

Papa, R.; Nanni, L.; Sicard, D.; Rau, D. and Attene, G. 2006.The evolution of genetic diversity in *Phaseolus vulgaris*. *In*: Motley, T. J.; Zerega, N. and Cross, H. (Eds.) Darwin's Harvest: new approaches to the origins, evolution and conservation of crops. Columbia University Press, New York, USA, pp. 121-142.

Pereira, P. A. A. and Souza, C. R. B. 1992.Tipos de faseolinaemraçascrioulas de feijão no Brasil.*PesquisaAgropecuariaBrasileira*, 27: 1219-1221.

Pifiero, D. and Eguiarte, L. 1988. The origin and biosystematics status of *Phaseolus coccineus* sub sp. *polyanthus*: Electrophoretic evidence. *Euphytica*, 37: 199-203.

Poryazov, I.; Zagorcheva, L.; Nikolova, V. and Rodeva, V. 1988. Results of the hybridization between *P. vulgaris* L. and *P. coccineus*L. or *P. acutifolius*A. Gray, *Proceedings of the Symposium on Interspecific Hybridisation and Methods for Transfer of Alien Genetic Material*, Sofia, 29-30: 296-300.

Pushpa, G.; Mahishi, D. M.; Satyan, B. A. and Kulkarni, R. N. 1979. Karyomorphology of three species of *Phaseolus*. *Mysore Journal of Agricultural Sciences*, 13: 147–151.

Santalla, M.; Rodiño, A. P. and De Ron, A. M. 2002. Allozyme evidence supporting southwester Europe as a secondary center of genetic diversity for common bean. *Theoretical and Applied Genetics*, 104: 934– 944.

Santalla, M.; Sevillano, M. C. M.; Monteagudo, A. B. and De Ron, A. M. 2004. Genetic diversity of Argentinean common bean and its evolution during domestication. *Euphytica*, 135: 75-87.

Sarbhoy, R .K. 1980. Karyological studies in the genus *Phaseolus* Linn. *Cytologia*, 45: 363–373.

Sarbhoy, R. K. 1977. Cytogenetical studies in the genus *Phaseolus* Linn. III. Evolution in the genus *Phaseolus*. *Cytologia*, 42: 401–414.

Sarbhoy, R. K. 1978. Cytogenetical studies in genus *Phaseolus* I and II: Somatic and meiotic studies in fifteen species of *Phaseolus*. *Cytologia*, 43: 161–180.

Schmit, V.; Muñoz, J. E.; du Jardin, P.; Baudoin, J. P. and Debouck, D. G. 1995. Phylogenetic studies of some *Phaseolus* taxa on the basis of chloroplast DNA polymorphisms.*In*: *Phaseolus* Beans Advanced Biotechnology Research Network. Roca, W. M.; Mayer, J. E.; Pastor- Corrales, M. A. and Tohme, J. (Eds.). Proceedings of the Second International Scientific Meeting, 7-10 September 1993, CIAT (Cali, Colombia), pp. 69-75.

Schmit, V.; du Jardin, P.; Baudoin, J. P. and Debouck, D. G. 1993.Use of chloroplast DNA polymorphism for the phylogenetic study of seven *Phaseolus* taxa including *P. vulgaris* and *P. coccineus*.*Theoretical and Applied Genetics*, 87: 506-516.

Sharma, P. N.; D´iaz, L. M. and Blair, M. W. 2013. Genetic diversity of Indian common beans elucidated with two germplasm collections and by morpho- logical and microsatellite markers. *Plant Genetic Resources: Characterisation and Utilisation*, 11: 121–130.

Singh, A. K.; Singh, K. P. and Singh, B. K. 2007. Genetic variability, heritability and genetic advance in French bean (*Phaseolus vulgaris* L.). *Haryana Journal of Horticultural Sciences*, 36(3/4): 352- 353.

Singh, S. P. 2001. Broadening the genetic base of common bean cultivars: A review. *Crop Science*, 41: 1659-1675.

Singh, S. P.; Gepts, P. and Debouck, D. G. 1991. Races of common bean (*Phaseolus vulgaris*, Fabaceae). *Economic Botany*, 45: 379-396.

Sinha, S. S. N. and Roy, H. 1979a. Cytological studies in the genus *Phaseolus*: Mitotic analysis of fourteen species. *Cytologia*, 44: 191–199.

Sinha, S. S. N. and Roy, H. 1979b. Cytological studies in the genus *Phaseolus* II. Meiotic analysis of sixteen species.*Cytologia*, 44: 201–209.

Smartt, J. 1979. Interspecific hybridization in the grain legumes - a review. *Economic Botany*, 33:329-337.

Smartt, J. 1990. Grain Legumes: Evolution and Genetic Resources. Cambridge University Press, Cambridge, UK.

Smýkal, P.; Coyne, C. J.; Ambrose, M. J.; Maxted, N.; Schaefer, H.; Blair, M. W.; Berger, J.; Greene, S. L.; Nelson, M. N.; Besharat, N.; Vymyslický, T.; Toker, C.; Saxena, R. K.; Roorkiwal, M.; Pandey, M. K.; Hu, J.; Li, Y. H.; Wang, L. X.; Guo, Y.; Qiu, L. J.; Redden,R. J. andVarshney,R. K. 2015. Legume Crops Phylogeny and Genetic Diversity for Science and Breeding. *Critical Reviews in Plant Sciences*, 34(1-3): 43-104.

Solano, J. P. L. 2005. Patterns of phaseolins and RAPD analysis in domesticated species of *Phaseolus*. *RevistaFitotecnia Mexicana*, 28: 195-202.

Sturtevant, E. L. 1919. *Sturtevant's Edible Plants of the World.(ManĝeblaPlantoj de sturtevant de la Mondo).* Hedrick U. P. (Ed.) J. B. Lyon Company, New York (State) Dept. of Agriculture. *27th Annual Report*, 1918/19, Report of the New York Agricultural Experiment Station, 2 (II): 696.

Tomlekova, N. B. 2012. Genetic Diversity of Bulgarian *Phaseolus vulgaris* L. Germplasm Collection ThroughPhaseolin and Isozyme Markers. (In:) *The Molecular Basis of Plant Genetic Diversity*, Prof. Mahmut Caliskan (Ed.), *In*: Tech (Open Science: Open Minds), pp. 181-230.

Toro, C. O.; Lareo, L. and Debouck, D. G. 1993. Observations on a noteworthy wild Lima bean, *Phaseolus lunatus*L., from Columbia. *Annual Report Bean Improvement Cooperative*, 36: 53–54.

Van Schoonhoven, A. and Voysest, O. 1991. Common beans: research for crop improvement. CIAT, Cali, Colômbia.

Vavilov, N. I. 1951. The origin, variation, immunity and breeding of cultivated plants. *Chronica Botany*, 13 (1-6):1949-1950.

Weinstein, A. I. 1926. Cytological Studies on *Phaseolus vulgaris*. *American Journal of Botany*, 13 (4): 248–263.

Wortmann, C. S.; Kirkby, R. A.; Eledu, C. A. and Allen, D.J. 2004. Atlas of Common bean (*Phaseolus vulgaris*-L.) production in Africa. CIAT Publication, pp. 297.

Yarnell, S. H. 1965. Cytogenetics of vegetable crops. IV. Legumes. *Botanical Review*, 31(3): 247-330.

Zeven, A. C. 1997. The introduction of the common bean (*Phaseolus vulgaris* L.) into Western Europe and the phenotypic variation of dry beans collected in The Netherlands in 1946. *Euphytica*, 94: 319-328.

Zhang, X.; Blair, M. W. and Wang, S. 2008. Genetic diversity of Chinese common bean (*Phaseolus vulgaris* L.) landraces assessed with simple sequence repeats markers. *Theoretical and Applied Genetics*, 117: 629 -664.

Zohary, D. and Hopf, M. 2000.*Domestication of Plants in the Old World*. Oxford University Press, Oxford.

9

Garlic

R K Singh, R P Gupta and V Mahajan

Garlic *(Allium sativum* L.*)* is one of the important bulbous crops grown and used as a spice or a condiment throughout India. The genus *Allium* is a large and diverse one containing over 1,250 species. Its close relatives include elephant garlic, chives, onion, leek and shallot. Garlic has been used throughout recorded history for culinary, medicinal use and health benefits. Currently, the interest in garlic is highly increasing due to nutritional and pharmaceutical value including high blood pressure and cholesterol, atherosclerosis and cancer. The use of garlic is as condiment, garlic oil as insecticide, garlic paste as bio-fungicide, garlic residue with antibacterial properties. Garlic reproduces almost exclusively by means of underground cloves or vegetative bulbils (topsets) in the inflorescence and is mostly sterile.

Variation in plant type, bulb size, bulb weight, colour, coat layer, leaf length and width, growth habit, stress resistance, number of leaves, ability to flower, adaptation to different environmental conditions has been reported in garlic (Senula and Keller, 2000 and Volk and Stern, 2009) and this has been attributed to its apomictic nature which leads to the existence of extensive somatic mutations (Ata, 2005), chromosomal aberrations and genome plasticity. Since garlic has been propagated asexually for many generations, an accumulation of chromosome aberrations such as aneuploidy and translocations and/or inversions could also significantly reduce the incidence of balanced gametes. Therefore variation in garlic occurs only through random or induced mutation (Burba, 1993) and/

or soma-clonal variation (Novak, 1990) and new cultivars are bred by clonal selection, induced mutations, soma-clonal variation or genetic engineering (Jones and Mann 1963; Rubatzky and Yamaguchi 1997; Robinson 2007).

Taxonomy and Botany of *Allium sativum* L.

Garlic (*Allium sativum* L.) is a diploid species ($2n=2x=16$) in the subgenus *Allium* of the Alliaceae (formerly in the Liliaceae, and then the Amaryllidaceae). The edible underground stem is the composite bulb made of numerous smaller bulbs called cloves. The other cultivated plants in this subgenus are leek, usually tetraploid, and elephant garlic, usually hexaploid ($2n=2x=48$) (both *A. ampeloprasum* L.). Garlic (hard-neck type) produce inflorescences having topsets with/without flowers and usually no seed, except in few recently discovered fertile garlic clones. Elephant garlic produces a large leek like inflorescence but, seed produced are sterile and it rarely forms topsets (bulbils) in the inflorescence. If the plant of elephant garlic does not flower, the bulb consists of a single, large clove, termed as 'round'. Elephant garlic and garlic form a bulb, but leek does not. Elephant garlic is related to leek (*A. ampeloprasum* L.), but forms cloves resembling those of garlic, although appearance and flavor predominantly resemble leek (Fritsch and Friesen 2002). Elephant garlic bulbs consist of 2 to 6 large cloves and 2-10 corms (bulblets) while garlic bulbs usually have more cloves of a relatively consistent size, especially for bolting types.

Early History

Garlic has been cultivated for thousands of years in India and other parts of world. It is among the most ancient cultivated spice/vegetables giving pungency of the genus *Allium*. Original abode of garlic is said to be Central Asia and Southern Europe especially Mediterranean region. Some authorities consider that *Allium longicuspis* Regal which is endemic to Central Asia is the wild ancestor and spread in ancient times to the Mediterranean region. It is known in Egypt in pre-dynastic times, before 3000 BC and also to ancient Greeks and Romans. It has long been grown in India and China. Garlic was carried to the Western hemisphere by the Spanish, Portuguese and French. It was used in England as early as first half of the 16th century. The early domestication of garlic took quite different turn from that of the seed propagated leek and onion. Garlic becomes exclusively vegetatively propagated by cloves or bulbils. Most of the garlic from Central Asia belongs to the rather diverse *longicuspis* group (large bolting plants, many small topsets, to some extent still fertile cultivars). They might have been the genetic pool from which the other cultivar groups developed-the *subtropical* and *Pikinense* subgroups (smaller plants, few large topsets)-which possibly developed under the special climatic conditions of South, South-East and East Asia; the Mediterranean sativum group (bolting and non-bolting types, large topsets); and the *Ophioscorodon* group from Central and East Europe (long coiling scapes, few large topsets).

Recent History

Not much work was done on garlic improvement until National Horticultural Research and Development Foundation started work on development of varieties three and half decades ago. The work was also initiated by different State Agricultural Universities and

ICAR Institutes subsequently. Major work on garlic varieties is being taken up at NHRDF, DOGR Rajgurunagar, HAU Hissar, MPKV Rahuri, PAU Ludhiana, VPKAS Almora, ARU Almora, IARI, New Delhi, GAU Junagarh etc. It is seen that garlic grown near its centre of origin is bigger in size with bold cloves. As it advances towards sub tropics and tropics the size become small with more number of small cloves. The yield potential of such adopted types is less as compared to long day types. The short day garlic grown in plains of northern India, western India and hills of Nilgiris suffer from degeneration effects, small size of cloves, and susceptibility to diseases, pests and finally low yield. Improvement over these aspects is thus a continued process.

Crop Improvement in Garlic

Garlic is a vegetatively propagated crop. It is a neglected crop in regards to crop improvement. Varietal improvement work was started under All India Co-ordinated Vegetable Improvement Project. National Horticultural Research and Development Foundation took the lead and developed several varieties. Directorate of Onion and Garlic Research, Rajgurunagar exclusively working on onion and garlic improvement and developed two garlic varieties at national level. Still mostly local strains are grown in different garlic areas. Plant sterility usually hinders crop improvement by means of cross hybridization. Clonal selection thus remains a major breeding method for garlic as it is exclusively propagated vegetatively. The varieties released so far have been developed through clonal selections from local material.

Garlic grown near its centre of origin is bigger in size with bold cloves. The size becomes small with more number of small cloves as it advances towards sub-tropics and tropics. The yield potential of such adapted type is less as compared to long day types. The short day garlic grown in plains of North India, western India and hills of Nilgiris suffer from degeneration effects, small size of clove and susceptibility to diseases, pests and finally low yield. Therefore, improvement is needed to solve these problems.

Being an asexually propagated crop, methods of improvement through cross- pollination are not viable in garlic. Most of the varieties developed are through introductions and clonal selection. Based on temperature and day-length response, garlic has been classified as having long-day and short day varieties. It has also been classified as having hard neck and soft neck varieties. Hard-neck varieties bolt and flower but these flowers are usually sterile, while soft-neck varieties do not flower at all. Hard neck varieties cannot be braided for storage whereas soft-neck varieties can be braided and stored. Hard neck (long- day varieties) is characterized by big bulbs, less number of cloves (10-15), ease of peeling and, generally, have low storage life (Peter and Pradeep kumar, 2008; Ram, 1998; Batchvarov, 1993). Typical examples are Agrifound Parvati, Agrifound Parvati-2 and Chinese garlic. Because of big size, their productivity is higher and these fetch a good price in local and international markets. Soft-neck (short-day) varieties are characterized by small bulbs, more number of cloves (20-45), more aroma and are, generally, good storers *e.g.*, Indian garlic varieties G-41, G-1, G-50, G-282, G-323, G-189, G-384 and G-386 *etc.* On the basis of consumption, area and production statistics, garlic is an important commodity in the Indian market, yet, public or private research on this crop has been less than encouraging. The main reason for this may be its asexual nature which limits breeding methods and area under its cultivation. At the international front, there are a few reports of flowering and seed production, but even now garlic is considered a sexually sterile species. Breeding methods for development of garlic are limited to clonal selection and mutagenesis among conventional methods, and somaclonal variation among biotechnological approaches. In India, most varieties have been developed through clonal selection and one or two through introduction. National Horticultural Research and Development Foundation (NHRDF) has been at the forefront of garlic research (with maximum number of varieties developed under their research programmes), followed by agricultural universities, *viz.*, Gujarat Agricultural University (GAU), Punjab Agricultural University (PAU), MPKV, Rahuri, *etc.* (Mahajan and Lawande, 2011). The NHRDF initiated its research and development work with garlic germplasm collections from indigenous and abroad. The germplasm collections at the institute were considerably augmented by introducing 475 accessions from exotic and indigenous sources in garlic. The significant progress has taken place during the past three and half decades in development of ten high yielding varieties. During the course of evaluation/investigation/study of germplasm several genotypes selected for further recommendation for varieties development (Pandey, *et al.*1995, Gupta *et al.*, 1998, Singh, *et al.* 2011, Dubey and Singh, 2012, Singh and Gupta, 2012, Singh *et. al.*, 2013, Singh and Gupta, 2014).

Clonal Selection

Before the development of techniques to produce garlic seed in relatively large quantities, garlic breeding was not a realistic possibility. In fact, there is no evidence indicating that sexual reproduction and selection were ever utilized by garlic growers throughout history so that, while garlic has been one of the longest cultivated vegetable crops, breeding the crop has just begun. Clonal selection has been successful in altering some traits in garlic such as clove number and earliness (Burba, 1997) and routine treatments to reduce or eliminate viruses clearly improve production (Van Dijk 1994; Verbeek *et al.* 1995; Salomon 2002), but without sexual reproduction, desired traits found in different clones cannot be combined. With the possibility of seed production, garlic breeding can commence.

Due to non-flowering and sterility, almost all the varieties of garlic are through introduction and clonal selection. There is urgent need to induce flowering in garlic for varietal improvement. Clonal selection has been the most widely used method for generating new garlic material. It is based on the variability existing in populations. The development of new garlic clones for asexual propagation will follow the same process of evaluation and utilization as is currently being used to evaluate new asexually propagated clones that have been acquired from abroad and are being tested for performance. Selection for yield in garlic is difficult because the size of planting material *i.e.* garlic clove and also the conditions in which it was produced, *e.g.* plant density, affect its yield potential. Simple mass selection of the largest, best-looking bulbs gives little improvement (Messiaen *et al.* 1993). Instead, by planting cloves from selected individual garlic bulbs, a number of clonal families were developed. Subsequently these families were planted in comparative trials and the highest yielding families were selected as those with the highest mean values of the product (weight of harvested bulb x weight multiplication between planting and harvest). The weight multiplication is simply the weight of harvested bulb divided by the weight of the planted clove. The process of adequately testing of a new high yielding clone will take several years, since the increase of a single plant to produce an adequate supply of bulbs is necessary to perform replicated field trials, storage trials, and evaluation of added-value characteristics such as phytonutrient content or soluble solids. New cultivars developed this way must fit all existing parameters to meet grower and processor needs, but also exhibit enough superior traits or unique new combinations of traits to warrant considering replacing existing cultivars with them. New clonally propagated cultivars have likely been gradually replacing older ones in Central Asia throughout history, although this process has never been documented.

Systematic evaluation of naturally available genetic diversity needs to be undertaken for improving garlic. Since the plant is an exotic introduction, though an old one, genetic diversity in Indian cultivars is inadequate to meet the demands of breeders. This has to be compensated by raising collections from the centre of origin which lies somewhere in or around Afghanistan. As mentioned before, clonal variation exists in respect of bulb characteristics such as the number of cloves per bulb, size of individual clove, order *etc.* Since these characteristics are of direct importance, a well conceived selection programme is likely to yield good dividends.

Garlic Varieties Developed for Different Regions

NHRDF, Nashik: Agrifound White (G-41), Yamuna Safed (G-1), Yamuna Safed-2 (G-50), Yamuna Safed-3 (G-282), Yamuna Safed-4 (G-323), Yamuna Safed-5 (G-189), Yamuna Safed-8 (G-384), Yamuna Safed-9 (G-386), Agrifound Parvati (G-313) and Agrifound Parvati-2 (G-408). DOGR, Pune: Bhima Purple and Bhima Omkar. MPKV, Rahuri: Godavari, Sweta, GG-4. TNAU, Tamil Nadu: Ooty-1. VPKAS, Almora: VL Garlic-1 and VL Lahsun-2. CITH, Srinagar: CITH-G-1 and CITH-G-2. Besides these, varieties selected by farmers over the years are also available in the market. The distinct types namely 'Fawari and Rajalle Gaddi with slightly bigger bulbs are grown in the Bellary district of South India. Besides, Madrasi, Tabiti, Creole, Eknalia, T-56-4, Jamnagar Local, Ooty Local, Jeur Local *etc.* are known clones.

Germplasm Resources

Mountain regions of the former Soviet Union republics in Central Asia such as Kirgstan, Uzbekistan and others, and the pre-forest regions of the Caucasus and the Mediterranean possess the maximum variety of species, subspecies and genetic forms of cultivated garlic. International Plant Genetic Resources Institute, Rome is the nodal agency for collection and conservation of garlic germplasm. Global collections for conservation of *Allium* are the Centre for Genetic Resources, Wageningen, Netherlands; Horticultural Research International, Wellesbourne, UK; National Plant Germplasm System-USA and Pacific, for some species; Research Centre for Agrobotany, Tapioszele, Hungary (for South and East European countries) and National Institute of Agrobiological Research, Tsukuba, Japan. In India, NBPGR, New Delhi is the nodal agency for collection and conservation of garlic germplasm. Besides, National Horticultural Research and Development Foundation, Nasik; Directorate of Onion and Garlic Research, Rajgurunagar; MPKV, Rahuri; VPKAS, Almora; and Junagadh Agricultural University, Junagadh also collect and conserve garlic germplasm.

Garlic Breeding

Traditional garlic-breeding research has been limited to evaluation for yield and other morphological characters to identify the best genotypes (Figliuolo *et al*, 2001; Khar *et al*, 2005a, 2005b). Genetic studies have revealed positive interaction between plant-height, bulb-weight, bulb-diameter and mean clove-weight (Zhila, 1981). Significant positive correlation between clove and bulb mean-weight, negative correlation between clove mean-weight and clove-number has also been reported (Baghalian *et al.*, 2005). Variation in yield is explained by leaf number and bulb mean-weight. Therefore, these important characteristics could help in garlic selection programme and yield improvement (Baghalian *et al*, 2006). Although garlic is propagated vegetatively, considerable variation has been found in morphological traits (Shashidhar and Dharmatti, 2005; Khar *et al*, 2006). Major characters found to contribute to genetic diversity are bulb weight, diameter, yield, number of cloves per bulb, maturity, plant height, number of green leaves and bulbing period. Diversity assessment on the basis of morphological (Panthee *et al*, 2006; Baghalian *et al*, 2005), physical-chemical, productive and molecular characteristics, allicin content (Baghalian *et al.*, 2006), productive and qualitative characteristics (Resende *et al*, 2003) and chemotaxonomic classification (Storsberg *et al*, 2003) have been studied. In diversity assessment, Baghalian *et al*, (2005) did not detect any significant relationship between genetic diversity and geographical origin. Therefore, probably, genetic factors have more influence than ecology. Allicin is a major chemical constituent of garlic and is use in harmaceuticals. Multiple factors, *viz.*, genotype, environment, S fertilization and light spectrum (Huchette *et al*, 2005), relative water content, soil type and harvesting date (Yang *et al*, 2005) have been found to influence allicin content in garlic bulbs, whereas, Baghalian (2005) found no significant correlation between ecological condition and allicin content. Production of true seed in garlic (*Allium sativum* and *A. longicuspis*) has been known for several years. It was with the discovery of fertile clones by Etoh (1986) that efforts were started to induce flowering and seeds in garlic. With the advent of flowering garlic, Jenderek and Hannan (2004) were able to evaluate reproductive characteristics and true seed production in garlic germplasm and were successful at producing S1 bulbs in a few fertile clones. This represented valuable material for studies on garlic genetics (Jenderek 2004). Jenderek and Zewdie (2005) studied within and between family variability for important bulb and plant traits and observed that bulb weight, number of cloves, and clove weight were the main factors contributing to yield, and concluded that vegetative propagation of garlic over the centuries had produced highly heterozygous plants. Koul *et. al* (1979) studied prospects for garlic improvement in the light of its genetic and breeding systems and Simon and Jenderek (2004) made a comprehensive review about flowering, seed production and genesis of garlic breeding. Cultivated garlic, being non-flowering, has limited variability. Breeders depend upon natural clonal mutations and selection of superior clones from the germplasm. Induced mutations and somaclonal variation are the best way to broaden germplasm.

Mutation Breeding

On account of absolute apomixes, specific defects in commercial varieties of garlic might be rectified using mutation breeding. Even for the purpose of inducing resistance to common fungal and insect pests, mutation breeding needs to be fully tapped so that some otherwise good varieties can be developed for commercial cultivation. Apomixis has a factor that is extremely favourable for the improvement of this species in the hands of a mutation breeder. The mutations are fixed more rigidly by apomixes than amphimixis. Although mutations may be a source of variability, they are rather limited; therefore, breeding using this strategy has not resulted in significant progress (Etoh and Simon, 2002).

Experiment conducted in cold desert of trans-Himalayan region of Ladakh showed that frost induced mutant in garlic produced normal clove sized aerial bulbils and significantly higher bulb yield than normal clove planting (Mathew *et al.*, 2007). Among the chemical mutagens assessed for plant growth and yield characteristics in garlic cv. G-41, mutagens EMS (0.826%), colchicines (0.124%) and sodium azide (0.095%) at low dose for 6 hours treatment was found suitable for the creation of variability (Mahajan *et al.*, 2014).

Polyploidy and Somaclonal Variations

The practice of vegetative propagation and the desirability of improvement in the vegetative organs, particularly the bulb, constitute special advantages in pursuing polyploidy breeding in this crop. The advantage of polyploidy, if any, could readily be exploited since maintenance and propagation of polyploids pose no problem. Since no natural polyploids or aneuploids are known in garlic, the polyploidy breeding programme has to start with induced polyploidy. The evaluation of yield and other qualities will have to be done through the assessment of a diverse range of genotypes with doubled chromosomes.

Various experiments have been undertaken in garlic to create somaclonal variants to create irreversible variability to improve the original plants. Novak (1983) induced polyploidy by treating meristems with colchicines found to be effective in regenerating tetraploid plants (35%). Koch and Salomon (1994) cultured basal plates on BDS medium with 2,4-D and kinetin for callus inductions and further transferred to medium containing kinetin and IAA for development of adventitious shoots. Badria and Ali (1999) cultured root meristems on MS medium with kinetin, IAA and 2,4-D for callus induction and indentified somaclonal variants and after four cycle exhibited significant difference in bulb characteristics with three time increases in *allicine* content of variants (14.5mg/g) than original plant (3.8mg/g), but with no change in there chromosome number.

A somaclone derived from cultivar Rosado, in Argentina, had more agronomical desirable traits (taller, thicker pseudostem length, bulb with less but bigger size cloves) compare to original plant (Ordonez *et al.*, 2002). Mukhopadhyay *et al.*, (2005) observed chromosomal stability among the plants regenerated through callus of variety *Rossete* and frequency of aneuploid cells was increased with callus aging.

Interspecific Hybridization

The discovery of fertile garlic clone opened the way for producing interspecific hybrids between fertile garlic and other *Allium* species. Ohsumi *et al.* (1993) generated this interspecific hybrid using garlic as the female and pollinating with onion. The ovules were extracted and cultured to rescue the embryo. The interspecific hybrid was intermediate in phenotype between garlic and onion. This plant possesses unique combinations of the flavor compounds from onion and garlic. This was a very significant and interesting accomplishment, since the two species belong to sections *Cepa* and *Allium*.

The hybrid plants had only 2% pollen viability and did not produce seeds. However, there is a possibility of introducing or introgressing the garlic genome more broadly into the genus *Allium* by backcrossing, using onion, followed by embryo rescue. Interestingly, combinations of thiosulfinates from various alliums may show more health-enhancing effects than those from individual alliums (Morimitsu *et al.* 1992). The interspecific cross between *A. longicuspis* and garlic was successfully accomplished, just after the discovery of first fertile clone (Etoh, 1984), by pollination of sterile *A. longicuspis* with pollen from fertile garlic. The resulting hybrids, however, were also sterile. Offspring of crosses between *A. sativum* and *A. longicuspis* made by Pooler and Simon (1994) and by Jenderek (1998) were indistinguishable from those resulting from garlic x garlic crosses. *Allium longicuspis* plants tend to have higher flowering rates, smaller topsets, purple anthers and better seed production than garlic plants.

Another successful interspecific hybridization was performed between leek (*A. ampeloprasum)* (Female) and fertile garlic (*A. sativum*) (Sugimoto *et al.* 1991). The two species both differ in ploidy level. Leek is a tetraploid plant with 32 chromosomes, while garlic is diploid plant with 16 chromosomes. Hence, several interspecific triploids with 24 chromosomes and near-triploid aneuploids were recovered. Tetraploids and diploids were also obtained, but these may not be hybrids. Other interspecific hybrids between fertile garlic and more closely related species, such as *A. tuncelianum*, described by Mathew (1996) may be interesting for future studies.

Biotechnological Improvements

Tissue Culture

Regeneration via somatic embryogenesis in garlic was first reported from calli of bulb leaf discs and stem tips by Abo El-Nil (1977). Regeneration has also been reported from basal plate (Al-Zahim *et al.* 1999), leaf (Wang *et al.* 1994; Zheng *et al.* 1998), receptacle (Xue *et al.* 1991), and flower buds (Suh and Park 1988), shoot tip or stem disc explants (Myers and Simon 1999; Kondo *et al.* 2000; Hasegawa *et al.* 2002) and root tips (Haque *et al.* 1999, Barandiaran *et al.* 1999, Robledo *et al.* 2000). Myers and Simon (1998) reported on a continuous callus production and regeneration system from garlic root segments including root tips and this system was further refined by using both apical and non-apical garlic root segments (Zheng *et al.* 2003). Subsequently, Fereol *et al.* (2005) used young leaf explants from four European garlic cultivars to produce embryogenic calli and obtained efficient regeneration via somatic embryos. Metwally *et al.* (2012) studied the field performance of tissue-cultured garlic plants and concluded that it takes four vegetative generations (four years) for the micro-propagated plants to reach commercial size and these developed plantlets were considered to be new source for breeding and improvement of garlic.

Meristem Tip Culture

Normally garlic cultivars are propagated vegetatively because of their inability to produce seed. This vegetative propagation leads to accumulation of viruses and it is well established that garlic is susceptible to accumulation of a complex of viruses, notably members of the genera *Potyvirus*, *Carlavirus*, *Allexivirus* and *Potexvirus* (King *et al.*,

2012) that are spread from (vegetative) generation to generation through the bulbs. Losses in yield and deterioration in quality are the well established problems associated with virus infections. Control of these viruses is problematic and involves the production of virus-free plants by meristem-tip culture and subsequent multiplication of plants under aphid-free conditions. Production of virus free garlic plants has been attained through shoot tip culture (Pena-Iglesias and Ayuso, 1982), scape-tip culture (Ma *et al.* 1994), small inflorescence bulbils culture (Ebi *et al.* 2000), "stem disc dome culture" (Ayabe and Sumi, 2001), meristem tip culture (Wei and We, 1992). Attempts to obtain virus free garlic through thermotherapy (Conci and Nome, 1991; Ucman *et al.* 1998), combination of mersitem tip culture and thermotherapy (Robert *et al.* 1998) and use of chemotherapy (Ramirez *et al.* 2006) has been reported. It has also been concluded that virus free garlic yields better and have better quality than the virus infected plants (Ramirez *et al.* 2006).

Genetic Transformation

Introduction of alien DNA into plant cells can be achieved by using the bacterium *Agrobaterium tumefaciens* (indirect method) or biolistic method (direct method) as a vehicle. There are a few reports of garlic transformation using biolistic particle delivery or mediated by *Agrobacterium tumefaciens*. In biolistic approaches, Barandiaran *et al.* (1998) were first to attempt garlic transformation using biolistic approach to transfer and detect the transient expression of *uid*A gene into different garlic tissues, including regenerable calli using nuclease inhibitor aurintricarboxylic acid. Later, Ferrer *et al.* (2000) introduced, by biolistic method, reporting gene *uidA* and selection gene *bar* in leaf tissue, basal plate disc and embryogenic calli and reported maximum expression of *uidA* gen in calli and leaves. Sawahel (2002) showed that biolistic transformation can lead to the expression and stable integration of a DNA fragment into immature cloves whereas Park *et al.* (2002) established an effective biolistic transformation procedure for obtaining chlorsulfuron resistant transgenic plants by incorporating *ALS* gene coding for acetolactate synthase. Later Robledo-Paz *et al.* (2004) were able to introduce DNA into embryogenic garlic callus and produce stably transformed garlic plants.

Use of *Agrobacterium* mediated transformation was initiated by Kondo *et al.* (2000) and they were able to develop a stable transformation system of garlic using highly regenerative calli. Zheng *et al.* (2004) developed a reliable transformation system to produce garlic plants containing *Bt* resistance genes which conferred resistance to beet armyworm (*Spodoptera exigua*). Khar *et al.* (2005) studied the transitory expression of the reporter gene *gusA* in two garlic cultivars after infecting them with *A. tumefaciens* whereas Eady *et al.* (2005) recovered garlic transgenic plants from immature embryos using *A. tumefaciens* containing the vector pBIN *mgfp-ER* which includes the modified *gfp* reporter gene and the *nptII* selectable marker gene. Later, Kenel *et al.* (2010) developed a method for garlic transformation from immature leaves containing the *mgfp*-ER reporter gene and *hpt* selectable gene. Regenerated transgenic plants survived in the glasshouse and matured into healthy plants. Transgenic garlic plants stably integrated and expressed the phosphinothricin acetyltransferase (PAT) gene and they demonstrated that transgenic plants conferred herbicide resistance, whilst non transgenic plants and weeds died. Quality and yield of garlic is diminished due to white rot disease (*Sclerotium cepivorum* Berk).

Molecular Markers and Diversity

Garlic has been cultivated for millennia, but the taxonomic origins of this domestication process have not been identified. Modern taxonomy subdivides the world's garlic germplasm into five distinct groups: Sativum, Ophioscordon, Longicuspis, Subtropical and Pekinense (Fritsch and Friesen, 2002). The Longicuspis group from central Asia is recognized as the most primitive, the one from which the other group were derived (MaaB and Klaas, 1995; Etoh and Simon, 2002; Fritsch and Friesen, 2002). Central Asia was hypothesized to be the primary centre of garlic evolution and diversity (Fritsch and Friesen 2002), and recent studies on primitive garlic types in the Tien-Shan mountains strongly support this assumption (Etoh, 1986; Kamenetsky *et al*, 2003). A wide range of morphological diversity has been observed in garlic including flowering ability, leaf traits, bulb traits, plant maturity, bulbing response to temperature and photoperiod, cold hardiness, bulbil traits and flower traits (Simon and Jenderek, 2003). MaaB and Klaas (1995) included subtropical and Pekinense clones in their study, and suggested that the subtropical clones were clearly separated from all other types, while the Pekinense subgroup was relatively similar to the stalking type. Molecular markers are being used extensively for determination of genetic diversity because of their neutral nature, reproducibility of results across labs and no environmental effect on their expression. Genetic diversity of garlic has been assessed by isozymes (Pooler and Simon, 1993). RAPD techniques have been mostly reported for characterization of garlic germplasm from different researchers all over the world. RAPDs have been used for characterization of Australian genotypes (Bradley *et al*, 1996). Further various researchers viz. Maaß and Klaas, 1995 and Ipek *et al.* 2003, Taiwanese (Hsu *et al*, 2006), Brazilian (Buso *et al*, 2008), Chinese (Xu *et al*, 2005), Chilean (Paredes *et al*, 2008), Guatemalan (Rosales *et al*, 2007) and Indian garlic (Khar *et al*, 2008) also used RAPD for characterization. Inter-simple sequence repeat (ISSR) markers (Jabbes *et al.* 2011), combination of RAPD and ISSR markers (Shaaf *et al.* 2014), sequence related amplified polymorphism (SRAP) markers (Chen *et al.* 2013) were also studied. In addition to this, AFLP (Amplified Fragment Length Polymorphism) technique has also been used to characterize garlic (Ipek *et al*, 2003; Lampasona *et al*, 2003; Volk *et al*, 2004; Ipek *et al*, 2005; Garcia-Lampasona *et al.* 2012; Volk *et al.* 2004) and locus specific markers (Ipek *et al.* 2008). Ipek *et al* (2003) compared AFLPs, RAPD and isozymes for diversity assessment of garlic and detection of putative duplicates in germplasm collections and concluded that there was good correlation between the markers and demonstrated that genetic diversity among closely-related clones, which could not be differentiated with RAPD markers and isozymes, was detected by AFLPs. Therefore, AFLP is an additional tool for fingerprinting and detailed assessment of genetic relationships in garlic. Most of the reports have concluded that diversity assessment is not correlated with geographical location though a few studies reported correlation between geographical locations and the diversity (Lampasona *et al*, 2003). Volk *et al* (2004) reported that 64% of the U.S. National Plant Germplasm System's garlic collection, held at the Western Regional Plant Introduction Station in Pullman, Washington, USA, and 41% of commercial garlic collections, were duplicates. Rapid characterization of garlic accession is important for avoiding duplicate genotypes. For this purpose, Ipek *et al* (2008) developed several locus-specific polymerase chain reaction (PCR) based DNA markers and tested them for characterization of garlic clones and concluded that locus specific markers could be used as another tool for rapid characterization of garlic germplasm collection. Markers have also been used to clarify the taxonomic status of other well-characterized locally grown garlic (Ipek *et al*, 2008; Figliuolo and Stefano, 2007). Genetic fidelity of micro-propagated crops (Al Zahim *et al*, 1997, 1999), traits like pollen fertility (Etoh *et al*, 2001) and marker related to white rot (Nabulski *et al*, 2001) have also been reported. A wide range of morphological diversity has been observed in garlic including flowering ability, leaf traits, bulb traits, plant maturity, bulbing response to temperature and photoperiod, cold hardiness, bulbil traits and flower traits (Simon and Jenderek, 2004). With the reporting of flowering garlic, linkage maps have been developed (Ipek *et al*, 2005; Zewdie *et al*, 2005) which will help tag important genes in future.

But most of these markers are not ideal choice nowadays because of instability in RAPD markers, less and complex conditions of temperature and other parameters for amplification of ISSR markers, cumbersome procedures, lack of mechanization for AFLP markers and stability/reproducibility of markers across the laboratories using different chemicals, PCR machines and other conditions. Today, microsatellite markers or SSRs are the markers of choice for a broad number of genetic studies because of their high polymorphism, codominance, genomic abundance and transferability among various laboratories. The only criterion for using SSR markers is that prior knowledge about the genomic sequence for development of markers is needed.

Despite being cultivated since ages and known for its culinary and medical benefits (Kik and Gebhardt, 2001), genomic resources in garlic have not increased exponentially. Estimation of garlic diversity using microsatellite markers was first reported by Ma *et al.* (2009) wherein they were able to develop a SSR enriched library and finally reported eight SSRs for diversity estimation. The same eight SSRs were used by Zhao *et al.* (2011) for molecular genetic diversity, population structure and core collection estimation followed by Jo *et al.* (2012) who classified genetic variation in 120 accessions from five different countries using the seven primers out of the same eight SSRs reported earlier. Cunha *et al.* (2012) reported a new set of 16 SSR markers using (CT)8-and (GT)8- enriched library and found 10 markers to be polymorphic whereas Chen *et al.* (2014) used the same set of markers and found eight to be polymorphic. Khar, (2012) used ninety nine SSRs comprising 30 onion genomic (AMS1-30; Fischer and Bachmann, 2000), 30 onion EST-SSR microsatellites (McCallum *et al.* 2008), 8 garlic microsatellites (Ma *et al.* 2009) and 31 primers mined from garlic EST database (Kim *et al.* 2009) reported 18 polymorphic SSRs for estimation of genetic diversity in garlic. Recently, Cunha *et al.* (2014) were able to assess the genetic diversity and structure of Brazilian accessions using 17 SSR markers developed by Ma *et al.* (2009) and Cunha *et al.* (2012).

Prospects

Garlic is a widely recognized and appreciated crop with a long history of asexual propagation. The productivity of garlic is very low due to low yield potential of the available garlic varieties and non-availability of virus-free planting material. Way-forward is to combine conventional and biotechnological approaches to generate genetic variability and produce healthy planting material on commercial scale. Varieties with high yield, good storability and carrying tolerance/resistance to major pests and diseases need to be developed.

Fertility restoration will enable in the near future the development of new and improved cultivars propagated from seed. Garlic being a cross pollinated crop is highly heterozygous i.e. it has better seedlings survival and plant vigour. Hence, modern breeding should aim at seed propagated F_1 hybrids, thus eliminating the main ailments of clonal propagation, including carryover of pests from one generation to another, low propagation rate, voluminous storage of bulbs, rotting and sprouting, and spatial position of the transplanted cloves. The use of true seeds will save the costs of vegetative propagation and spare the need for virus elimination.

References

Abo, El-Nil M. M. 1977. Organogenesis and embryogenesis in callus cultures of garlic (*Allium sativum* L.). *Plant Science Letters.* 9: 259-264.

Al Zahim, M., Newbury, H.J. and Ford Lloyd, B.V. 1997. Classification of genetic variation in garlic (*Allium sativum* L.) revealed by RAPD. *Hort. Sci.,* 32:1102-1104.

Al-Zahim M. A, Ford Lloyd, B. V and Newbury, H. J. 1999. Detection of somaclonal variation in garlic (*Allium sativum* L.) using RAPD and cytological analysis. *Plant Cell Reports.* 18: 473-477.

Ata, A. M. 2005. Constitutive heterochromatin diversification of two *Allium* species cultivated in Egypt. Proceedings of the 7th African crop science society conference, Kampala, Uganda, 5-9 Dec. 225-231.

Ayabe, M. and Sumi, S. 2001. A novel and efficient tissue culture method "stem-disc dome culture" for producing virus-free garlic (*Allium sativum* L.). *Plant Cell Reports.* 20: 503-507.

Badria, F.A. and Ali, A.A. 1999. Chemical and Genetic Evaluation of Somaclonal Variants of Egyptian Garlic (*Allium sativum* L.). *J. Med Food* 2: 39-43.

Baghalian, K., Sanei, M.R., Naghavi, M.R., Khalighi, A. and Badi, H.A.N. 2005. Post-culture evaluation of morphological divergence in Iranian garlic ecotypes. *Acta Hort.,* 688:123-128.

Baghalian, K., Naghavi, M.R., Ziai, S.A. and Badi, H.N. 2006. Post-planting evaluation of morphological characters and allicin content in Iranian garlic (*Allium sativum* L.) ecotypes. *Scientia Hort.,* 107:405-410.

Barandiaran, X, Martin, N., Rodriguez, Conde, M. F., Di Pietro, A. and Martin, J. 1999. Genetic variability in callus formation and regeneration of garlic (*Allium sativum* L.). *Plant Cell Reports.* 18: 434-437.

Barandiaran, X., Pietro, A. D. and Martin, J. 1998. Biolistic transfer and expression of a *uidA* reporter gene in different tissues of *Allium sativum* L. *Plant Cell Reports.* 17: 737-741.

Batchvarov S. 1993. Garlic (*Allium sativum* L.). (In): G. Kalloo and B.O. Bergh (eds), Genetic improvement of vegetable crops. Pergamon Press Ltd. pp. 15-28.

Bradley, K.F., Rieger, M.A. and Collins, G.G. 1996. Classification of Australian garlic cultivars by DNA fingerprinting. *Aust. J. Exptl. Agri.,* 36:613-618.

Burba, J. L. 1993. Producción de "Semilla" de Ajo. Asociación Cooperadora EEA, La Consulta, Argentina.

Burba, J. L. 1997. Obtencion de nuevas cultivar de ajo. 50 Temas Sobre Prod. de Ajo 2:49-53. CAB Internacional, Wallingford, U.K.

Buso, G.S.C., Paiva, M.R., Torres, A.C., Resende, F.V., Ferreira, M.A., Buso, J.A. and Dusi, A.N. 2008. Genetic diversity studies of Brazilian garlic cultivars and quality control of garlic clover production. *Genet. and Mole. Res.*, 7:534-541.

Conci, V. and Nome, S. 1991. Virus Free Garlic (*Allium sativum* L.) Plants Obtained by Thermotherapy and Meristem-Tip Culture. *J. Phytopath.* 132: 186-192.

Cunha, C. P., Hoogerheide, E. S. S., Zucchi, M. I., Monteiro, M. and Pinheiro, J. B. 2012. New microsatellite markers for garlic, *Allium sativum* (Alliaceae). *Am. J. Bot. 99* : E17–E19.

Cunha, C. P., Resende, F. V., Zucchi, M. I. and Pinheiro, J. B. 2014. SSR-based genetic diversity and structure of garlic accessions from Brazil.*Genetica*, 142 : 419-431.

Chen, S., Chen, W., Shen, X., Yang, Y., Qi, F., Liu, Y. and Meng, H. 2014. Analysis of the genetic diversity of garlic (*Allium sativum*) by simple sequence repeat and inter simple sequence repeat analysis and agro-morphological traits. *Biochem. Syst. Ecol.* 55 : 260-267.

Chen, S., Zhou, J., Chen, Q., Chang, Y., Du, J. and Meng, H. 2013. Analysis of the genetic diversity of garlic (*Allium sativum* L.) germplasm by SRAP. *Biochem. Syst. Ecol.* 50 : 139–146.

Dubey, B. K. and Singh, R. K. 2012. Selection of planting genotypes for yield quality as well as storage in garlic (*Allium sativum* L). *Ind. J. Hort.* 69(1):125-128.

Eady, C. C., Davis, S., Catanach, A., Kenel, F. and Hunger, S. 2005. *Agrobacterium tumefaciens*-mediated transformation of leek (*Allium porrum*) and garlic (*Allium sativum*). *Plant Cell Reports.* 24 : 209-215.

Ebi, M., Kasai, N. and Masuda, K. 2000. Small inflorescence bulbils are best for micro propagation and virus elimination in garlic. *Hort. Science* 35: 735-737.

Etoh, T. 1984. Germination of seeds obtained from a clone of garlic, *Allium sativum* L. Proc. Jpn. Acad. 59, Ser. B, 4:83-87.

Etoh, T. 1986. Fertility of the garlic clones collected in Soviet Central Asia. *J. Jpn. Soc. Hort. Sci.*, 55:312-319.

Etoh, T., Watanabe, H. and Iwai, S. 2001. RAPD variation of garlic clones in the center of origin and the westernmost area of distribution. *Mem. Fac. Agric. Kagoshima University*, 37:21-27.

Etoh, T., and P. W. Simon. 2002. Diversity, fertility, and seed production of garlic. p. 101–117. (In:) H. Rabinowitch and L. Currah (eds.), Allium crop science: Recent advances. CABI Publ., New York.

Fereol, L., Chovelon, V., Causse, S., Triaire, D., Arnault, I., Auger, J. and Kahane, R. 2005. Establishment of embryogenic cell suspension cultures of garlic (*Allium sativum* L.), plant regeneration and biochemical analyses. *Plant Cell Reports.* 24 : 319-325.

Ferrer, E., Linares, C. and Gonzalez, J. M. 2000. Efficient transient expression of the beta-glucuronidase reporter gene in garlic (*Allium sativum* L.). *Agronomie* 20 : 869-874.

Figliuolo, G; Candido, V., Logozzo, G., Miccolis, V. and Zeuli, P.L.S. 2001. Genetic evaluation of cultivated garlic germplasm (*Allium sativum* L. and *A. ampeloprasum* L.). *Euphytica,* 121:325-334.

Figliuolo, G. and Stefano, D.di. 2007. Is single bulb producing garlic *Allium sativum* or *Allium ampeloprasum Scienti. Hort.,* 114:243-249.

Fischer, D. and Bachmann K. 2000. Onion microsatellites for germplasm analysis and their use in assessing intra- and interspecific relatedness within the subgenus *Rhizirideum. Theor. Appl. Genet.* 101 : 153-164.

Fritsch, R. M. and Friesen, N. 2002. Evolution, domestication and taxonomy. p. 5-30. (In:) H.D. Rabinowitch and L. Currah (Eds.), *Allium* crop sciences: recent advances, CAB Int., Wallingford, UK.

Garcıa-Lampasona, S., Asprelli, P. and Burba, J. L. 2012. Genetic analysis of garlic (*Allium sativum* L.) germplasm collection from *Argentina. Sci. Hortic.* 138 : 183–189.

Gupta, R. P. and Singh, D. K. 1998. Studies of the performance of different advance lines in garlic. *NHRDF Newsletter* 18 (3): 13-18.

Haque, M. S., Wada, T. and Hattori, K. 1999. Anatomical changes during in vitro direct formation of shoot bud from root tips in garlic (*Allium sativum* L.). *Plant Production Science* 2: 146-153.

Hasegawa, H., Sato, M. and Suzuki, M. 2002. Efficient plant regeneration from protoplasts isolated from long-term, shoot primordia-derived calluses of garlic (*Allium sativum*). *Journal of Plant Physiology* 159: 449-452.

Hsu H.C., Hwu, K., Deng, T and Tsao, S. 2006. Study on genetic relationship among Taiwan garlic clones by RAPD markers. *J. Taiwan Soc. Hort. Sci.,* 52:27-36.

Huchette, O., Kahane, R., Auger, J., Arnault, I. and Bellamy, C. 2005. Influence of environmental and genetic 113 factors on the alliin content of garlic bulbs. *Acta Hort.* 688:93-99.

Ipek, M., Ipek, A., Almquist, S. G and Simon, P. W. 2005. Demonstration of linkage and development of the first low-density genetic map of garlic, based on AFLP markers. *Theoretical and Applied genetics,* 110: 228-236.

Ipek, M., Ipek, A. and Simon, P. W. 2003. Comparison of AFLPs, RAPD markers, and isozymes for diversity assessment of garlic and detection of putative duplicates in germplasm collections. *J. Am. Soc. Hortic. Sci.* 128, 246–252.

Ipek, M., Ipek, A. and Simon, P. W. 2008. Rapid characterization of garlic clones with locus-specific DNA markers. *Turk. J. Agric. For.* 32: 357–362.

Jabbes, N., Geoffriau, E., Le Clerc, V., Dridi, B. and Hannechi C. 2011. Inter simple sequence repeat fingerprints for assess genetic diversity of Tunisian garlic populations. *J. Agric. Sci.* 3: 77–85.

Jenderek, M.M. 2004. Development of S1 families in garlic. *Acta Hort.*, 637:203-206.

Jenderek, M. M. 1998. Generative reproduction of garlic (*Allium sativum* L.) (in Polish). Zeszyty Naukowe Akademii Rolniczej im. *H. Kollataja w Krakowie* 57:141–145.

Jenderek, M.M. and Hannan, R.M. 2004. Variation in reproductive characteristics and seed production in the USDA garlic germplasm collection. *Hort. Sci.*, 39:485-488.

Jenderek, M.M. and Zewdie, Y. 2005. Within and between family variability for important bulb and plant traits among sexually derived progenies of garlic. *Hort. Sci.*, 40:1234-1236.

Jo, M., Ham, I., Moe, K., Kwon, S., Lu, F., Park, Y., Kim, W., Won, M., Kim, T. and Lee, E. 2012. Classification of genetic variation in garlic (*Allium sativum* L.) using SSR markers. *Aust. J. Crop. Sci.* 6: 625–631.

Jones, H. A., and L. K. Mann. 1963. Onions and their allies. Leonard Hill Books, London, UK.

Kamentsky, R., London Shafir, I., Bizerman, M., Khassanov, F., Kik, C. and *Rabi*nowitch, H.D. 2003. Garlic (*Allium sativum* L.) and its wild relative from Central Asia: evolution for fertility potential. *Proceeding of XXVIth International Horticultural Congress*, Toronto, Canada. *Acta Hort..*, 673:83-91.

Kenel, F., Eady, C. and Brinch, S. 2010. Efficient *Agrobacterium tumefaciens*-mediated transformation and regeneration of garlic (*Allium sativum*) immature leaf tissue. *Plant Cell Rep.* 29: 223-230.

Khar, A., Mahajan,V., Devi, A.A. and Lawande, K.E. 2005a. Genetical studies in elite lines of garlic (*Allium sativum* L.). *J. Maha. Agri. Univ.*, 30:277-280

Khar, A., Asha Devi, A., Mahajan, V. and Lawande, K.E. 2005b. Genotype X environment interactions and stability analysis in elite lines of garlic (*Allium sativum* L.). *J. Spices Arom. Crops*, 14:21-27.

Khar, A., Asha Devi, A., Mahajan, V. and Lawande, K.E. 2006. Genetic divergence analysis in elite lines of garlic (*Allium sativum* L.). *J. Maha. Agri. Univ.*, 31:52-55

Khar, A., Asha, Devi A. and Lawande, K. E. 2008. Analysis of genetic relationships among Indian garlic (*Allium sativum* L.) cultivars and breeding lines using RAPD markers. *Ind. J. Gen.* 68: 52-57.

Khar, A., Yadav, R. C., Yadav, N. and Bhutáni, R. D. 2005. Transient *gus* Expression Studies in Onion (*Allium cepa* L.) and Garlic (*Allium sativum* L.). Akdeniz Universitesi Ziraat Fakultesi Dergisi., 18 : 301-304.

Khar, A. 2012. Cross amplification of onion derived microsatellites and mining of garlic ESTdatabase for assessment of genetic diversity in garlic. *Acta Hort.* 969: 289-295.

Kik, C. K. R. and Gebhardt, R. 2001. Garlic and health. *Nutr. Metab. Cardiovasc. Dis.* 11: 57–65.

Kim, D. W., Jung, T. S., Nam, S. H., Kwon, H. R., Kim, A., Chae, S. H., Choi, S. H., Kim, D. W., Kim, R. N. and Park, H. S. 2009. Garlic ESTdb : an online database and mining tool for garlic EST sequences. *BMC Plant Biology* 6: 1-6.

King, A. M., Adams, M. J., Lefkowitz, E. J. and Carstens, E. B (Eds). 2012. Virus taxonomy: classification and nomenclature of viruses: Ninth Report of the International Committee on Taxonomy of Viruses. Elsevier.

Koch, M. and Salomon, R. 1994. Improvement of garlic via somaclonal variation and viruses' elimination. *Acta. Hort.* 358:211-214

Kondo, T., Hasegawa, H. and Suzuki, M. 2000. Transformation and regeneration of garlic (*Allium sativum* L.) by *Agrobacterium*-mediated gene transfer. *Plant Cell Reports* 19: 989-993.

Koul, A.K., Gohil, R.N. and Langer,A. 1979. Prospects of breeding improved garlic in the light of its genetic and breeding systems. *Euphytica*, 28:457-464

Lampasona,S. G., Martinez, L., and Burba, J.L. 2003. Genetic diversity among selected Argentinean garlic clones (*Allium sativum* L.) using AFLP (Amplified Fragment Length Polymorphism). *Euphytica,* 132: 115-119

Ma, K. H., Kwag, J. G., Zhao, W. G., Dixit, A., Lee, G. A., Kim, H. H., Chung, I. M., Kim, N. S., Lee, J. S., Jun, J. J., Kim, T. S. and Park, Y. J. 2009. Isolation and characteristics of eight novel polymorphic microsatellite loci from the genome of garlic (*Allium sativum* L.). *Sci. Hortic.* 122: 355-361.

Ma, Y., Wang, H. L., Cun-Jin, Z., Zhang, C. J. and Kang, Y. Q. 1994. High Rate of Virus-Free Plantlet Regeneration via Garlic Scape-Tip Culture. *Plant Cell Rep.*, 14: 65-68.

Maaß, H. I. and Klaas, M. 1995. Infraspecific differentiation of garlic (*Allium sativum*) by isozyme and RAPD markers. *Theor. Appl. Genet.* 91: 89–97.

Mahajan, V., Asha Devi, Anil Khar and K.E. Lawande, 2014, Studies on mutagenesis in garlic using chemical mutagens to determine lethal dose (LD_{50}) and create variability in garlic (*Allium sativum* L.). *Indian Journal of Horticulture, 72(2): 289-292.*

Mahajan, V., and Lawande K.E. 2011. Genetic diversity and crop improvement in onion and garlic. *Souvenir: Exploiting spices production potential of the Deccan region*, SYMSAC-VI, Indian Society for Spices, 8-10 Dec., 2011, Dharwad, 19-40.

Mathew, B .1996. *A Review of Allium Section* Allium. Royal Botanic Garden, Kwe, Richmond, UK, pp.176.

Mathew, D., Zakwan Ahmed and N. Singh 2007 Formulation of flowering index, morphological relationships and yield prediction system in true garlic aerial seed bulbil production. *Hort. Science.* 40(7):2036-2039.

McCallum, J., Thomson, S., Pither-Joyce, M., Kenel, F., Clarke, A. and Havey, M. J. 2008. Genetic diversity analysis and single-nucleotide polymorphism marker development in cultivated bulb onion based on expressed sequence tag–simple sequence repeat markers. *J. Amer. Soc. Hort. Sci.* 133: 810–818.

Messiaen, C. M., Cohat, J., Leroux, J. P., Pichon, M. and Beyries, A. 1993. *Les Allium Alimentaires Reproducts par Voie Vegetative*. INRA, Paris, pp. 230.

Metwally, EI., El-Denary, M. E., Omar, A. M. K., Naidoo, Y. and Dewir, Y. H. 2012. Bulb and vegetative characteristics of garlic (*Allium sativum* L.) from *in vitro* culture through acclimatization and field production. *African Journal of Agricultural Research*, 7: 5792-5795.

Morimitsu,Y.,Y. Morioka, and S. Kawakishi. 1992. Inhibitors of platelet aggregation generated by mixtures of *Allium* species and/or S-alk (en) nyl-L-cysteine sulfoxides. *J. Agr. Food Chem.* 40:368–372.

Mukhopadhyay, M.J., Sengupta, P., Mukhopadhyay, S. and Sen, S. 2005. *In vitro* Stable regeneration from Onion and Garlic from Suspension Culture and Chromosomal Instability in Solid Callus Culture. *Sci. Hort* 104(1):1-9

Myers, J. M. and Simon, P. W. 1998. Continuous callus production and regeneration of garlic (*Allium sativum* L.) using root segments from shoot tip-derived plantlets. *Plant Cell Reports* 17: 726-730.

Myers, J. M. and Simon, P. W. 1999. Regeneration of garlic callus as affected by clonal variation, plant growth regulators and culture conditions over time. *Plant Cell Reports* 19: 32-36.

Nabulski, I., Safadi A.I., Mit, B., Ali, N. and A*rabi*, M.I.E. 2001. Evaluation of some garlic (*Allium sativum* L.) mutants resistant to white rot disease by RAPD analysis. *Ann. Appl. Biol.*, 138:197-202.

Novak, F.J. 1983: Production of Garlic (*Allium sativum* L.) Tetraploids in shoot-tip in vitro culture. Z.*Pflanzen* 91: 329-333.

Novak, F. J. 1990. *Allium* Tissue Culture. (In): *Onions and Allied Crops*. Rabinowitch, H. D., and Brewster, J. L., (Eds)., pp. 233-250, Vol. II, CRC Press, Boca Raton, Fl., U.S.A.

Ohsumi, C., A. Kojima, K. Hinata, T. Etoh, and T. Hayashi. 1993. Interspecific hybrid between *Allium cepa* and *Allium sativum*. *Theor. Appl. Genet.* 85:969–975.

Ordonez, A., Torres, L.E., Hidalgo, M.G. and Munoz, J.O. 2002. Analisis citologia de una variante genetica somatica de ajo (*Allium sativum* l.) Tipo rosado. *Agriscientia* 19: 37-43

Pandey, U. B., Bhonde, S. R., Chouhan, K. P. S and Singh, D. P. 1995. Evaluation of garlic varieties for storage performance. *Allium Improvement News letter* 5:52-54.

Panthee, D.R., Regmi, H.N., Subedi, P.P., Bhattarai, S. and Dhakal, J. 2006. Diversity analysis of garlic (*Allium sativum* L.) germplasm available in Nepal based on morphological characters. *Genet. Res. Crop Evol.*, 53:205-212.

Paredes, C.M., Becerra, V.V. and Gonzalez, A.M.I. 2008. Low genetic diversity among garlic (*Allium sativum* L.) accessions detected using random amplified polymorphic DNA (RAPD). *Chilean J. Agril. Res.*, 68:3-12.

Park, M. Y., Yi, N. R., Lee, H. Y., Kim, S. T., Kim, M., Park, J. H., Kim, J. K., Lee, J. S., Cheong, J. J. and Choi, Y. D. 2002. Generation of chlorsulfuron-resistant transgenic transgenic garlic plants (*Allium sativum* L.) by particle bombardment. *Molecular Breeding* 9: 171-181.

Peña-Iglesias, A. and Ayuso, P. 1982. Characterization of Spanish Garlic Viruses and Their Elimination by *in vitro* Shoot Apex Culture. *Acta. Hort.* 127: 183-193.

Pooler, M. R. and Simon, P. W. 1993. Characterization and classification of isozyme and morphological variation in a diverse collection of garlic clones. *Euphytica.* 68: 121–130.

Pooler, M. R., and Simon, P. W 1994. True seed production in garlic. *Sexual Plant Reprod.* 7: 282- 286.

Peter, K.V. and Pradeep Kumar, T. 2008. Genetics and breeding of vegetable crops. ICAR, New Delhi, pp. 48-67.

Ram, H. H. 1998. Vegetable Breeding-Principles and Practices, Kalyani Publishers, New Delhi. pp. 309-321.

Ramírez-Malagón, R., Pérez-Moreno, L., Borodanenko, A., Salinas-González, G. J. and Ochoa-Alejo N. 2006. Differential organ infection studies, potyvirus elimination, and field performance of virus-free garlic plants produced by tissue culture. *Plant cell, tissue and organ culture, 86:* 103-110.

Resende, G.M-de, Chagas, S.J-de R, Pereira, L.V., 2003. Productive and qualitative characteristics of garlic cultivars. *Horticultura-Brasileira*, 21:686-689.

Robert, U., Zel, J. and Ravnikar, M. 1998. Thermotherapy in virus elimination from garlic: influences on shoot multiplication from meristems and bulb formation *in vitro. Sci Hort.* 73:193–202.

Robinson, R. A. 2007. *Self-Organizing Agro-ecosystems.* Share books Publishing, ISBN 698-0- 9783634-1-3, Available: Sharebooks e-book.

Robledo Paz, A., Cabrera Ponce, J. L., Villalobos Arámbula, V. M., Herrera Estrella, L. and Jofre Garfias, A. E. 2004. Genetic Transformation of Garlic (*Allium sativum* L.) by Particle Bombardment. *Hort. Sci.* 39: 1208-1211.

Robledo Paz, A., Villalobos Arambula, V. M. and Jofre Garfias, A. E. 2000. Efficient plant regeneration of garlic (*Allium sativum* L.) by root-tip culture. *In Vitro Cellular and Developmental Biology – Plant* 36: 416-419.

Rosales-Longo, F.U, Molina-Monterroso, L.G., 2007. Genetic diversity of the garlic (*Allium sativum* L.) grown in Guatemala, revealed by DNA markers. *Agronomia-Mesoamericana,* 18: 85-92.

Rubatzky, V. E., and M.Yamaguchi 1997. World vegetables: Principles, Production and Nutritive Values, 2nd ed., Chapman and Hall, New York.

Salomon, R. 2002. Virus diseases in garlic and the propagation of virus-free plants. p. 311-327. In: H. Rabinowitch and L. Currah (Eds.), *Allium* Crop Science: Recent advances. CABI Publ., New York.

Sawahel, W. A. 2002. Stable genetic transformation of garlic plants using particle bombardment. *Cellular and Molecular Biology Letters* 7: 49-59.

Senula, A., Keller, E. R. J. and Leseman, D. E. 2000 Elimination of viruses through meristem culture and thermotherapy for the establishment of an *in vitro* collection of garlic (*Allium sativum*). *Acta Horticulturae*, 530:121-128.

Shaaf, S., Sharma, R., Kilian, B., Walther, A., Özkan, H., Karami, E. and Mohammadi, B. 2014. Genetic structure and eco-geographical adaptation of garlic landraces (*Allium sativum* L.) in Iran. Genet Resour. *Crop Evol. doi*:10.1007/s10722-014-0131-4.

Shashidhar, T.R., and Dharmatti, P.R. 2005. Genetic divergence studies in garlic. *Karnataka J. Hort.*, 1:12-15.

Simon, P.W. and Jenderek, M.M, 2004. Flowering, seed production, and the genesis of garlic breeding. *Pl. Breed. Rev.*, 23:211-244.

Singh, R. K., Dubey, B. K. Bhonde, S. R. and Gupta, R. P. 2011. Correlation and path coefficient studies in garlic (*A. sativum* L). *J. Spices and Aromatic Crops,* 20 (2):81-85.

Singh, R. K., Dubey, B. K. and Gupta, R. P. 2012. Studies on variability and genetic divergence in garlic (*Allium sativum* L). *J. Spices and Aromatic Crops,* 21 (2): 136-144.

Singh, R. K., Dubey, B. K. and Gupta, R. P. 2013. Intra and Inter cluster studies for quantitative traits in garlic (*Allium sativum* L). *SARK Journal of Agricultural* 11 (2): 61-67.

Singh, R. K., Gupta, R. P. and Dubey, B. K. 2014. "G-389 (IC-0596521; INGR14009), a garlic (*Allium sativum* L) germplasm for earliness. *Ind. J. Plant Gent. Resources*: 27 (2): 187-189.

Storsberg, J, Schulz, H. and Keller, E.R.J. 2003. Chemotaxonomic classification of some Allium wild species on the basis of their volatile sulphur compounds. *J. Appl. Bot.*, 77:160-162.

Sugimoto, H., Tsuneyoshi, T., Tsukamoto, M., Uragami, Y. and Ehoh, T. 1991. Embryo-cultured hybrid between garlic and leek. *Allium Improvement Newsletter* 1: 67-68.

Suh, S. and Park, H. G. 1988. Somatic embryogenesis and plant regeneration from flower organ culture of garlic (*Allium sativum* L.). *Korean Journal of Plant Tissue Culture* 15: 121-132.

Ucman, R., Zel, J. and Ravnikar, M. 1998. Thermotherapy in Virus Elimination from Garlic: Influences on Shoot Multiplication from Meristems and Bulb Formation *in vitro*. *Sci. Hort.*, Vol. 73, pp. 193-202.

Van Dijk, P. 1994. Virus diseases of *Allium* species and prospects for their control. *Acta. Hort.* 358: 299-306.

Verbeek, M. P., Van Dijk, and Van Well, P. M. A. 1995. Efficiency of eradication of four viruses from garlic (*Allium sativum*) by meristem-tip culture. *Eur. J. Plant Path.* 101:231-239.

Volk, G. M., Henk, A. D. and Richards, C. M. 2004. Genetic diversity among U. S. garlic clones as detected using AFLP methods. *J. Am. Soc. Hort. Sci.* 129: 559-569.

Volk, G. M. and Stern, D. 2009. Phenotypic characteristics of ten garlic cultivars grown at different North American locations. *Hort. Sci.* 44: 1238-1247.

Wang, H. L., Kang, Y. Q. and Zhang, C. J. 1994. Embryogenesis via culture of garlic sprout leaf. *Acta Agriculturae Boreali Sinica* 9: 92-94.

Wei, N. S. and We, Y. F. 1992. Identification of virus diseases and virus free meristem culture of garlic. *Acta Universitatis Agriculturalis Boreali Occidentalis*, 20:76-81.

Xu, P., Yang, C., Qu, S. and Yang, C.Y. 2005. A preliminary study on genetic analysis and purity assessment of the garlic germplasm and seed bulbs by the "fingerprinting" technique. *Acta. Hort.*, 688:29-33.

Xue, H. M., Araki, H., Shi, L. and Yakuwa, T. 1991. Somatic embryogenesis and plant regeneration in basal plate and receptacle derived-callus cultures of garlic (*Allium sativum* L.). *Journal of the Japanese Society for Horticultural Science* 60: 627-634.

Yang, F., Liu, S. and Wang, X. 2005. Effects of boron in soil on the physiological-biochemical characteristics, yield and quality of garlic. *Sci. Agricultura Sinica.* 38:1011-1016.

Zewdie, Y. and Jenderek, M. M. 2005. Within and Between family variability for important bulb and plant traits among sexually derived progenies of garlic. *Hort. Sci.,* 40:1234-1236.

Zhao, W. G., Chung, J. W., Lee, G. A., Ma, K. H., Kim, H. H., Kim, K. T., Chung, M. I., Lee, J. K., Kim, N. S., Kim, S. M. and Park, Y. J. 2011. Molecular genetic diversity and population structure of a selected core set in garlic and its relatives using novel SSR markers. *Plant Breed.* 130: 46-54.

Zheng, H. R., Shen, M. J., Zhong, W. J., Zhang, Z. Q. and Zhou, Y. 1998. Induction and utilization of globular bodies on calli from garlic (*Allium sativum* L.) leaf explants. Study of cellular histology in morphogenesis. *Acta Agriculturae Shanghai* 14: 33-38.

Zheng, S. J., Henken, B., Ahn, Y. K., Krens, F. A. and Kik, C. 2004. The development of a reproducible *Agrobacterium tumefaciens* transformation system for garlic (*Allium sativum* L.) and the production of transgenic garlic resistant to beet armyworm (*Spodoptera exigua* Hübner). *Molecular Breeding* 14: 293-307.

Zheng, S. J., Henken, B., Krens, F. A. and Kik, C. 2003. The development of an efficient cultivar independent plant regeneration system from callus derived from both apical and non-apical root segments of garlic (*Allium sativum* L.). *In Vitro Cellular and Developmental Biology- Plant* 39: 288-292.

Zhila, E.D. 1981. Correlations between phenotypic traits in garlic of the bolting type. *Tsitologiya-i-Genetika*, 15:46-48.

10

Hyacinth Bean

Alok Nandi

Hyacinth bean(*Lablab purpureus*), also known as field bean or dolichos bean is a popular legume vegetable in South Asia, China, Japan, West Africa, and the Caribbean. It is also popular as a nitrogen-fixing green manure to contribute to soil N and improve soil quality. Lablab is a popular choice as a cover crop on infertile, acidic soils, and it is drought tolerant once established. Like other legumes, it can be incorporated into a grazing rotation.

It has long been cultivated in India. It is grown as a field crop in Tamil Nadu, Andhra Pradesh, Karnataka, Madhya Pradesh and Maharashtra. In Kerala, Odisha, West Bengal and other states the photo-sensitive pole types are grown in homesteads by trailing to bower for its tender fruits which are used as cooked vegetable.

Although originally a perennial, hyacinth bean is often grown as an annual. It is typically a twining plant, 1.5 to 6 m tall, but there are bushy forms as well. The trifoliate (3-leafleted) leaves are tinged purple. The white, pink, or purple flowers give rise to 5 to 15 cm pods that are flat, glossy, and suffused with purple. These pods contain three to six seeds, which may be white, cream, buff, reddish, brown, or black. Mature and dried seeds are reportedly toxic due to high levels of cyanogenic glucosides (more specifically, glucosides) and should be boiled in two changes of water before eating to remove the toxins. The tap-root is well developed with many laterals and well developed adventitious roots. Probably no other legume shows such variation in form and habit.

Cytotaxonomic Background

The biosystematics of hyacinth bean and its relatives were reviewed and revised (Magness *et al.*,1971). Formerly, *Lablab* was included in the genus *Dolichos* following Linnaeus, but is now assigned to the monotypic genus *Lablab*. Three subspecies are recognized in *L. purpureus* ; ssp. *uncinatus*: the wild ancestral form distributed mainly in East Africa with small, scimitar-shaped pods of about 40 mm x15 mm; ssp. *purpureus*, cultivated as a pulse crop, has larger, scimitar-shaped pods, 100 mm x 40 mm; includes commercial varieties; and ssp. *bengalensis*, Asiatic origin, has linear-oblong shaped pods, longer than other subspecies, up to 140 mm x 10-25 mm. Although pod shape is a significant morphological difference, it is widely believed that ssp. *bengalensis* and ssp. *purpureus* are genetically very similar. Although most domesticated material is either ssp. *purpureus* or ssp. *bengalensis*, ssp. *uncinatus* has been domesticated in Ethiopia. Studies in lablab have shown that the perennial types have considerable genetic and morphological diversity. Lablab is predominantly self-fertilizing. Chromosome number is $2n = 22$.

Botanical name:	*Lablab purpureus* (L.) *Sweet*
Order:	Fabales
Family:	Fabaceae
Subfamily:	Faboideae
Species:	Lablab purpureus
Tribe:	Phaseoleae
Sub-tribe:	Faboideae
Genus:	Lablab

Synonyms

Dolichos bengalensis Jacq. ;*Dolichos lablab* L. ;*Dolichos purpureus* L.;*Lablab niger* Medikus; *Lablab purpurea* (L.) Sweet; *Lablab vulgaris* (L.) Savi; *Vigna aristata* Piper.

Common Names

Hyacinth bean, lablab bean, field bean, pig-ears, rongai dolichos, lab-lab bean, poor man's bean, Tonga bean (English); dolique lab-lab, dolique d'Egypte, frijol jacinto, quiquaqua, caroata chwata, poroto de Egipto, chicarros, frijol caballo, gallinita, frijol de adorno, carmelita, frijol caballero, pois nourrice (Spanish); faselbohne, helmbohne, schlangenbohne, batao, wal, sem, lubia (the Sudan); fiwi bean (Zambia); antaque, banner bean (Caribbean); wal (India); batao (Philippines); natoba, toba (Fiji); pois Antaque; pois de Senteur, tapirucusu.

Linnaeus (1754) described the species under *Dolichos* L. Adanson (1763) phrased the name *Lablab* for *Dolichos* L.(Gowda, 2012). The first of these species combined in *Lablab adans* was *L. niger* of Medikus (1787), based on *Dolichos lablab* L. A new epithet was required as *Lablab* would be tantonymous. Since *Lablab niger* is synonymous with *D. lablab*, the two genera are homotypic by lectotypification.

Dolichos described in Flora India by Roxburgh (1832) contained seven varieties, of which five were cultivated and the rest were wild(Gowda, 2012). The former was further divided into two categories: (a) *Dolichos lablab* var. *typicus* and (b) *Dolichos lablab* var. *lignosus*(Barker, 1911). It was seen that while many marked differences existed between the two, clear-cut separation between the two was not possible. The distinction between them on the basis of pod tips(Roxburgh, 1832) and on the basis of the disposition of the long axis of the seeds to the sutures(Prain, 1897) is not clear-cut enough to serve as reliable guides to their separation, especially in view of the wealth of blended forms.

Purseglove (1968) has recognized two types of *Lablab* in India and are sometimes considered as distinct species. They are:

Dolichos lablab var. *typicus* Prain: It is commonly called Lablab bean, Bonavist bean, Hyacinth bean, Indian butter bean (Hindi-Sem; Bengali-Shim; Gujarathi-Val; Marathi-Pavta; Telugu-Chikkudu; Tamil-Avarai; Kannada-Chapparadavare; Malayalam- Avara.)

A perennial twining herb, cultivated mostly as an annual, distributed throughout the tropical and temperate regions of Asia, Africa and America. In India, it is grown as a garden crop. Several types differ in colour of flowers, size, shape and texture of pods and size and colour of seeds. A type with showy purple flowers is cultivated as an ornamental plant in temperate regions. The pods are white, green or purple-margined. Seeds white, yellow, brownish, purple or black.

Dolichos lablab var. *lignosus* Prain: It is commonly known as Australian pea, Field Bean. (Hindi - Ballar; Gujarati - Val; Telugu - Anumulu; Tamil - Mochai; Kannada - Avare; Malayalam - Mochakotta).

It is a semi-erect, bushy, perennial herb, cultivated as an annual. It shows little or no tendency to climb. Leaflets innately trifoliate, smaller than those of var. *typicus*. Flowers borne on a straight upright stalk, often a foot high on which they open in succession. Pods oblong, flat and broad, firm-walled and fibrous, contain 4-6 seeds with their long axis at right angles to the suture. Seeds almost rounded white, brown or black. The plant emits a characteristic odour.

Early History

Lablab purpureus L. (Sweet) is considered to be native to India or South-East Asia. Summarizing archaeological findings, Fuller (2003) mentioned that *Lablab* bean is an ancient crop in India with the earliest find from before 3500 BC (Gowda, 2012). Its wild forms are found in India. The crop was perhaps taken to tropical Africa and since then it might have been distributed to many countries like, Malaysia, Indonesia, Philippines, Mainland China, Papua New Guinea, Egypt, Sudan, East and West Africa, the Caribbean, Central and South America, etc. Hoshikawa (1981) reported that it was introduced into Japan from China in 1654 where it is called "Fujimame" and young pods are used as vegetables.

But Maass *et al.* (2005) did not find any evidence through AFLP studies to show that it had originated from elsewhere other than eastern and or southern Africa. Additionally, the wild forms collected from India and analysed through molecular marker studies were found genetically placed intermediate between wild and cultivated forms. And that these forms did not exist among the African collections led Maass *et al.* (2003) to hypothesise that the pattern of domestication and distribution of *Lablab purpureus* was from Africa to Asia. This was further affirmed by Maass *et al.* (2007) through studies relating to changes in seed characteristics between wild and cultivated forms. Still there is a school of thought, which believes in the dual centers of origin–Africa and Asia.

The nature of the pod texture constitutes the basic economic difference between the two varieties(Ayyangar and Nambiar, 1935). The field variety has tough firm-walled parchmented pods. In this variety the seeds alone are consumed. The garden variety has tender edible pods, some of them so soft as to wrinkle on drying. The marked difference in appearance between the field and garden varieties consists in the bushy habit of the former and the ready twining habit of the latter.

The extent of physiological changes introduced into the garden variety consequent upon intense domestication, so much so that years of habit have tended to stabilise differences between the field and garden varieties. The evolution tending to the production of garden varieties seems to have been along lines that softened the pods and eliminated the oil from it in addition to giving the vine a marked twining habit, drawn out flowering and a more continuous supply of edible pods (Ayyangar and Nambiar, 1935).

A second direction in which the garden varieties have been evolved from the field ones is in the production of drooping pods(Ayyangar and Nambiar, 1935). Most of the field varieties have erect pods, the pods being perpendicular to or making an upward acute angle with the peduncle. Some field varieties have drooping pods. All the garden varieties have drooping pods. This drooping condition is a consequence of pod softening, in contrast to the erectness of tough pods.

A third line of evolution leading towards common domestic varieties is a widening or narrowing of the medium pod of the field varieties (Ayyangar and Nambiar, 1935). There is a wealth of pod sizes and shapes characterising the garden varieties. In the field variety there is only one common pod shape. This is medium in width and a little broader towards the tip.

A fourth line of evolution(which arose through an absence of the parchment layer and a softening of the pod tissue) has been a disappearance in some narrow pods of the septate condition and its replacement by the bloated non-septate condition(Ayyangar and Nambiar, 1935). One collateral effect of this erasure of the septa in narrow pods is a realignment in the disposition of the seed. In the septate pod it is perpendicular to the pod suture. With the disappearance of the septa and the attendant bloating, the seed tends towards having a disposition parallel to the suture.

Further division of the cultivated ones was proposed as short-day varieties with a photoperiod of 10–11 hours and the others relatively unaffected by day length (Rivals, 1953).

In 1965, Verdcourt proposed the rejection of *Dolichos* L. and in its place conservation of *Dolichos* Lam. with *D. uniflorus* Lam. as a type, because *Dolichos* L. with lectotype *D. lablab* L. would restrict the generic name *Dolichos* to *D. lablab* and possibly a few allies.

However, the committee for Spermatophyta (1968) rejected the proposal for 3 specified reasons: (1) The species proposed as the type is not one of those originally included by Linnaeus; (2) the illegitimate name *Lablab* Adanson would need conserving; (3) botanists not agreeing with the segregation of *Lablab* and other generic splits could not use the name *Dolichos* L. for the combined genus.

Three years later, Verdcourt (1968) again proposed to retypify *Dolichos* L. this time by *D. trilobus* L. The committee for Spermatophyta (1970) found this proposal nomenclaturally preferable to the earlier one for the following seasons: (1) the proposed lectotype is one of the original Linnaean species. (2) the historical type of *Dolichos* Lam *D. lablab* is not separated from the genus as it was in the former proposal to conserve *Dolichos* Lam., so those choosing to include D. lablab as an inclusive *Dolichos* L. may still do so. (3) If *D. lablab* is held to be genetically distinct from *Dolichos* L. with the type *D. trilobus*, the name *Lablab adons* is available. Because, *Lablab adons*, was typified by *Dolichos lablab* L., it is not illegitimate unless the latter species is the type of *Dolichos.* (4) The proposal implies that the number of necessary new combinations will be reduced to about 30. However, the committee's report also noted: "there is already a well-established type of the name *Dolichos*, namely *D. Lablab.* Why not simply accept this and make all the necessary new combinations and go ahead on the basis of priority?" But the renewed proposal was rejected.

However, Verdcourt (1970 a) put forth a new argument, which clearly demonstrated that the delimitation of *Dolichos* L. is by no means settled, but is still very much a matter of opinion, and that, when a future complete and detailed revision becomes available, impredictable segregations may prove necessary. The committee for Spermatophyte (1972) finally accepted the proposal, "over the objections of a strongly dissenting minority".

Recent History

Verdcourt (1970) recognized three sub-species: (a) *unicinatus*, (b) *purpureus* and (c) *bengalensis.* He also mentioned the chromosome number of the species as 2n =22 or 24. Among the three subspecies, ssp. *uncinatus* was the ancestral form distributed mainly in East–Africa with small pods (40mm x 15mm); ssp. *purpureus.* with large pods (100mm x 400mm) was the cultivated one with commercial varieties and ssp. *bengalensis* had linear oblong shaped pods (140mm x 10–25mm) and found widely spread in Asia.

Although there were significant differences in respect of pod shape, it was supposed that ssp. *purpureus* and ssp. *bengalensis* were genetically very similar and most of the domesticated material in India belongs either to ssp. *purpureus* or ssp. *bengalensis.* ssp *uncinatus* has been domesticated only in Ethiopia (Magness *et al.*, 1971).

Verdcourt (1980) finally assigned it to monotypic genus Lablab and now widely known as *Lablab purpureus* L. (Sweet) although some, still, refer to it as *Dolichos lablab.* Sometimes, it is also referred to as *Lablab niger* Medik, seldom though. The name "*Lablab*" is an Arabic or Egyptian name describing the dull rattle of the seeds inside the dry-pod.

Maass *et al.*(2010) stated that Crop improvement in South Asia, Australia and the USA is based on limited genetic diversity. Although, lablab is widespread in many Asian countries, it is not among the mainstream crops and its role in diets continues to decline in many societies. For example, lablab is disappearing from southern Japan and, while the entire reasons for this may be quite complex, it is undoubtedly related to declining relationship between traditional crops and communities, and the failure of younger generations to continue the tradition of growing heirloom varieties. Conversely, lablab is a highly popular crop in South Asia, where it is especially cultivated under drought-prone conditions. Genetic studies and plant improvement were conducted at Tamil Nadu Agricultural University in Coimbatore, India since the 1930s resulting in a number of cultivars being developed. At the University of Agricultural Sciences (UAS) Bangalore, India, several improved cultivars have been developed by selecting for desirable traits in segregating populations. The most widely distributed Australian cultivars, however, resulted from research towards forage improvement. This program also endeavoured to combine traits of the successful, wide-spread forage cv. Rongai with an African wild, perennial germplasm accession, which led to the release of cv. Endurance. While the Australian improvement program has been discontinued, a lablab forage cultivar, cv. Rio Verde, has recently been released in Texas, USA. The current most active lablab improvement programs for food, however, probably exist in India and Bangladesh. More than 30 improved lablab varieties have been produced at various Indian institutions since breeding began. In Coimbatore, India, the photo-insensitive, ultra-short-duration vegetable lablab, CoLT 22/1, commences flowering at 40–42 days after sowing, and enables the harvesting of green vegetable pods as early as 48 days after sowing so enabling use throughout the year. Byre Gowda from UAS Bangalore, India, recently succeeded in breeding the high-yielding, photo-insensitive cultivar HA-4.

There are also major plant breeding programs at Bangladesh Agricultural University in Mymensingh and at Bangabandhu Sheikh Mujibur Rahman Agricultural University in Gazipur. Two photo-insensitive lines have been bred and released as year-round cultivars, cvs. IPSA Seam-1 and IPSA Seam-2. These can be grown during both the hot/dry and humid/cooler seasons with flowering in both cultivars commencing between 45 and 60 days after sowing. This compares positively with the traditional photo-sensitive landraces that flower 2–6 months after sowing and are only grown during the hot/dry winter season. Lablab plant improvement programs are also underway in China. The aforementioned plant breeding successes illustrate most Asian activities targeting vegetable or pulse improvement, where the most important goals appear to be the development of short-duration, determinate, bushy types that are insensitive to photoperiod, fairly homogeneous and pest-resistant. To support these and other plant improvement objectives, genetic maps containing molecular and morphological markers have been constructed from different mapping populations or improved by comparative mapping with mungbean (*Vigna radiata*). As a consequence of the historical interest in the crop and the focus on plant improvement, relatively large collections of landraces and other genetic stocks have been assembled in a number of institutions. Several of these collections have provided the basis for various molecular studies of diversity that have helped to understand the genetic patterns of diversity of this crop. Despite the impressive agro-morphological variation available, particularly in southern Asia, it appears that lablab in that region is genetically less diverse than material from Africa.

The recognition of limited genetic diversity available in Indian indigenous landraces and in Indian plant improvement programs has led Indian scientists to initiate the addition of 'exotic' germplasm sourced from the USDA-held collection to their breeding program at UAS Bangalore.

Maass *et al.*(2010) again confirmed that in contrast to South Asia, wild, undomesticated lablab has been found to occur naturally in several African countries, and both the wild types and domesticated African landraces have been shown to be genetically diverse. Lablab has been a traditional crop in eastern Africa although its use has dramatically decreased in recent years. Despite this decline, however, lablab appears to persist as a garden crop (rather than a field crop) in eastern and southern Africa. There is almost no ongoing lablab research in Africa, except for programs focusing on improving soil properties by using green-manure/forage crops, such as in maize- based systems of Kenya, Malawi and Nigeria, or sorghum- and millet-based systems in the semi-arid tropics of Mali. Almost all of these African initiatives have included and continue to include one popular late-maturing forage cultivar, cv. Rongai and, as a result, the potential role of the species as a pulse or vegetable in Africa is likely to be severely underestimated. Only recent work at ILRI in Ethiopia and CSIRO in Australia and, subsequently, in southern Africa explored a much larger range of accessions for feed and food and identified germplasm, which was well adapted to drier climates and crop use. Similarly, at IITA in Nigeria germplasm accessions almost largely acquired from USDA were evaluated for the dual purpose of feed and food. They identified accessions with good grain and forage yields of high potential for use in the cereal-legume livestock systems in the moist savannah zone of West Africa.

To improve understanding of diversity of *Lablab purpureus* and establish relationships among 103 germplasm accessions collected from diverse geographic origins, amplified fragment length polymorphism markers were used (Maass *et al.*, 2005). Four primer sets selected out of 16 produced 289 clear, repeatable polymorphisms. UPGMA analysis of similarity data clustered the accessions according to their subspecific taxonomic organization, i.e., subsp. *purpureus* and subsp. *uncinatus*, as well as to cultivated and wild forms. The well-represented landraces from Africa and Asia, belonging predominantly to subsp. *purpureus*, displayed moderate genetic diversity. Wild forms from Africa showed far greater levels of diversity that would justify taxonomic re-assessment of the wild subsp. *uncinatus*. The molecular analysis identified forms that were collected in the wild in India but were genetically placed intermediate between wild and cultivated forms. As these plant types did not exist among the African accessions, it is suggested that they might represent escapes from early attempts of domestication. These results support the suggested pathway of domestication and distribution of *L. purpureus* from Africa to Asia.

Molecular diversity in *Lablab purpureus* was assessed using amplified fragment length polymorphism markers on fifty Kenyan lablab accessions obtained from farmers' fields and the Kenya National gene bank (Kimani *et al.*,2012). One hundred and eighty polymorphic bands were revealed using fifteen selective primer pairs. The overall mean expected heterozygosity (He) for the five populations was 0.189. Estimates of components of molecular variance revealed that most of the genetic variation resided within populations (99%) and only 1% variance was among the populations, while Principal Coordinate Analysis showed an overlap between accessions from different geographic origins. The UPGMA cluster analysis generated from the distance matrix of the 50 assayed accessions, revealed low diversity among most of the accessions. The low diversity observed may be due to the narrow genetic base for breeding stocks, and extensive exchange of germplasm among smallholder farmers across the country.

Prospects

It was stated by Maass *et al.*(2010) that the probability of occurrence and potential use of lablab throughout eastern and southern Africa can be estimated by using passport data from a range of germplasm accessions and herbarium specimens as input into the FloraMap® program and by subsequently calculating and mapping the distribution probability based on climatic similarities. The resulting map is in contrast to that generated by a database on tropical forages and derived primarily from known adaptation of widespread forage cultivars, such as cvs. Rongai or Highworth. While the tropical forages mapping suggests that large parts of eastern Africa are only marginally suited to the species, the analysis using passport data indicated that many accessions were collected from precisely some of those regions defined as "marginally suited". Such divergent outcomes from GIS analyses can be expected when different ecotypes of a species are included, as is the case here. The integrity of the divergence between the adaptation maps based on different ecotypes is supported by on-the-ground adaptation studies in southern Africa, where cvs. Rongai and Highworth were deemed to be unsuitable due to long vegetative phases that made them prone to early frosts or drought and impeded seed production. Overall, the two maps demonstrate the wide potential range of adaptation for this species in Africa, depending both on specific genotypes (cultivars) and its purpose of use(s). Few germplasm or herbarium collections originated from West and Central Africa. However, it would be expected that the species be adapted over a large proportion of that region given its rainfall and latitude. Widespread adaptation in West and Central Africa is supported by the forage accession analyses' prediction of adaptation in the region.

Future Opportunities

While the current initiatives to develop this underutilized plant for food and feed in Africa continue to be founded on a very narrow genetic base, and often on plant types selected for forage rather than pulses or vegetables, there is every reason for concern that known and existing biodiversity, which may be more suited to Africa and its range of climates, is being overlooked.

Adebisi and Bosch (2004) summarized that lablab has considerable promise as a crop species because its grain yields can be higher than those of cowpea and its spectrum of adaptability to differing ecological conditions is wider than for any other leguminous plant. However, not enough is known about adaptation to drought across the species (Ewansiha and Singh, 2006). Even less information exists about the physiological mechanisms of that adaptation. The history, success and failure of lablab evaluation in various African environments, either as a crop or forage, have clearly been based on narrow genetic diversity and a few commercial cultivars, some of which were initially selected for their forage value only. Basing decisions on the potential value of the crop in Africa on results from this limited genetic base will undoubtedly lead to the risk that the species as a whole may be discarded (Maass *et al.* 2003). A wider range of diversity exists and is available from the world's genebanks, including African indigenous materials. And this wider range of germplasm needs to be evaluated in future work. This may not only provide new insights into the potential role of lablab, in semi-arid regions in particular, but such an approach may also aid in ensuring that indigenous germplasm is conserved.

Due to its drought tolerance, lablab might offer comparable opportunities for African agriculture in view of global change. The mutual benefit and potential collaborations arising from the exchange of materials and knowledge between African and Asian researchers will hopefully have been stimulated during the First International Lablab Meeting near Arusha, Tanzania, in March 2008. Improved, high-yielding cultivars from drought-prone areas in India could contribute to African food security in regions with similar climates. Genetically distant African landraces may offer to the South Asian breeding programs new sources for pest and disease resistance together with other important traits. Collecting special purpose germplasm from semi-arid regions, such as in Namibia, where the wild *L. purpureus* subsp. uncinatus var. rhomboïdeus has been recorded (Verdcourt, 1970) and which has never been included in any screening program, may add needed traits for future breeding programs.

L. purpureus can grow in a variety of soils, from sand to clay, in a pH range of 4.5–7.5 (Cook *et al.*, 2005). It does not grow well in saline or poorly-drained soils, but it grows better than most legumes under acidic conditions (Valenzuela and Smith, 2002). It can continue to grow in drought or shady conditions, and will grow in areas with an average annual rainfall is 25–120 in. (Cook *et al.*, 2005). It is more drought resistant than other similar legumes like common beans (*Phaseolus vulgaris*) and cowpea (Maass *et al.*, 2010), and can access soil water 6 feet deep (Cook *et al.*, 2005). It is better adapted to cold than other warm-season forages such as velvet bean (*Mucuna pruriens*) or cowpea. It yields 5–10 t/ha of green matter, which can be used as fodder or green manure. It improves the soil condition and is also a good cover crop. As forage, it is very palatable, either as green fodder or as silage. It has been used for centuries in India and China as an edible pod and animal forage but it is primarily an ornamental annual vine in the US. The distinctive long–lasting pods are suitable for cut–stems for the cut–flower industry. In addition, the pods are so unique that they could be used for decoration or harvested for Chinese food wholesalers. Some local and regional florists enjoyed the special colour and texture that hyacinth beans offered their arrangements while others had little interest in use of these stems. With yields as 70 cut–stems per linear meter, the potential returns could be quite high. But the availability of the established market determines the overall economy of the crop (Anderson, *et al.*, 1996).

Lablab purpureus is grown as a pulse crop (crop harvested for dry seed) in Africa, Asia, and the Caribbean. It is also consumed as a green vegetable (green bean, pod, leaf). Protein isolate from the bean can be used as a food additive for improving cake quality (Maass *et al.*, 2010).

L. purpureus is used as forage, hay, and silage. As forage, it is often sown with sorghum or millet. The leaf is very palatable but the stem is not. The seeds are moderately palatable. Overall, it is one of the most palatable legumes for animals (Valenzuela and Smith, 2002). The leaf has crude protein of 21 to 38% and the seed contains 20 to 28% crude protein (Cook *et al.*, 2005). The seeds contain large amounts of various vitamins and minerals, but contain tannins and trypsin inhibitors so must be soaked or cooked before human consumption.

The leaves make excellent hay for cattle and goats, but the stem is difficult to dry, and must be mechanically conditioned through crushing (FAO, 2012). Silage made from a mix of *L. purpureus* and *Sorghum* sp. raised the protein content of sorghum by roughly 11% with a 2:1 mixture of lablab: sorghum (FAO, 2012).

L. purpureus is used as a nitrogen-fixing green manure to improve soil quality. It often produces more dry matter than cowpea (*Vigna unguiculata*), especially during drought, and can produce roughly 1,750 lb of leaf matter (Cook *et al.*, 2005) or 2.5 tons of total biomass per acre (Valenzuela and Smith, 2002). Each ton of biomass produced 50 lb of nitrogen (Valenzuela and Smith, 2002). It not only produces nitrogen through fixation, but returns nitrogen through leaf decay (FAO, 2012). Initially growth is slow, but once established, it competes well with weeds. It has an extensive root system that improves the physical condition and function of the soil.

Currently the most common use of *L. purpureus* in the United States is as an ornamental crop in the cut flower industry (Stevens, 2012). It is valued for its late summer flowers and colorful, purple peapods. Depending on the weather in late summer, harvest yields can be up to 55 cut stems per plant (Anderson *et al.*, 1996).

L. purpureus has been used in the Philippines and China as a stimulant, to reduce fever, to reduce flatulence, to stimulate digestion, and as an antispasmodic (Stuart, 2011). In Namibia, the root has been used to treat heart conditions (Pennacchio *et al.*, 2010). The hyacinth bean, is used throughout Asia and Africa for human food and medicine with potential to be a source of phytochemicals, nutraceuticals and pharmaceuticals. Hyacinth bean seeds and pods could be used as famine food worldwide where humans suffer from malnourishment and disease (Bradley,2009).

References

Adebisi, A. A. and Bosch C.H. 2004. *Lablab purpureus* (L.) Sweet. (In:) Grubben GJH, Denton OA (eds) Plant Resources of Tropical Africa (PROTA), no. 2, vegetables. PROTA Foundation, Wageningen, The Netherlands / Backhuys, Leiden, The Netherlands/ CTA, Wageningen, The Netherlands, pp. 343–348.

Anderson, R.G., Bale, S. and Jia, W. 1996. Hyacinth bean: stems for the cut flower market. In: J. Janick and J.E. Simon, editors, Progress in new crops. ASHS Press, Arlington, VA. pp. 540–542.

Ayyangar, G.N.R. and Nambiar, K.K.K. 1935. Studies in *Dolichos lablab* (Roxb.) and L.- The Indian field and garden bean. I. *Proc. Ind. Acad. Sci.,B,*vol.1, pp. 857-867.

Bradley, M.J. 2009. Morphological and reproductive characterization in hyacinth bean, *Lablab purpureus* (L.) sweet germplasm with clinically proven nutraceutical and pharmaceutical traits for use as a medicinal food. *Journal of dietary supplements* 6 (3): 263-279.

Cook, B.G., Pengelly, B.C., Brown, S.D., Donnelly, J.L., Eagles, D.A., Franco, M.A., Hanson, J., Mullen, B.F., Partridge, I.J., Peters, M. and Schultze-Kraft, R. 2005. Tropical forages: an interactive selection tool. *Lablab purpureus*. CSIRO, DPI&F(Qld), CIAT, and ILRI, Brisbane, Australia.

Ewansiha, S.S. and Singh, B.B. 2006. Relative drought tolerance of important herbaceous legumes and cereals in the moist and semi-arid regions of West Africa. *J. Food Agr Environ.* 4(2):188–190.

FAO. 2012. Grassland species index. *Lablab purpureus.* http://www.fao.org/ag/AGP/AGPC/doc/Gbase/DATA/Pf000047.HTM.

Gowda, M. B. 2012. www.lablablab.org/html/origin-distribution.html

Hoshikawa,K. 1981. Fuji mame (Hyacinth bean). in "Shokuyou Sakumotu" (Food Crops). Yoken-do, Tokyo. (in Japanese) pp.540-542.

Kimani, E. N., Wachira, F. N. and Kinyua, M. G. 2012. Molecular Diversity of Kenyan Lablab Bean (*Lablab purpureus* (L.) Sweet) Accessions Using Amplified Fragment Length Polymorphism Markers, *American Journal of Plant Sciences*, 3 (3):313-321.

Maass, B. L. , Jamnadass, R. H., Hanson, J. and Pengelly, B.C. 2005. Determining Sources of Diversity in Cultivated and Wild *Lablab purpureus* Related to Provenance of Germplasm by using Amplified Fragment Length Polymorphism. *Genetic Resources and Crop Evolution* 52(6) :683-695.

Maass, B.L., Knox, M.R., Venkatesha, S.C., Angessa, T.T., Ramme, S. and Pengelly, B.C. 2010. *Lablab purpureus*-a crop lost for Africa? *Trop. Plant Biol.* 3(3):123–135.

Magness, J.R., Markle, G. M. and Compton C. C. 1971. Food and feed crops of the United States. Interregional Research Project IR-4, IR Bul. 1, (Bul. 828 New Jersey Agric. Expt. Sta.)

Mullen, C.L., Holland, J.F. and Heuke, L. 2003. Cowpea, lablab, and pigeon pea. Agfact P4.2.21. NSW Agriculture, Orange, New South Wales. http://www.dpi.nsw.gov.au/__data/assets/pdf_file/0006/157488/cowpea-lablab-pigeon-pea.pdf

Murphy, A.M. and Colucci, P.E. 1999. A tropical forage solution to poor quality ruminant diets: a review of *Lablab purpureus. Liv. Res. Rur. Dev.* 11(2) http://ftp.sunet.se/wmirror/www.cipav.org.co/lrrd/lrrd11/2/cont112.htm

Murtagh, G.J. 1972. Seedbed requirements for *Dolichos lablab. Australian J. Exp. Agr. Anim. Husb.* 12 (56):288–292.

Pennacchio, M., Jefferson, L.V. and Havens, K. 2010. Uses and abuses of plant-derived smoke: its ethnobotany as hallucinogen, perfume, incense, and medicine. Oxford Univ. Press Inc., New York.

Purseglove, J. W. 1968. Tropical Crops: Dicotyledons. Wiley, New York, 2 vols., xx + 719 pp.

Stevens, J.M. 2012. Bean, hyacinth— *Dolichos lablab* L., or *Lablab purpureus* (L.) Sweet. Publication #HS552. Institute of Food and Agricultural Sciences (IFAS), Univ. of Florida Extension. http://edis.ifas.ufl.edu/mv019.

Stuart, G. 2011. Stuartxchange- Philippine alternative medicine. http://www.stuartxchange.org/AltMed.html.

Valenzuela, H. and Smith, J. 2002. Sustainable agriculture green manure crops. SA-GM-7. Cooperative Extension Service, College of Tropical Agric. and Human Resources, Univ. of Hawaii at Manoa. http://www.ctahr.hawaii.edu/oc/freepubs/pdf/GreenM.

Verdcourt, B. 1970. Studies in the Leguminosae-Papilionoideae for the 'Flora of Tropical East Africa': II*. *Kew Bulletin* Formerly the Bulletin of Miscellaneous Information 3 published for The Royal Botanic Gardens, Kew Volume 24, No. 2.

Verdcourt, B. 1980. The classification of *Dolichos* L. emend. Verdc., *Lablab* Adans., *Phaseolus* L., *Vigna* Savi and their allies. In "Advances in Legume Science" eds.R.J.Summerfield and A.H.Bunting. Kew Royal Botanic Gardens. pp.45-48.

11

Kokum

P M Hadankar, Y R Parulekar and S M Sawratkar

Kokum [*Garcinia indica* (Dupetit-Thouars) Choisy] is a perennial, tropical, endemic and potential, under exploited, wild and semi domesticated spice crop currently gaining much commercial and medicinal importance. It is also known as *kokum* in Hindi, *kokan* in Gujarati, *brindan* in Goa, *bhirond, kokum, katambi, ratamba* in Marathi and Konkani, *murugal* in Kannada and *parampuli* in Malyalam (Sullivan and Triscari, 1977). It is slender, evergreen tree reaching to height of 10-15 m with drooping or spreading branches; young leaves are red; while mature leaves are dark green above and pale beneath, ovate or oblong or elliptic, lanceolate. It is dioecious in nature but bisexual types are also reported (Mathew *et al.*, 2001).

The fruit is of commercial importance which poses agreeable flavour and sweetish acid taste. It is anthelmintic, cardiotonic and used in treating piles, dysentery, tumors, pains and heart complaints. Fruit juice is given in bilious affections. It is used both as digestive tonic and to cope with paralysis. Kokum juice is also effective against allergies due to bee bites and other insect bites and sun exposure related symptoms as well as acidity. The fruit contain rich amount of antioxidants that bind with free radicals prevent oxidative damage to body cells. It also promotes cell regeneration and repair (Haldankar *et al.*, 2012).

Kokum juice is especially popular during scorching summer months as it has a cooling effect on the body and shields the body against dehydration and sunstroke. It also helps in bringing down the fever and allergic reactions. It is known to strengthen the cardiovascular system and stabilizes liver function. The hydroxy citric acid present in the fruit fights cholesterol and curbs lipogenesis which helps for weight loss. The oil extracted from seed remains solid at room temperature which is popularly known as kokum butter. It is considered nutritive, demulcent, astringent and emollient. It is suitable for ointments, suppositories and other pharmaceutical purposes. It is used for local application to ulcerations and fissures of lips, hands, etc. (Haldankar *et al.* 2012). Use of *Garcinia* fruit to interrupt the synthesis of various lipids, fatty acids, LDL cholesterol and triglycerides is reported. The principal acid in the fruit rinds of *Garcinia* is identified as (-) hydroxycitric acid, HCA (1,2- dihydroxypropane-1,2,3- tricarboxylic acid) (Lewis & Neelakandan, 1965). Lowenstein (1971) found that (-) HCA strongly inhibited fatty acid synthesis in living systems. HCA the biological compound is known to inhibit the synthesis of lipid and fatty acids. *Garcinia* also contains a high amount of Vitamin C and is been used as heart tonic.

The seeds of the fruit yield (23-26% on the weight of seeds; c. 44% on the weight of kernels) a valuable edible fat known in commerce as kokum butter, (The wealth of India, Volume IV F- G pp 102). In a seedling population, 30-50 kg yield per plant is obtained.

Cytotaxonomical Background

Kokum belongs to genus *Garcinia*. It is hypothesized that the genus *Garcinia* originated before the continental drift, followed by separate diversification of Canters in the Afro-Madagaskar and Indo-Malayan area (Seetharam, 2006). *Garcinia* are native to old world tropics, especially Asia, Australia, tropical and southern Africa and polynesia, consists of over 250 species worldwide (Stevens, 2001). The maximum concentration of *Garcinia* species is in the Asian Continent. In India, Western Ghats region is hot spot of biological diversity with a very high endermism. About 7 species of *Garcinia* are endemic to this region of which five species are being commercially exploited. Kokum is the most important among them. The chromosome number of kokum is reported as 2n =54 by Krishnaswamy and Raman in 1949 and as 2n=48 by Thomare in 1964.

Kokum is dioecious in nature and about 11 different types of sex expression have been reported. These types could be broadly classified into staminate (male), pistillate (female) and hermaphrodite (bisexual) types. This feature of sex expression encourages cross pollination and sunbsequent generation of heterogeneous population of kokum. Besides this, the sexual mode of propagation (seedling origin) has resulted in heterozygosity in the genetic make up of tree. This renders each and every individual tree to be genetically different from each other (Rawat and Bhatnagar, 2005).

The sexual system in the genus *Garcinia* is highly diverse and includes dioecious, gynodioecious, androdioecious, polygamodioecious, monoecious and andromonoecious species. *Garcinia indica* Choisy (Kokum tree) is reported to be polygamodioecious or gynodioecious. Four basic kinds of flowers namely, male flowers, male flowers with pistillode, female flower with staminodes and bisexual flowers are reported in kokum. These basic kinds of flowers occurred individually on a particular tree, producing unisexual individuals, or in combination of two, forming cosexual individuals. Thus, this species is trioecious, although the percentage of cosexual trees varied in different geographical locations. Male flowers had larger petal and female flowers were in general smaller than the other kinds. However, bisexual flowers were not always intermediate in size between male and female flowers (Joseph and Murthy, 2015).

History

Though kokum is indigenous, its documented evolution history is very limited. Its improvement and domestication got attention only recently. Besides its several medicinal and domestic applications, the crop remained neglected till end of twentieth century.

The improvement of this crop till today is exclusively depended upon selection of chance variant from sexually originated seedlings. In India, the attempts were made to conserve the elite types of kokum at various locations. While identifying the promising variants, emphasis was given mainly on the yield and yield contributing characters such as earliness in bearing, size, weight and colour of fruit, fruit fly damage, rind thickness, rind weight, yield etc. In Kerala, seventy-six accessions of kokum collected from Karnataka, Maharashtra, Goa and Kerala are maintained at NBPGR Station, Trissur (Abraham *et al.*, 2010). In Karnataka, total of 40 promising accessions were collected with their passport data including that of global positioning system information, tree morphology, leaf, flower, fruit and kernel characters (Patil *et al.*,2010). In Maharashtra, 320 Kokum accessions are maintained which are mostly from Konkan region (Raorane, 2003).

Up till now, only two varieties of Kokum are developed in India by Dr. Balasaheb Sawant Konkan Krishi Vidyapeeth, Dapoli. First one, Konkan Amruta which is early, high yielding (135 kg/ tree) with an average fruit weight of 34 g and Konkan Hatis having bold size fruits (fruit weight 91 g) and high yielding type (250 kg/ plant) (Chavan *et al.*, 2010).

The progress in kokum improvement is relatively slow as compared to other commercial horticultural crops. Till now, hybridization is not attempted in this crop. Furthermore, the precise information on inheritance pattern is yet to be known.

Prospects

Kokum is at focus of attention recently. The demand from farmers for supply of quality planting material of superior type is increasing day by day. At present, the raw material supply to processing industry differ greatly with respect to quality, which is further reflected in variation in quality of finished product. Development of superior variety will certainly help to overcome this constraint faced by the processors.

At present, enormous variability exists among the kokum genotypes in Western Ghats. Though, certain attempts were made to study and conserve the elite genotypes, still there is vast scope for survey of unexplored kokum variants. At the moment, selection and hybridisation are the key approaches for crop improvement in kokum.

Apart from yield, plant canopy and quality parameters with respect to rind and seed in future the focused goal on Hydroxy Citric Acid content and colour would help to develop superior Kokum varieties in future. Kokum seed kernel contains oil which is known as Kokum butter. Oil content in kernel ranges from 29.51 to 46.98 percent (Abraham *et al.*, 2010). The attention needs to be given on developing the varieties having high kernel oil content (>40%).

References

Abraham Z., S. K. Malik, M. Latha, Rekha chaudhury, N. Mangayarkarassi and Sharma S.K. 2010. Genetic Resources management of *Garcinia* species in India. *National symposium on Garcinia genetic resources: Linking Diversity, Livelihood and Management.* 8-9 May, 2010.

Abraham Z., Malik, S. K., Rao, G. E., Narayanan, S. L. and Biju, S. 2006. Collection, maintenance and characterization of Malabar tamrind (*Garcinia cambogia*) (Gaertn. Desr.). *Genetic Resources and Crop Evolution.* 53:401-406.

Abraham, Z., Sangita Yadav, M. Latha, S. Mani and Mangayarkkarasi N. 2010. Genetic Resources of *Garcinia indica* (Dupetit- Thouars) Choisy. National bureau of Plant Genetic Resources Regional Station. *National symposium on Garcinia genetic resources: Linking Diversity, Livelihood and Management.* 8-9 May, 2010.

Chavan S. A., B. R. Salvi and Dalvi N. V. 2010. Collection, maintenance of *Garcinia* Germplasm at Regional Fruit Research Station, vengurla. *National Bureau of Plant Genetic Resources Regional Station. National Symposium on Garcinia Genetic Resources: Linking Diversity, Livelihood and Management. 8-9 May, 2010.*

Haldankar, P. M; Y. R. Parulekar and Haldavnekar P. C. 2012. A Monograph on Kokum (*Garcinia indica* choicy)

Joseph, K. S. and Murthy H. N. 2015. Sexual system of *Garcinia indica* Choisy: geographical variation in trioecy and sexual dimorphism in floral traits. *Plant Systematics and Evolution.* Vol 301 (3) pp. 1065-1071.

Krishnaswamy, N and Raman V. S. 1949. A Note on the chromosome number of some economic plants in India. *Curr. Sci.*, Vol 18:376-378.

Lowenstein J. M. 1971. (-) Hydroxy citrate on fatty acid synthesis by rat liver *in viva J. Bio. Chem.* 246: 629-632.

Lewis, Y. S. and Neelakandan, S 1965. (-) Hydroxy citric acid- The principal acid in the fruits of *Garcinia cambogia. Phytochemistry* 4: 619.

Mathew, P. A., Krishnamorthy B. and Rema J. 2001. Collection, maintenance and Conservation of Garcinia Germplasm at Indian Institute of Spices Research, Calicut. *Proceeding of first National Seminar on Kokum.*, 5-8.

Patil S. K, G. O. Manjunatha, Harsha Hegde and Channabasappa K.S. 2010. Diversity in Germplasm of *Garcinia indica* (Choisy) in Karnataka. *National Symposium on Garcinia Genetic Resources: Linking Diversity, Livelihood and Management.* 8-9 May, 2010.

Rawat, R., and Bhatnagar A. K. 2005. Flowering and pollination in *Garcinia indica. Acta Biol. Cracoviensia Series Botanica.*, 47. Suppl. 1. p.45.

Raorane G. P 2003. Studies on growth, flowering, fruiting and some aspects of post harvest handling of Kokum (*Garcinia indica*). A Ph. D thesis submitted to Dr. Balasaheb Sawant Konkan Krishi Vidyapeeth, Dapoli.

Seetharam, Y. N. 2006. Diversity of androecium and pollen grains in the genus *Garcinia* L. and its bearing on geographical distribution and evolution. *Proceedings of the Indian Academy of Science, Plant Sciences.*

Stevens, P. F. 2001. Onwards. Angiosperm phylogeny website. Version 6.

Sullivan A. C and Triscari J 1977. Metabolic regulation as acontrol for lipid disorders. Influence of (-) hydroxyl citrate on experimentally induced obesity in the rodent. *Am. J. Clin. Nutr.* 30:767-776.

Thomare M. V. 1964. Studies in *Garcinia indica* choisy *sci. cult.*, 30:453-454.

The Wealth of India- Vol. IV (F - G) pp. 99-108.

12

Luffa Gourds

Amish Kumar Sureja and Anilabh D Munshi

The immature fruits of *L. acutangula* (ridge gourd) and *L. cylindica* (sponge gourd) are edible and are consumed as fresh slices, in soups similar to okra, or cooked in the fashion of squash (Behera *et al.*, 2012). In India, they are eaten boiled or in curry (mixed with potato or solo), where fruits mixed with *khus* (poppy seed) seeds is often highly sought after. In Japan, the immature fruits are sliced, dried, and then kept at room temperature for future use. While the young insipid leaves of sponge gourd are consumed in Malaysia (Porterfield, 1955), the leaves in Africa are also used as a leafy vegetable and the seeds are employed in soup and sauce preparations (Adebooye, 2009).

The genus *Luffa* (2n = 2x = 26) is economically important to tropical agriculture, where some of its members have high nutritive value and possess important medicinal attributes (Desai and Musmade, 1998). Well known members of the genus include *Luffa cylindrica* (L.) M.J. Roemer (synom. *L. aegyptiaca* Miller; 2n = 2x = 26), commonly called sponge, smooth, loofa, vegetable sponge, bath sponge or dish cloth gourd, and *Luffa acutangula* (L.) Roxburgh (2n = 2x = 26), commonly called ridge, ribbed gourd or angled loofah (Swarup, 2006). These species are popular vegetables in tropical and subtropical regions of the World, and are widely cultivated in India, Southeast Asia, China, Japan, and parts of Africa (e.g., Egypt) (Swarup, 2006). The genus derives its name from the product 'loofah', which is used as bath sponges, scrubbing pads, door mats, pillows, mattresses and cleaning utensils (Swarup, 2006). Both the species contain a gelatinous amorphous bitter alkaloid called luffein (Swarup, 2006).

Luffa agricultural species are typically cultivated on a small scale in isolated rural areas (Behera *et al.*, 2012), so precise information on cultivated area, production, and yield are limited. Nevertheless, it is an important component of crop rotation during the "*pre-kharif*" and "*kharif*" (i.e., monsoon) seasons in north Indian where it is cultivated commercially, in home gardens, and as mixed cropping along river beds (Choudhury, 1996). River bed cultivation of *Luffa* during winter and early spring or summer seasons is a specialized form of off-season cultivation in north India (Swarup, 2006). In tropical south India, *Luffa* species are cultivated throughout the year (Seshadri and More, 2009). Cultural practices for growing *L. aegyptiaca* and *L. cylindrica* have been standardized for production in more temperate climates (Davis, 1994). This review provides information relating to the biology, genetics, breeding, taxonomy, and horticultural aspects of this crop species.

Origin, Distribution, and Historic Uses

Luffa is an Old World genus, which includes two commercially important, domesticated species *L. aegyptiaca/cylindrica* and *L. acutangula*, whose young fruits possess culinary attributes that are particularly sought after in Africa and southeast Asia (Jeffrey, 1980a, Heiser and Schilling, 1988). The earliest records of *Luffa* are mentioned in Kautilya's Arthshastra, ca. 350-300 B.C. in India (Swarup, 2006, Seshadri and More, 2009). In fact, the Sanskrit name 'Koshataki' indicates that the early cultivation of *Luffa* took place in India (Seshadri and More, 2009). Since sponge gourd characteristics were described in Egyptian writings, the name 'Luffa' or 'Loofah' is of Arabic origin (Seshadri and More, 2009). Likewise, early Chinese literature describes 'Szkua'-dish cloth gourd or towel gourd (Seshadri and More, 2009). Although Bretschneider (1878) introduced sponge gourd into Europe and North America, he never collected wild specimens. Regardless of the place of origin and domestication, it is clear that loofah has existed in India, China, Japan, Malaysia and the Middle East since ancient times (Swarup, 2006, Seshadri and More, 2009, Behera *et al.*, 2012).

The genus *Luffa* consists of eight species. Four are native to the Old world [*Luffa cylindrica* (L.) M.J. Roem., *L. acutangula* (L.) Roxb., *L. echinata* Roxb., and *L. graveolens* Roxb.]. Three are native to the new world [*L. operculata* (L.) Cogn., *L. quinquefida* (Hook. & Arn.) Seem. and *L. astorii* Svens]. One is native exclusively to Australia [*Luffa saccata* F. Muell.ex I.Telford sp. nov.] (Chakravarty, 1959, 1982, Dutt and Roy, 1990, Heiser and Schilling, 1990, Filipowicz *et al.*, 2014). India is rich in genetic resources of *L. acutangula, L. cylindrica, L. echinata*, and *L. graveolens* (Ram *et al.* 2007, Seshadri and More, 2009, Behera *et al.*, 2012, Pandey *et al.*, 2014). The species *Luffa echinata* var. *longistylis* (Edur.) Clarke is widely distributed in the northwest Himalayan Mountains and the upper Gangetic plains (Swarup, 2006). *Luffa graveolens* Roxb. occurs in the northeast plains of India, extending south to Tamil Nadu, and sporadically appears in Eastern Himalaya (Swarup, 2006). *Luffa acutangula* var. *amara* (Roxb.) Clarke is mainly distributed in peninsular India (Swarup, 2006). Another species, *L. umbellata* (Klein) M.J. Roem. is endemic to the southwest Ghats of the Kerala state in India (Chakravarty, 1959, 1982) and is a rare and threatened species (http://www.keralabiodiversity.org/index.php?). *Luffa operculata* occurs from Mexico (Gulf of California) to Nicaragua, *L. quinquefida* is present from Panama to southern Brazil, and *L. astorii* is indigenous to coastal Venezuela to Ecuador, and Peru (Heiser and Schilling, 1990, Filipowicz *et al.*, 2014).

Luffa echinata is considered the most primitive species of this genus, and the wild *L. graveolens* is the most probable progenitor species of cultivated *L. acutangula* and *L. cylindrical* market types (Swarup, 2006). Moreover, the wild dioecious species, *L. echinata* and monoecious *L. operculata* are likely to have evolved from *L. graveolens* (Swarup, 2006). DeCandolle (1959) suggested that *L. acutangula* was domesticated in Asia. More recent information, however, refines the site of domestication to India (Zeven and Zhukovsky, 1975; Heiser and Schilling, 1990), where the bitter wild or feral forms of *L.acutangula* var. *amara* (Roxb.) C.B. Clarke are endemic and are potential progenitors of commercial *Luffa*.

Sponge gourd often occurs as small wild or feral populations ranging from southeast Asia to the Philippines, and south to northeastern Australia and the South Pacific islands (Paris and Maynard, 2008). Ridge gourd is extensively grown in India, and to a lesser extent in southeast Asia, Indonesia, the Philippines and islands in the Caribbean Sea. These gourds originated in subtropical regions of Asia, and most probably have a primary center of origin and domestication in India (Seshadri and More, 2009). Their centres of diversity are considered to be south and southeast Asia (Seshadri and More, 2009).

Luffa species were used in commerce in Japan between 1890 and 1895 (Porterfield, 1955). Gourds were produced initially because their porous mesocarp could be used as filters in marine steam and diesel engines, which historically was of particular interest to the United States Navy (Porterfield, 1955, Heiser, 1979). In fact, at one time the entire supply of such filters came exclusively from Japan, which stopped at the beginning of World War II. This led to an order by the U.S. War Production Board on April 8, 1942 forbidding delivery, sale, or use of loofah sponges, in order to conserve the country's stockpile. Thus, the value of sponge gourd was officially recognized, and a wartime government program encouraged fresh market production of gourds. Although temperate region gourd production was not successful, tropical production in Mexico, Cuba, Haiti, Dominican Republic, El Salvador, and Guatemala proved commercially viable during the early to middle part of the 20th Century (Porterfield, 1955). Similarly, the United Kingdom started production programs through its Colonial Products Advisory Bureau in Jamaica, Cyprus, Gambia, Nigeria, Nyasaland, Southern Rhodesia, Sierra Leone, Tanganyika, British Guiana, Uganda, St. Kitts, British Honduras, the Gold Coast, and Antigua. However, production in those regions did not match theyield and quality of Japanese loofah gourd production. With the reduction in ship building after World War II and more competitive prices in Japan, the production demand for sponge gourd in the United States and United Kingdom markets fell dramatically.

Taxonomy

The genus *Luffa* is a member of the subfamily Cucurbitoideae, tribe Benincaseae, sub-tribe Luffinae C. Jeffr. in the family Cucurbitaceae (Jeffrey, 1962; Jeffrey, 1980a). It is the only taxon in this sub-tribe that has both Old and New World species. In some respects, species of this sub-tribe morphologically resemble members of the Cyclantherinae, an entirely New World sub-tribe of the Siceyeae. *Luffa* thus may evolutionarily represent a genetic bridge between the Benincaseae and Sicyeae tribes (Heiser and Schilling, 1990).

In a comprehensive taxonomic treatment of *Luffa,* Cogniaux and Harms (1924) recognized eight species. One of these species *L. variegata* Cogn., has since been reclassified as *Lemurosieyos* (Keraudren, 1965), and two others, *L. forskalii* Schwein. Ex Harms and

L. umbellata (Klein) M.J. Roem. are either synonymous with, or a botanical variety of *L. acutangula* (L.) Roxb. (Heiser and Schilling, 1988; Jeffrey, 1980a). Some reports have suggested that four *Luffa* species are of Old World and one is of New World origin (Heiser and Schilling, 1988 and Jeffrey, 1980b). Heiser *et al.* (1988) proposed that, in fact, three *Luffa* species (*L. operculata* (L.) Cogn., *L. quinquefida* (Hook. & Arn.) Seem and *L. astorii* Svens) originated in the New World. Sex expression is not considered a critical taxonomic discriminator in *Luffa* since all species are monoecious, except *L. echinata*, which is dioecious (Dutt and Saran, 1998). Likewise, although *L. acutangula* is commonly considered monoecious, it does have a hermaphroditic form found in the Indian and Nepal landrace cultivar Satputia, which bears small fruits in clusters (Swarup, 2006). All *Luffa* species possess a vinning habit and bear solitary pistillate flowers and racemes of male flowers.

In 1963, Singh and Bhandari (1963) formally described another species, *L. hermaphrodita* Singh and Bhandari, which bears hermaphrodite flowers and fruits in clusters, and is cultivated in the Indian state of Bihar, West Bengal and parts of north-central India. However, *L. hermaphrodita* is taxonomically distinct from ridge gourd 'Satputia' [Jeffrey (1980b) and Heiser and Schilling (1990)], and is commonly regarded as a cultigen of *L. acutangula* (Seshadri and More, 2009, Prakash *et al.*, 2013).

Chakravarty (1959, 1982) reported that *L. tuberosa* Roxb.,is a perennial vinning plant that occurs wild in the Indian states of Andhra Pradesh, Karnataka, Madhya Pradesh, Maharashtra and Tamil Nadu, as well as in tropical Africa in mostly dry habitats (Pandey and Prakash, 2013). On the basis of characters of endosperm haustorium and morphology of fruit that resembles *L. acutangula* (L.) Roxb. var. *amara* (Roxb.) C.B. Clarke, and the absence of cystoliths in leaf and absence of foliaceous bracts, *L. tuberosa* was placed in the genus *Luffa* (Singh, 1964; Chakravarty, 1982). However, there still exists some ambiguity with regard to the taxonomy of *L. tuberosa*. For instance, seed coat anatomy (Singh and Dathan, 1990), seed fat characteristics (conjugated triene acid) (Azeemoddin and Rao, 1966), morpho-anatomy (Kumar *et al.*, 2011), phytochemistry (Rao *et al.*, 1999; 2001; Shantha *et al.*, 2009) and species ecogeography (Joseph John, 2005) supports its placement in the genus *Momordica*. Nevertheless, molecular phylogenetic assessment of this species (i.e., nrDNA ITS sequences data analysis) supports the reclassification of *L. tuberosa* to *Mormodica tuberosa* (Roxb.) Cogn. (Ali *et al.*, 2010). Based on comparative morphological assessments of *L. tuberosa* with other taxa of *Luffa* and *Momordica*, Pandey and Prakash (2013) also supported its reclassification into the genus *Momordica* as *M. tuberosa* (Roxb.) Cogn. (syn. *M. cymbalaria* Hk. f).

Chakravarty (1959) and Chandra (1995) have provided detailed treatments of New and Old World *Luffa* species that allows for its current taxonomic classification. They recognized seven species, which are described below. In addition to this taxonomic classification, one New World species *L. saccata* F. Muell. ex I.Telford recognized by Telford *et al.* (2011) and one rare and threatened Old World species, *L. umbellata* M Roem. are also described to recognize their taxonomic significance to the genus (Chakravarty, 1959, 1982).

Old World Species

Luffa echinata Roxb.: This species is found in both Asia and Africa and is morphologically distinct from all the other Luffa species given its unique dioecious character (Chakravarty,

1959, Heiser and Schilling, 1990). Nevertheless, the complete extent of morphological and sexual differentiation between Asian and African types is unknown. Cogniaux and Harms (1924) and Keraudren (1965) stated that petals of *Luffa* are white, but some African floras indicate the petals can be yellow (Hooker, 1871; Hutchinson and Dalziel, 1954) or various hues of yellow to white (Andrews, 1950). Prakash *et al.* (2013) stated that in *L. echinata*, leaves are sparsely hairy, shallowly five angled, reniform-suborbicular and sparsely puberulent on both surfaces. The fibrous fruit wall is loosely knitted, and fruit are pale brown, round-ovoid, with the wall densely covered with echinate long bristles, seed coat verrucose (coarsely pitted), yellowish-brown to black and seeds extremely bitter with pungent smell (Pandey *et al.*, 2014).

Luffa acutangula (L.) Roxb.: Ridge gourd or ribbed gourd is one of the two most domesticated species of *Luffa*, and its fruit possess conspicuous ridges (Chakravarty, 1959, Heiser and Schilling, 1990). The stems are acutely five-angled, tendrils are hairy and commonly trifid. Leaves are medium in size, light to dark green in color, with 5-7 angled or shallow lobes. Male flowers are borne on a raceme with 5-20 flowers/raceme, and female flowers are borne solitary in the leaf axils. The staminate flowers appear first followed by pistillate flowers. The flowers are cream to light yellow in color, and mildly fragrant. Flowers open in the evening, where stamens appear as group of three. Fruits are club-shaped and angled, having 10 ribs, 15-50 cm x 5-10 cm in length and width, respectively, in which immature fruits are crowned by enlarged calyx and style. Seeds are ellipsoid, 0.4-1.3 cm x 0.7-0.9 cm in length and width, respectively, and are pitted and black, without wing-like margins (Chandra, 1995).

Four botanical varieties of *L. acutangula* are recognized: 1) var. *acutangula* (L.) Roxb., which is the large-fruited cultivated form, whose origin is likely India; 2) var. *amara* (Roxb.) C.B. Clarke, which is a wild or feral form with extremely bitter fruits and is confined to India; 3) var. *forskalii* (Harms) Heiser and Schilling, which is also a wild form but found exclusively in Yemen and likely originating as an escape from cultivation based on fruit flavonoid analyses (Schilling and Heiser, 1981), and; 4) var. *hermaphrodita* Singh & Bhandari, which is common to northeastern Uttar Pradesh and the Bihar state of India and is espoused as the most taxonomically primitive varietal form as it bears small- to medium-sized hermaphrodite fruits on racemes (Chandra, 1995).

Luffa aegyptiaca Mill. *or L. cylindrica* (L.) M.J. Roem.: This is the most extensively commercialized species of loofah and is commonly known as sponge gourd or smooth gourd (Chandra, 1995). Its broadly ovate to reniform leaves are comparatively large and dark green. While staminate and pistillate flowers are borne separately on the same vine, staminate flowers appear first in clusters of 5-20 and pistillate flowers are borne solitarily. Its deep yellow flowers expand in the morning hours as anthesis takes place. Flowers are pentamerous and deep yellow, housing 3 or 5 stamens. Mature fruits are sub-cylindrical, smooth and devoid of conspicuous surface ridges, 30-60 cm long and immature fruits are crowned by enlarged calyx and style. Seeds are broadly ellipsoid, between 1-1.5 cm long, and are smooth, black or sometimes white, having a narrow wing-like margin (Chandra, 1995).

Domesticated *L. aegyptiaca* species are self-compatible (Sinnot and Blotch, 1943) and there are no crossing barriers with wild *L. acutangula* (Seshadri and More, 2009). However, time to anthesis between these species often requires artificial hand pollination to ensure successful hybridization. As a consequence of flavaniod analyses, breeding system, and leaf, flower, and fruit morphology, Heiser and Schilling (1990) grouped *L. acutangula* and *L. aegyptiaca* into a single taxonomic group, apart from the other five *Luffa* species (Heiser and Schilling, 1990). However, this grouping is not entirely supported by chloroplast DNA maker analysis (Chung *et al.*, 2003).

Even though var. *aegyptiaca* is commonly found in India and neighboring countries, cultivar-dependent production practices are used to produce fruits as fresh vegetables or as sponges (Chandra, 1995, Seshadri and More, 2009). Fruits of var. *aegyptiaca* are more deeply furrowed, less bitter and larger than those of var. *leiocarpa* (Heiser and Schilling, 1990). In contrast to var. *aegyptiaca*, wild forms of var. *leiocarpa* occur from Myanmar to the Philippines, and south to northeastern Australia and Tahiti (Heiser and Schilling, 1990).

Luffa graveolens Roxb.: This species is found in India and northeastern Australia and is considered a possible progenitor of domesticated *Luffa* species (Singh, 1990, Swarup, 2006). Fruits are tuberculate with projections less than 1 mm long that are scattered over the fruit surface (Heiser and Schilling, 1990). Cogniaux and Harms (1924) determined that *L. graveolens* is endemic to Samoa based on a specimen collected by Whitmee (number 82; Heiser and Schilling, 1990). Although Parham (1972) did not include this *Luffa* species in his taxonomic treatment of the flora of the Samoa islands, he did describe a *Luffa* variant as *L. aegyptiaca* var. *leiocarpa* (Heiser and Schilling, 1990), thus agreeing with Heiser and Schilling (1990). Leaves of *L. graveolens* are reniform without evident lobbing and with a scabrid leaf surface. Its fibrous fruit wall is loosely knitted (Prakash *et al.*, 2013), and fruit are grey, ovoid, sparsely tuberculate-spinose, possessing seed that is white to ash in color with a smooth (finely pitted) and bitter (no pungent smell) seed coat (Pandey *et al.*, 2014).

Luffa umbellata (Klein) Roem.: This annual species possesses elongate, sulcate, and glabrousstems. Its broad, 3-5 cm long leaves are alternate and ovate-cordate, which are slightly lobed (3-5), acute, subulate, and dentate. The petioles are 6-12 cm long, slender, glabrous or puberulous. Tendrils are puberulous and bifid and flowers are white. Staminate flowers have 15-28 cm longpeduncles with three stamens with bifid filaments, and are usually slender, glabrous with 0.5-2.0 cm longpedicels. In contrast, pistillate flowers have 0.5-1.0 cm long peduncles. Fruit are short, ecostate, densely long spinous with densely woollyspines (Chakravarty, 1959; http://indiabiodiversity.org/species/show).

New World Species

L. operculata (L.) Cogn.: This is the most widespread species in South America, ranging from Panama to Peru and east to Brazil to include the archipelago of Fernando de Noronha (Heiser *et al.*, 1988, Heiser and Schilling, 1990). The species name is derived from the word, *operculum*, meaning "little lid." When the fruits mature and the seeds are ready to be dispersed, a small part of the bottom tip (blossom-end) opens up and the seeds fall to the ground. The species possses shallow-lobed (3-5) to nearly entire leaves, where sinus lobes are rounded to acute. The corolla is typically pale yellow and fruiting peduncles are 7-20 mm long. Fruits are usually ovoid tooblong, tuberculate, less than 10 cm long, and usually densely and finely pubescent. Rugose seeds are 7.5-10.0 mm long and 5.0-5.5 mm wide (Heiser *et al.*, 1988, Heiser and Schilling, 1990). Fruits contain an inner rather dense network of fibers in which the seeds are borne on a loose outer network. In *L. operculata*, fruit fibers are mostly connected, although a few fibers can occur as free and bifid elements at the tips of fruit (Heiser *et al.*, 1988).

L. astorii Svens: This species is reported from Venezuela (single collection), western Ecuador including the Galapagos Islands and northwestern Peru (Heiser *et al.*, 1988, Heiser and Schilling, 1990). The leaves of *L. astorii* are shallowly to moderately deeply lobed, and its corolla is deep yellow in color with lobes, where staminate flowers can be over 12 mm long. Its fruits are nearly ovoid and lightly pubescent to subglabrous (Heiser *et al.*, 1988, Heiser and Schilling, 1990). The fruits house both an inner rather dense network of fibers in which the seeds are borne and a loose outer fiber network. The outer fiber network in *L. astorii* is somewhat denser than that observed in the fruit of *L. operculata* and *L. quinquefida* (Heiser *et al.*, 1988). The fibres of outer network are soley anastomosed (the connection of separate parts of a branching system to form a network) in *L. astorii* (Heiser *et al.*, 1988).

L. quinquefida (Hook & Arn.) Seem.: Although this species was initially included in *L. operculata* (L.) Cogn. (Jeffrey, 1992), it is geographically isolated and is morphologically distinguishable from all other New World species (Heiser *et al.*, 1988, Heiser and Schilling, 1990). The species' range extends from western Mexico to Nicaragua (Heiser *et al.*, 1988, Heiser and Schilling, 1990). Its leaves are deeply lobed, and its flowering corolla is light to deep yellow in color (Heiser *et al.*, 1988, Heiser and Schilling, 1990). Fruits are usually ovoid to oblong and are densely pubescent (Heiser *et al.*, 1988, Heiser and Schilling, 1990). In *L. quinquefida*, many of its fruit fibres are free and bifid at the fruit tip (Heiser *et al.*, 1988).

Luffa saccata F. Muell.ex I.Telford sp. nov.: This species is relatively recent in its taxonomic description and is endemic to Australia (Telford *et al.*, 2011). Leaves of this species are ovate to broadly ovate (3–14 cm long and 2.5–13 cm wide), with 3 to 5 broad rounded or obtuse lobes. Inflorescences are usually unisexual and only rarely uniquely staminate or pistillate. Yellow flowers are usually solitary but are sometimes paired in the leaf axils. Fruits are ovoid, between 2.5 to 4.5 cm long and between 2 to 4 cm in diameter.

Cytogenetics and Species Cross Compatibility

Cytogenetic analyses of the Old World *Luffa* has revealed that all species are uniformly $2n = 2x = 26$, where their basic chromosome number is 13 ($x = 13$) (Naithani and Das, 1947, Dutt and Roy, 1969, Roy *et al.*, 1970). Cytological appraisal of the New World species *L. operculata*, and *L. quinquefida* also detected the presence of numerous globular inclusions in the pollen mother cells. However, it was determined by assessment of species hybrids, that chromosome number in these species was $n = 13$ (Heiser *et al.*, 1988, Heiser and Schilling, 1990).

Comparative morphological and chromosome pairing analysis in interspecific hybrids of the Old World wild and cultivated species indicates that *L. graveolens* is the principal species which gave rise to the two cultivated monoecious species, *L. acutangula* and *L. cylindrica* (Dutt and Roy, 1971). Dutt and Roy (1971) assessed pairing relationships in two interspecific hybrids of Luffa: *L. acutangula* × *L. graveolens* and *L. echinata* × *L. graveolens*, where various types of chromosome associations from univalents to quadrivalents were observed. The hybrid *L. echinata* × *L. gvaveolens* showed a higher frequency of univalents compared to *L. acutangula* × *L. graveolens* progeny, but chromosome pairing data of both hybrid progeny indicated partial homology exists among most of the chromosomes of the three parental species examined. The study of these F_1 hybrids has been utilized to establish phylogenetic relationship among various *Luffa* species and to elucidate evolution pathways in the genus (Dutt and Roy, 1990, Heiser and Schilling, 1990). Such studies indicate that two wild species, *L. graveolens* and *L. echinata*, are closely related, whereas, *L. graveolens* is more distantly related to *L. acutangula* than to *L. echinata*. Moreover, data also indicate that dioecism in *L. echinata* is a derived feature, established after its putative derivation from the monoecious more primative species, *L. graveolens*.

Pathak and Singh (1949), similarly, documented the cytological behavior of interspecific hybrid progeny between *L. acutangula* × *L. cylindrica*, and found reduced hybrid fertility when compared to their parents. Fertility, however, could be restored when the F_1 progeny were backcrossed to either parent. Further cytological studies revealed considerable species homology along with some non-homologous segments in such F_1 species hybrids (Dutt and Roy, 1990). The interspecific hybridizations among *L. graveolens*, *L. aegyptiaca/cylindrica*, *L. acutangula*, and *L. echinata* resulted in hybrid sterility (Dutt and Roy, 1969 and 1971). The frequent occurrence of univalents and bivalents at diakinesis and Metaphase I was also detected among hybrids between *L. acutangula* and *L. graveolens* and their amphidiploids (2n = 4x = 52), which resulted in lower fertility (Dutt and Roy, 1976). The number of univalents ranged from 0.0-18.0, where the mean was 5.8 per pollen mother cell (PMC). Rod bivalents varied from 0.0-14.0, whereas ring bivalents ranged between 8.0 to 20.0 per cell. The frequency of quadrivalent formation, however, was relatively low (<1%). At Anaphase I, regular disjunction of chromosomes along with spindle migration irregularities (e.g., unequal separation and lagging chromosomes) were also observed, which consequently resulted in a relatively high frequency of empty pollen formation and subsequent amphidiploid pollen sterility.

Autotetraploids (4x = 52) have been created in *L. acutangula* germplasm, but these products were considered to possess no real economic value (Roy and Dutt, 1971). Likewise, artificial hybrids made between Old and New World species (*i.e., L. aegyptiaca* × *L. astorii*, *L. aegyptiaca* × *L. operculata*, *L. echinata* × *L. operculata*, and *L. operculata* × *L. graveolens*) were sterile (Heiser and Schilling, 1988). When *L. quinquefida* was used as the female parent in crosses with *L. graveolens*, fruits matured but they were without seeds. In contrast, hybridizations among various New World species have been relatively successful (Heiser and Schilling, 1988, Heiser *et al.*, 1988). For instance, hybrids made within *L. astorii* and *L. operculata* are fully fertile, whereas successful hybridization within *L. quinquefida* was marked by a reduction in pollen stainabilty (46-71%) even though seed set was not substantially affected (i.e., equal to that of parents) (Heiser *et al.*, 1988). In those studies, the hybrids between *L. astorii* and *L. operculata* were obtained with great difficulty, where only one viable seed was retrieved and the resulting infertile plant possessed abnormally developed flowers without anthers. Hybrids between *L. quinquefida* and *L. astorii* were fairly readily obtained when *L. quinquefida* was used as the maternal parent. In this case, four hybrid progeny were grown to maturity, and cytological analysis revealed that only one plant produced 13 bivalents resulting in pollen with stainabilities ranging from 0.7 to 18%. In contrast, fertile hybrids are easily obtained from crosses between *L. operculata* and *L. quinquefida*, where *L. quinquefida* was used as the maternal parent. Hybrid plants from such crossing regularly possessed 13 bivalents and matured pollen with stainabilities ranging between 20 to 52%, with a mean of 37%. However, these hybrids typically produced only between 25-33% of the seeds typically retrieved from parental stocks

Prospects

Species in the genus *Luffa* are rapidly becoming indispensable for providing natural alternatives for human health and wellbeing as a healthy food source. Most of the genetic improvement in sponge and ridge grourd has been the result of rigorous phenotypic selection which has resulted in the release of many popular open-pollinated cultivars in Asia, particularly India principally by state agricultural universities and government agencies (e.g., the Indian Council of Agricultural Research, India) (Swarup, 2006, Ram *et al.*, 2007, Seshadri and More, 2009). Over the last two decades, heterosis breeding has been exploited in the improvement of *Luffa* sponges (Seshadri and More, 2009). In the future, strategically formulated classical hybridization programs are expected to create improved recombinants and transgressive segregants, which will likely be enhanced through the use of various biotechnologies (Behera *et al.*, 2012). In this regard, there continues to be a need for improved ridge and sponge gourd germplasm, which has been traditionally achieved through selection within landraces by farmers.

References

Adebooye, O.C. 2009. The properties of seed oil and protein of three underutilized edible Cucurbitaceae of Southwest Nigeria. *Acta Hort.* 806(1):347-354.

Ali, A.M., S. Karuppusamy, and Al-Hemaid Fahad M. 2010. Molecular phylogenetic study of *Luffa tuberosa* Roxb. (Cucurbitaceae) based on internal transcribed spacer (ITS) sequences of nuclear ribosomal DNA and its systematic implication. *Int. J. Bioinformatics Res.* 2:42-60.

Azeemoddin, G., and Rao, S.D.T. 1966. Seed fat of *M. tuberosa* or *L. tuberosa*. *Curr. Sci.* 36:100.

Behera, T.K., A.K. Sureja, S. Islam, A.D. Munshi, and Sidhu, A.S. 2012. Minor cucurbits. pp. 17-60. In: Y.H. Wang, T.K. Behera, and C. Kole (eds.), Genetics, *Genomics and Breeding of Cucurbits.* Science Publishers, Inc., Enfield, USA.

Bretschneider, E. 1878. History of European botanical discoveries in China.Vol. I, pp. 143-144.

Chakravarty, H.L. 1959. Monograph on Indian Cucurbitaceae. Rec. Bot. Survey India 17(1):73-83.

Chakravarty, H.L. 1982. Cucurbitaceae. *Fascicles of Flora of India.* No.11, Botanical Survey of India, Calcutta, West Bengal, India, pp. 85-116.

Chandra, U. 1995. Distribution, domestication and genetic diversity of *Luffa* gourd in Indian subcontinent. *Indian J. Plant Genet. Resour.* 8(2):189-196.

Choudhury, B. 1996. Vegetables. National Book Trust, New Delhi.

Chung, S.M., D.S. Decker Walters, and Staub, J.E. 2003. Genetic relationships within the Cucurbitaceae as assessed by consensus chloroplast simple sequence repeats (ccSSR) marker and sequence analyses. *Canadian J. Bot.* 81:814–832.

Cogniaux, A., and Harms, H. 1924. Cucurbitaceae-Cucurbiteae-Cucumerinae. pp. 1-246. In: A. Engler (ed.), Pflanzenreich Heft 88, IV.275.2. Wilhelm Engelmann, Leipzig.

Davis, J.M. 1994. Luffa sponge gourd production practices for temperate climates. *HortScience* 29(4):263-266.

DeCandolle, A. 1959. Origin of cultivated plants. (Reprint of the second edition, 1886). Hafner Pub. Co., New York.

Desai, U.T., and Musmade, A.M. 1998. Pumpkins, squashes and gourds. pp. 273-297. In: D. K. Salunkhe, and S.S. Kadam (eds.), Handbook of Vegetable Science and Technology: Production, Compostion, Storage and Processing. Marcel Dekker, Inc., New York.

Dutt, B., and Roy, R.P. 1969. Cytogenetical studies in the interspecific hybrid of *Luffa cylindrica* L. and *L. graveolens* Roxb. *Genetica* 40:7-18.

Dutt, B., and Roy, R.P. 1971. Cytogenetic investigation in Cucurbitaceae. I. Interspecific hybridization in *Luffa*. *Genetica* 42:139-156.

Dutt, B., and Roy, R.P. 1976. Cytogenetic studies in an experimental amphidiploid in *Luffa*. *Caryologia* 29(1):15-22.

Dutt, B., and Roy, R.P. 1990. Cytogenetics of the old world species of *Luffa*. pp. 134–140. In: D.M. Bates, R.W. Robinson, and C. Jeffrey (eds.), Biology and Utilization of the Cucurbitaceae. Cornell University Press, Ithaca, NY.

Dutt, B., and Saran, S. 1998. Cytogenetics. pp. 33-38. In: N.M. Nayar, and T.A. More (eds.), Cucurbits. Oxford & IBH Publishing Co. Pvt. Ltd., New Delhi.

Filipowicz, N., H. Schaefer, and Renner, S.S. 2014. Revisiting *Luffa* (Cucurbitaceae) 25 years after C. Heiser: Species boundaries and application of names tested with plastid and nuclear DNA sequences. *Syst. Bot.* 39(1):205-215.

Heiser, C.B. Jr. 1979. The Gourd Book. Univ. Oklahoma Press, Norman, USA, 248 p.

Heiser, C.B. Jr., and Schilling, E.E. 1988. Phylogeny and distribution of *Luffa* (Cucurbitaceae). *Biotropica* 20:185-191.

Heiser, C.B. Jr., and Schilling, E.E. 1990. The genus *Luffa*: A problem in phytogeography. pp. 120–133. In: D.M. Bates, R.W. Robinson, and C. Jeffrey (eds.), Biology and Utilization of the Cucurbitaceae. Cornell University Press, Ithaca, NY.

Heiser, C.B. Jr., E.E. Schilling, and Dutt, B. 1988. The American species of *Luffa* (Cucurbitaceae). *Syst. Bot.* 13:138-145.

Hooker, J.D. 1871. Cucurbitaceae. In: Flora of Tropical Africa (Ed. Oliver). Vol. 2, Reeve, London.

Hutchinson, J., and Dalziel, J.M. 1954. Flora of West Tropical Africa. 2nd ed. Vol. 1, Crown Agents, London.

Jeffrey, C. 1962. Notes on Cucurbitaceae, including a proposed new classification of the family. *Kew Bull.* 15:337-371.

Jeffrey, C. 1980a. A review of the Cucurbitaceae. *J. Linn. Soc. Bot.* 81:233-247.

Jeffrey, C. 1980b. Further notes on the Cucurbitaceae. V. The Cucurbitaceae of the Indian subcontinent. *Kew Bull.* 34:789-809.

Jeffrey, C. 1992. Names of indigenous Neotropical species of *Luffa* Mill. (Cucurbitaceae). Kew Bull. 47:741-742.

Joseph John, K. 2005. Studies on Ecogeography and Genetic Diversity of the Genus *Momordica* L. in India. Ph.D. thesis, Mahatma Gandhi University, Kottayam, Kerala, India.

Keraudren, M. 1965. Cucurbitaceae. (In): A. Aubreville (Ed.), Flore du Cameroun, Vol. 6. Museum National d' Histoire Naturelle, Paris.

Kumar, P., G. Devala Rao, B. Lakshmayya, and Setty, S.R. 2011. Morphoanatomy and phytochemical screening of entire fruits of *Momordica tuberosa* Cogn. (Cucurbitaceae). *Lat. Am. J. Pharm.* 30(3):593-598.

Naithani, S.P., and Das, P. 1947. Somatic chromosome numbers of cultivated cucurbits. *Curr. Sci.* 16:188-189.

Pandey, A., and Prakash, K. 2013. A note on debatable taxonomic identity of *Luffa tuberosa* Roxb. (Cucurbitaceae): A potential wild edible vegetable in India. *Indian J. Plant Genet. Resour.* 26(2):124-127.

Pandey, A., K. Pradheep, and Semwal, D.P. 2014. Notes on *Luffa* (Cucurbitaceae) genetic resources in India: Diversity distribution, germplasm collection, morphology and use. *Indian J. Plant Genet. Resour.* 27(1):47-53.

Parham, B. 1972. Plants of Samoa. New Zealand Dept. Sci. Indust. Res., Wellington.

Paris, H.S., and Maynard, D.N. 2008. *Luffa cylindrica* sponge gourd; *Luffa acutangula* angled luffa. p. 302–305. In: J. Janick and R.E. Paull (eds.), Encyclopedia of fruits & nuts. CAB International, Wallingford, Oxfordshire, UK.

Pathak, G.N., and Singh, S.N. 1949. Studies in the genus *Luffa*. I. Cytogenetic investigations in the interspecific hybrid *L. cylindrica*× *L. acutangula*. *Indian J. Genet.* 9:18-26.

Porterfield, W.M. 1955. Loofah - the sponge gourd. *Econ. Bot.* 9:211-223.

Prakash, K., A. Pandey, J. Radhamani, and Bisht, I.S. 2013. Morphological variability in cultivated and wild species of *Luffa* (Cucurbitaceae) from India. *Genet. Resour. Crop Evol.* 60:2319–2329.

Ram, D., M. Rai, and Singh, M. 2007. Temperate and subtropical vegetables. pp. 71-108. (In:) K.V. Peter and Z. Abraham (eds.), Biodiversity in Horticultural Crops, Vol 1. Daya Publishing House, New Delhi.

Rao, B.K., M.M. Kesavulu, R. Giri, and Appa Rao, C. 1999. Antidiabetic and hypolipidemic effects of *Momordica cymbalaria* Hk. fruit powder in alloxan-diabetic rats. *J. Ethnopharmacol.* 67:103-109.

Rao, B.K., M.M. Kesavulu, and Appa Rao, C. 2001. Antihyperglycemic activity of *Momordica cymbalaria* fruit in alloxan diabetic rats. *J. Ethnopharmacol.* 78:67-71.

Roy, R.P., and Dutt, B. 1971. Cytomorphological studies in induced polyploids in *Luffa acutangula* Roxb. *Nucleus* 15:17.

Roy, R.P., A.R. Mishra, R. Thakur, and Singh, A.K. 1970. Interspecific hybridization in the genus *Luffa*. *J. Cytol. Genet.* 5:16-20.

Schilling, E.E., and Heiser, C.B. 1981. Flavonoids and the systematics of *Luffa*. *Biochem. Syst. Ecol.* 9:263-265.

Seshadri, V.S., and More, T.A. 2009. Cucurbit Vegetables: Biology, Production and Utilization. Studium Press (India) Pvt. Ltd., New Delhi, 482 p.

Shantha, T.R., G. Venkateshwarlu, M.J. Indira Ammal, and Gopakumar, K. 2009. Pharmacognistical studies on fruits of *Momordica cymbalaria* Fenzl ex Nand. *J. Res. Educ. Indian Med.* 1:1-10.

Singh, A.K. 1990. Cytogenetics and evolution in the cucurbitaceae. pp. 10-28. In: D.M. Bates, R.W. Robinson, and C. Jeffrey (eds.), Biology and Utilization of the Cucurbitaceae. Cornell University Press, Ithaca, NY.

Singh, D. 1964. A further contribution to the endosperm of the Cucurbitaceae. *Proc. Indian Acad. Sci.* 60B:399-413.

Singh, D., and Dathan, A.S.R. 1990. Seed coat anatomy of Cucurbitaceae. pp. 225-238. In: D.M. Bates, R.W. Robinson and C. Jeffrey (eds.), Biology and Utilization of Cucurbitaceae. Cornell Univ. Press, Ithaca, USA.

Singh, D., and Bhandari, M.M. 1963. The identity of an imperfectly known hermaphrodite *Luffa*, with a note on related species. Baileya 11:132-141.

Sinnott, E.W., and Bloch, R. 1943. Luffa sponges, a new crop for the Americas. *J. New York Botanical Garden* 44:125-132.

Swarup, V. 2006. Vegetable Science and Technology in India. Kalyani Publishers, New Delhi. 656 p.

Telford, I.R.H., H. Schaefer, W. Greuter, and Renner, S.S. 2011. A new Australian species of *Luffa* (Cucurbitaceae) and typification of two Australian *Cucumis* names, all based on specimens collected by Ferdinand Mueller in 1856. *PhytoKeys* 5:21-29.

Zeven, A.C., and Zhukovsky, P.M. 1975. Dictionary of cultivated plants and their centres of diversity. Center of Agricultural Publishing and Documentation, Wageningen, The Netherlands.

13

Momordica

Amish Kumar Sureja and Anilabh D Munshi

Momordica (Family: Cucurbitaceae) is a genus of under-utilized and wild-gathered vegetables of importance as food and medicine and is a native of the Paleotropics (Robinson and Decker-Walters, 1997). The name *Momordica* derives from the Latin word 'mordeo' (means to bite) (Durry, 1864) or for the biting taste of the ripe fruits of *M. balsamina*, the type species (Genaust, 1996). The genus *Momordica* comprises 59 species (Schaefer and Renner, 2010) distributed in the warm tropics, mainly in Africa and with about 10 species in Southeast Asia (de Wilde and Duyfjes, 2002). Bitter gourd (*M. charantia*), teasel gourd (*M. subangulata* subsp., *renigera*) and spine gourds (*M. dioica* and *M. sahyadrica*) are nutritionally rich. Sweet gourd (*M. cochinchinensis*) is very rich source of β-carotene, the precursor of vitamin A and has longevity, vigour and vitality in enhancing properties.

Among the cultivated species of *Momordica*, bitter gourd (*Momordica charantia* L.), also known as *karela* in Hindi or balsam pear or bitter melon is the most popular and widely grown in India, Sri Lanka, Philippines, Thailand, Malaysia, China, Japan, Australia, tropical Africa, South America and the Caribbean. The fruits are cooked with other vegetables, stuffed or stir-fried or can be dehydrated, pickled, or canned. For some preparations, fruits are blanched, parboiled, or soaked in salt water before cooking to reduce the bitter taste. Young *Momordica* shoots and leaves are also cooked and eaten as leafy vegetables, and leaf and fruit extracts are used in the preparation of tea (Tindall, 1983; Reyes *et al.*, 1994). Bitter gourd is an important part of several Asian cuisines and has been used for centuries in ancient traditional Indian, Chinese, and African medicine. Bitter gourd extracts possess antioxidant, antimicrobial, antiviral, antihepatotoxic and antiulcerogenic properties while also having the ability to lower blood sugar (Raman and Lau, 1996). These medicinal activities are attributed to an array of biologically active plant chemicals, including triterpenes, pisteins and steroids (Grover and Yadav, 2004). *Momordica* spp. are extensively used in Indian systems of medicine including Ayurveda, Unani, Siddha, and Homeopathy. The extracted juice from leaf, fruit and even whole plant are used for treatment of wounds, ulcers, infections, parasites (*e.g.*, worms), measles, hepatitis, diabetes, and fevers (Beloin *et al.*, 2005; Behera *et al.*, 2010).

Bitter gourd fruits are a good source of carbohydrates, proteins, vitamins, and minerals and have the highest nutritive value among cucurbits (Behera *et al.*, 2007). The bitterness of bitter gourd is due to the cucurbitacin-like alkaloid momordicine and triterpene glycosides (momordicoside K and L) (Jeffrey, 1980; Okabe *et al.*, 1982). Unlike other cucurbitaceous vegetables, the bitter fruit flavour of *M. charantia* is considered desirable for consumption, and 'thus bitter flavour has been selected during domestication (Marr *et al.*, 2004).

Origin and Domestication

Momordica is a monophyletic genus that originated in tropical Africa and the Asian species are considered the result of one long-distance dispersal event that occurred about 19 million years ago (Schaefer *et al.*, 2009; Schaefer and Renner, 2010). Earlier *M. charantia* and *M. balsamina* were reported to be of Asian origin while a recent study using plastid and mitochondrial DNA-based markers (Schaefer and Renner, 2010) reported that these species are most likely of African origin and not Asian. Monoecious species evolved from dioecious species independently, always in Africa and mostly in the Savanna region. Among different species of *Momordica, M. charantia* is the only cultivated species which has been extensively studied in recent years and has a long history as a cultivated food and medicinal plant in Africa and Asia (Morton, 1967; Walters and Decker-Walters, 1988). It has two botanical varieties viz.; *M. charantia* var. *muricata* (syn. var. *abbreviata*) and *M. charantia* var. *charantia*, the former mostly wild and the latter cultivated (Bharathi and Joseph, 2013). The wild variety (*M. charantia* var. *muricata*) is considered as the progenitor of cultivated *M. charantia* var. *charantia* (Degner, 1947; Walters and Decker-Walters, 1988) which is found in tropical Asia and Africa.

The original place of domestication of bitter gourd is unknown or unclear (Li, 1970; Zeven and de Wet, 1982) for lack of credible archaeological evidences. The possible areas for domestication of bitter gourd proposed by various workers include southern China, eastern India (Sands, 1928; Degner, 1947; Walters and Decker-Walters, 1988; Raj *et al.*, 1993; Robinson and Decker-Walters, 1997; Marr *et al.*, 2004) and south-western India (Joseph, 2005). Both the domesticated and putative wild progenitors of bitter gourd are listed in floras of India, tropical Africa, and Asia as well as the New World tropics. From Africa, *M. charantia* is believed to have taken to Brazil via slave trade and thence to "Middle America" (Ames, 1939). Uncarbonized seed coat fragments have been tentatively identified from Spirit Cave in northern Thailand (Yen, 1977). However, there have been no archaeological reports of bitter gourd remains in China (Marr *et al.*, 2004) or Southeast Asia. The earliest written reference to *M. charantia* in China is from northern China and was reported in 1370 CE (Yang and Walters, 1992). A comprehensive compilation of plant remains from 124 Indian archaeological sites does not include bitter gourd (Kajale, 1991).

Based on both historical literature (Chakravarty, 1990; Raj *et al.*, 1993; Walters and Decker-Walters, 1988), and recent random amplified polymorphic DNA (RAPD; Dey *et al.*, 2006), intersimple sequence repeats (SSR; Singh *et al.*, 2007) and amplified fragment length polymorphisms (AFLP; Gaikwad *et al.*, 2008) molecular analyses, eastern India (includes the states of Odisha, West Bengal, Assam, Jharkhand and Bihar) may be considered as a probable primary center of diversity of bitter gourd, where a wild feral form *M. charantia* var. *muricata* (Chakravarty, 1990) currently exists.

M. charantia and *M. dioica* were originally described from Peninsular India by Linnaeus (1753) and Roxburgh (1832) based on van Rheede's *Hortus Malabaricus* (Van Rheede, 1688). Wild or small-fruited cultivated forms, however, are mentioned in Ayurvedic texts written in Indian Sanskrit writings from 2000 to 200 BCE by the Indo-Aryans who valued *Momordica* for use as medicine, food, containers, musical instruments, and literary metaphors (Decker-Walters, 1999), indicating an early cultivation of bitter gourd in India. The most prominent Sanskrit name for bitter gourd is *kdravalff* along with its various permutations and modern derivatives (e.g., *karilff, kariyalla, kareld, karelo*). The lack of a unique set of Indo-Aryan words indicates that the Aryans did not know bitter melon before entering India (Walters and Decker-Walters, 1988). The Indo-Aryan words for *M. charantia* may have been borrowed from the Dravidian (Turner, 1966) before the Indo-Aryans arrived, indicating an even earlier awareness of the plant (Decker-Walters, 1999). The most modern Hindi (Indian language) term "*karela*" may be of Dravidian origin (Turner, 1966).

Unlike *M. balsamina*, *M. dioica* has several names in modern Indo-Aryan (e.g., *kaksa, kakrol, jangli-karela*, etc.) and Dravidian (e.g., *agakral, hagal, karlikai, etc.*) languages (Decker-Walters, 1998), indicating its importance as a useful plant in recent centuries. *Kanchan-arak* is a Munda (aboriginal tribe from Central India) name for *M. dioica* (Chakravarty, 1982). *M. dioica* is indigenous to India and possibly evolved in Central India (Behera *et al.*, 2011).

The important characteristics most transformed in the process of domestication of bitter gourd were mainly those related with handling and uses, for example, relatively uniform germination and increase in the fruit size. The different degrees of variation in bitterness suggest a strong association of humans in selecting bitter types as both domesticate and the wild-type have bitter fruit (Marr *et al.*, 2004).

Distribution

Out of the ten species in Asia, seven species of *Momordica* viz. *charantia, balsamina, dioica, sahyadrica, subangulata, cochinchinensis* and *cymbalaria* are naturally distributed in India. *M. denudata* is naturally distributed in Sri Lanka. *M. denticulata, M. rumphii* and *M. clarkeana* are naturally distributed in Indonesia and Malaysia (Chakravarty, 1959; de Wilde and Duyfjes, 2002; Joseph, 2005; Bharathi and Joseph, 2013).

M. charantia is ubiquitous in distribution and is distributed in Tropical and subtropical Africa, South, East and SE Asia, Malesia, Australia and Pacific, India-Andaman and Nicobar Islands, South, North (except NW Himalayas), East and North-East, Pakistan, South-East Asia: Thailand, Indonesia (Sumatra, Java, Borneo), Malesia, Vietnam, Laos, Cambodia,

Myanmar, Sri Lanka, parts of China (Yunan) and Nepal (Chakravarty, 1959; de Wilde and Duyfjes, 2002). In India, it is distributed in the States of Tamil Nadu, Kerala, Andhra Pradesh, Karnataka, Odisha, Madhya Pradesh, Uttar Pradesh, Chhattisgarh, West Bengal, Assam, Rajasthan, Jharkhand, Mizoram and Andaman Islands (Joseph, 2005; Bharathi and Joseph, 2013).

M. dioica and *M. charantia* var. *muricata* had wider distribution spread all over India except the former in North–East and Northwest Himalayas. The major distribution areas of *M. dioica* are Peninsular India, extending beyond the Deccan Plateau to Central India. *M. subangulata* subsp. *renigera* is restricted to Northeast India and adjoining North Bengal hills. *M. subangulata* subsp. *renigera* is largely occur in Assam (Kokrajar and Kamrup districts), Meghalaya and Mizoram (Naga-Khasi-Jainthia and Garo hills) and occasionally in Arunachal Pradesh, Nagaland, Tripura and Sikkim. It has a preference for fertile loamy soils in humid high rainfall areas and do not tolerate climatic extremes. *M. cochinchinensis* has localised distribution in Andamans and a few localities in the Eastern and Northeastern states. *M. cochinchinensis* has a preference for heavy rainfall-high humidity regions. It occurs chiefly in low elevation coastal jungles in South, Middle and North Andamans, which is in continuation of its distribution in Vietnam, Thailand, Cambodia and Philippines to Australia. To a lesser extent, it occurs in Assam-Tripura-Manipur-Nagaland forests, which is in continuation of its distribution in Myanmar and Chittagong hill tracts to South China. *M. balsamina* is restricted to the arid belt comprising Rajasthan and Gujarat. *M. balsamina* has a preference for slightly high pH soils with hot days and cool nights. *M. sahyadrica* is endemic to the Western Ghats. The natural distribution of *M. sahyadrica* is restricted to low and mid-elevation range of Western Ghats from Kerala to Maharashtra through the Nilgiris and Konkan hills (Joseph, 2005; Bharathi and Joseph, 2013).

Taxonomy and Biosystematics

The genus *Momordica* belongs to the subtribe Thalidianthinae Pax, tribe Joliffieae Schrad., subfamily Cucurbitoideae of Cucurbitaceae (Jeffrey 1980; Jeffrey 1990; de Wilde and Duyfjes 2002). Generic and species descriptions (along with keys in some cases) are found in various monographic and floristic treatises (Willdenow, 1805; Blume, 1826; Seringe, 1828; Wight and Arnot, 1841; Thwaites, 1864; Hooker, 1871; Clarke, 1879; Keraudren, 1975; Jeffrey, 1980). Van Rheede's (1688) descriptions and illustrations of paval (=*Momordica charantia*) in the *Hortus Malabaricus* is the first printed record. Linnaeus (1753), de Candolle (1828), Roxburgh (1832), Clarke (1879), Cooke (1901), Gamble and Fischer (1919), Blatter (1919), Kanjilal *et al.* (1938), Santapau (1953), Saldhana and Nicholson (1976), Chakravarty (1959, 1982) and Mathew (1981, 1983) have extensively treated the systematics of the genus in their works.

Chakravarty's (1982) classification of *Momordica* in his Fascicles of Cucurbitaceae is by far the most relied upon in India. He has enumerated seven species from India including *M. denudata* from Kerala and *M. macrophylla* from the Assam–Manipur belt bordering Myanmar. He has also described a new variety, *i.e. M. charantia* var. *Muricata* based on Rheede's plate in *Hortus Malabaricus* as type. However, the reported occurrence of *M. denudata* in Kerala is based on wrong identification of *M. dioica* specimens (Joseph, 2005). Trimen (1893–1900) reported detailed technical description and key to the species of *Momordica* occurring in Sri Lanka. Backer and Brink (1963), Henderson (1974), and

Keraudren (1975) noted detailed floristic account of *Momordica* species in other South–East Asian countries. De Wilde and Duyfjes (2002) gave a detailed taxonomic treatment of *Momordica* genus in south and South–East Asia.

The genus *Momordica* is monophyletic and can be divided into 11 clades (Schaefer and Renner, 2010). The Asiatic species fall under three sects. monoecious species *M. charantia* and *M. balsamina* under the sect. *Momordica* and dioecious species like *M. cochinchinensis*, *M. dioica*, *M. sahyadrica*, *M. denticulata*, *M. denudata*, *M. clarkeana* and *M.subangulata* grouped under the sect. *Cochinchinensis*, and *M. cymbalaria* under the sect. Raphanocarpus (Schaefer and Renner, 2010).

Taxonomic position of *M. cymbalaria* has always been uncertain (Chakravarty, 1982; Joseph and Antony, 2010; Bharathi *et al.*, 2011). Joseph and Antony (2010) reported that species *cymbalaria* is more allied to genus *Luffa* than *Momordica*. However, recently, based on three genome (plastid, mitochondrial and nuclear DNA markers) phylogeny (Schaefer and Renner, 2010) the status of species *cymbalaria* under the sect. Raphanocarpus, genus *Momordica* is established.

The major diagnostic features of the genus *Momordica* are the presence of conspicuous floral bracts (male), calyx cup, entire petal, scales on corolla, pendulous, echinate or muricate fruits, sculptured seeds and viny habit. The minor diagnostic characters are flower colour, petal shape and size, petal markings, pubescence, bract shape, position, calyx cup colour, sepal shape, gland dottedness (petiole), floral scent, anthesis time, seed sculpture, shape, colour, pollinators, fruit surface ornamentation, *etc.* (Bharathi and Joseph, 2013).

Morphological Features of *Momordica* spp. Occurring in India

The genus *Momordica* has 45 species domesticated in Asia and Africa (Robinson and Decker-Walters, 1997). The genus *Momordica* has seven species in India. The monoecious taxa are *M. charantia* L. (var. *muricata* (Willd.) Chakrav. and var. *charantia* L.), *M. balsamina* L. and *M. cymbalaria*. The dioecious taxa are *M. dioica* Roxb., *M. sahyadrica* Joseph & Antony, *M. cochinchinensis* (Lour.) Spreng. and *M. subangulata* Blume subsp. *renigera* (G. Don) W. J. de Wilde (Joseph and Antony, 2010).

The annual monoecious species (*M. balsamina* and *M. charantia*) can be easily distinguished from each other. The male flower bract is positioned at the base/near the axis or below the middle of the flower stalk in *M. charantia*, whereas in *M. balsamina* it is situated in the upper middle or towards the tip of the peduncle. The anther filaments are fused to give a globose appearance in *M. charantia*, while it is split into lobes in *M. balsamina*. Both the monoecious species (*M. charantia* and *M. balsamina*) have evolved from a common ancestor and has diverged morphologically from the dioecious species (Bharathi and Joesph, 2013).

M. subangulata subsp. *renigera* has extra long fruit stalk when compared with other dioecious species. The flowers of *M. dioica* have smaller petals and do not have basal blotches in their petals which are the main distinguishing character from *M. cochinchinensis* and *M. subangulata* subsp. *renigera* (Bharathi *et al.*, 2009). Among the dioecious species, *M. dioica* and *M. sahyadrica* showed close similarities for most of the traits (except for anthesis time, flower size, calyx colour and fruit size) indicating close relationship between them. Although the calyx colour and fruit morphology of *M. sahyadrica* is closer to *M. subangulata* subsp. *renigera*, petal blotch was absent at the base of petals of *M. sahyadrica* (Joseph and Antony, 2007).

Bharathi (2010) studied the morphological variations in different accessions seven *Momordica* species and recognized three groups). The first group, containing *M. charantia* (var. *charantia*, var. *muricata*) and *M. balsamina* is characterised by n = 11, annual, monoecious, non-tuberous roots and muricate—tubercled fruit surface. The second group comprised *M. dioica*, *M. sahyadrica*, *M. subangulata* subsp. *renigera* and *M. cochinchinensis* which is characterised by n = 14, perennial, dioecious, tuberous tap roots and echinate—soft papillate fruit surface. The third group contained *M. cymbalaria* which is characterised by n = 9, perennial, monoecious, tuberous tap roots and ribbed fruit surface. *M. cymbalaria* differs from the other *Momordica* species in a large number of conspicuous characters including male flowers borne in short raceme, possession of milky white corollas, minute-scar-like bracts, dry dehiscent fruits, smooth and shiny seeds. In addition, there is no fruit set even after a number of attempts to cross with all the *Momordica* spp. available in India (Bharathi *et al.*, 2011).

The morphological features of *Momordica* species (Behera *et al.*, 2011; Bharathi and Joseph, 2013) occurring in India is described below.

Monoecious Species

1. ***M. balsamina* L.:** An annual, monoecious herbaceous climber with small lobed leaves and ovoid ellipsoid softly warted fleshy fruits. It is essentially an African species with distribution extending to Western India through West Asia. It has medicinal properties, has potential as an ornamental and is a wild gathered vegetable (leaves and tender fruits).

2. ***M. charantia* L.:** An annual monoecious herb with lobed leaves and ovoid-ellipsoid to elongate fruits varying in size from 5–500 g. Fruit surface has tubercles and murication.

Chakravarty (1990) classified Indian bitter gourd into two botanical varieties based on fruit size, shape, color, and surface texture: (1) *M. charantia* var. *charantia* has large fusiform fruits, which do not taper at both ends, and possess numerous triangular tubercles giving the appearance of a "crocodile's back". It shows wide variability for fruit size, shape and color and is cultivated in whole of South and South-East Asia; (2) *M. charantia* var. *muricata* (Wild), which develops small and round fruits with tubercles, more or less tapering at each end. It occurs wild in India and parts of Nepal and also cultivated to some extent in India for its fruits having medicinal and culinary properties.

Yang and Walters (1992) classified bitter gourd into three horticultural groups or types:

1. A small-fruited type where fruit are 10 to 20 cm long, 0.1 to 0.3 kg in weight, usually dark green, and very bitter.

2. A long-fruited type (most commonly grown commercially in China) where fruit are 30 to 60 cm long, 0.2 to 0.6 kg in weight, light green in color with medium-size protuberances, and are only slightly bitter.

3. A triangular-fruited type where cone-shape fruit are 9 to 12 cm long, 0.3 to 0.6 kg in weight, light to dark green with prominent tubercles, and moderately to strongly bitter.

3. ***M. cymbalaria* Fenzl ex Naudin:** A monoecious tuberous perennial with orbicular-reniform glabrous leaves. Male flowers appear in 2–5 flowered racemes with pale yellowish white corolla. Female flowers with fusiform ovary. Fruit 1.90–2.50 cm long, up to 3 g, pyriform or broadly fusiform, narrowed into a curved peduncle, fleshy, dark green, 8-ribbed, sparsely hairy, placenta spongy to fibrous, white.

Dioecious Species

1. ***M. cochinchinensis* (Lour.) Spreng:** A dioecious, stout perennial with tuberous roots and palmately lobed gland dotted leaves (petiole). Flowers light cream to white with purple-black bulls eye mark on three inner petals. Fruits very large weighing 350–600 g or more and is used as a vegetable (sweet gourd). It occurs wild in India (Andaman Islands, northeastern states), Philippines, Vietnam, and Thailand and is cultivated in Vietnam, Japan and other Asian countries for its fruits with medicinal properties.

2. ***M. dioica* Roxb.:** A dioecious tuberous perennial with ornamental lobed or unlobed small leaves. Flowers are small lemon yellow sweetly scented and open in the evening. Fruits are small, broadly ovoid-oblong clothed with soft spines. It is a wild gathered vegetable (tender fruits) and has potential for cultivation as a crop. It is native of peninsular India.

3. ***M. sahyadrica* Joseph and Antony:** A robust dioecious perennial, rainy season climber endemic to Western Ghats of India. Leaves are triangular-cordate and has large showy yellow flowers and fairy large softly spinicent (50 g) fruits. Fruits are used as vegetable and have potential for domestication as a high value vegetable.

4. ***M. subangulata* Blume subsp. *renigera* (G. Don) W. J. de Wilde and *M. subangulata* Blume subsp. *subangulata*:** A di oecious rainy season climber, perennating with both taproot and adventitious tubers. It has two subspecies; *M. subangulata* subsp. *subangulata* and *M. subangulata* subsp. *renigera*. The former is more robust and is cultivated and wild in sub mountain Himalayas, whereas the latter is a delicate herb restricted to Malaysia and South East Asia. Both have large showy creamy flowers with purple bull's eye blotch on three inner petals. Fruits are ridged prominently in the former and remnant in the latter. Fruits and tender leaves are used as vegetable.

Species Crossability

Bharathi *et al.* (2012) reported five major patterns of crossing behaviour among the *Momordica* species of Indian occurrence *viz.*

1. Cross-compatibility with fertile hybrids: between two varieties of *M. charantia* (var. *charantia* and var. *muricata*), also between *M. dioica* and *M. sahyadrica*;

2. Partial cross-compatibility with fertile hybrids: between *M. charantia* and *M. balsamina*;

3. Cross-compatibility with sterile hybrid: [between diploid species (*M. dioca*, *M. sahyadrica*, *M. cochinchinensis*) and tetraploid species (*M. subangulata* subsp. *renigera*)];

4. Partial cross-compatibility with sterile hybrids: between *M. dioica* and *M. cochinchinensis*; *M. sahyadrica* and *M. cochinchinensis*;

5. Cross-incompatibility: between monoecious and dioecious species.

Bharathi *et al.* (2014a) studied the potential for improving *Momordica* species through inter-specific hybridization followed by restoration of fertility by backcrossing. They produced and characterized a fertile female backcross progeny derived from an inter-specific hybrid [teasel gourd (*Momordica subangulata* Blume subsp. *renigera* (G. Don) WJJ de Wilde) × spine gourd (*Momordica dioica* Roxb)] by backcrossing the F_1 with the female parent. The F_1♀ BC bears more resemblance to maternal parent than the paternal parent and exhibited substantial heterosis for number of flowers and yield per plant. In F_1♀ BC, fruit set was significantly higher when pollinated with teasel gourd (84%) while the fruit set was very low upon sibbing (36%) and pollination with spine gourd (64%). The fruits of F_1♀ BC had little variation for shape from teasel gourd when pollinated with teasel gourd pollen, while fruits obtained from sibbing and dusting of spine gourd pollen were deformed and smaller in size. The proximate and phytochemical analysis revealed that the F_1♀ BC was intermediate between the original parents in its biochemical traits and antioxidant activity while mineral content was found to be higher or similar to its parents.

Bharathi *et al.* (2014b) developed a new synthetic species of *Momordica* (2n = 56) by crossing natural tetraploid *Momordica subangulata* subsp. *renigera* (2n = 56) with induced tetraploid *Momordica dioica* (2n = 4x = 56). The hybrid produced adventitious root tubers through which it perpetuates and propagates like its female parent and maintains its morphological characteristics in the progeny. The hybrid was naturally fertile and has the superior agronomic traits of both parents making it a good choice as a new vegetable crop.

Cytology and Cytogenetics

Cytological and cytogenetic studies help in the transfer of desirable traits from related wild species to the cultivated species. Cytogenetic investigations in *Momordica* species is difficult because the chromosomes are relatively small, are usually not well separated from each other and do not stain well (Roy *et al.*, 1966; Bhaduri and Bose, 1947); pollen mother cells of these species are not amenable for manipulation using conventional cytological techniques, cytoplasm also takes stain and it is very difficult to differentiate the chromosomes from the cytoplasm (Trivedi and Roy, 1972) and in dioecious *Momordica* spp. seed germination is very poor and the method to break dormancy is not yet standardised (Bharathi and Joseph, 2013).

The genus *Momordica* is considered to be dibasic with x = 11 and 14 (Coleman, 1982; Beevy and Kuriachan, 1996). However, *M. cymbalaria* has basic chromosome number of 9 (Bharathi *et al.*, 2011). *M. cymbalaria* belong to the section Raphanocarpus which diverged from the other two sections of *Momordica* occurring in Asia some 30 million years ago and may be the reason for the cytological differences (Schaefer and Renner, 2010). The chromosome number is 2n = 2x = 22 for *M. charantia* and *M. balsamina* (McKay, 1930; Whitaker, 1933; Bhaduri and Bose, 1947; Roy *et al.*, 1966; Trivedi and Roy, 1972; Varghese, 1973; Sen and Dutta, 1975; Bharathi *et al.*, 2011); 2n = 2x = 28 for *M. cochinchinensis, M. dioica* and *M. sahyadrica* (Jha *et al.*, 1989; Richharia and Ghosh, 1953; Bharathi *et al.*, 2011).

Bharathi *et al.* (2010a) observed a polyploidy chromosome number (2n = 4x = 56) in *M. subangulata* subsp. *renigera* in contrast to the earlier report 2n = 28 by Sarkar and Majumdar (1993). Based on meiotic analysis, Roy *et al.* (1966) also reported *M. subangulata* Blume subsp. *renigera* to be an autotetraploid of *M. dioica*. Polyploidy is normally associated vigorous growth with large size of vegetative and reproductive plant parts and some cases sterility as well. Bharathi *et al.* (2011) reported 2n = 18 in *M. cymbalaria* in contrast to earlier reports of 2n = 22 (Beevy and Kuriachan, 1996) and 2n = 16 (Mehetre and Thombre, 1980). In meiotic studies at diakinesis/ metaphase I regular formation of 11 bivalents and regular anaphase I with equal segregation of 11 chromosomes to each pole have been noticed in *M. charantia* and *M. balsamina* (Roy *et al.*, 1966; Trivedi and Roy, 1972; Sen and Dutta, 1975; Jha and Trivedi, 1989); 14 chromosomes to each pole in *M. dioica* (Richharia and Ghosh, 1953; Jha and Trivedi, 1989; Trivedi and Roy, 1972; Bharathi *et al.*, 2011) and in *M. cochinchinensis* (Sen and Dutta, 1975; Bharathi *et al.*, 2011).

Chromosome size was found to be considerably uniform for the annual monoecious and perennial dioecious group with the exception of *M. cymbalaria*. Polyploidy is observed in *M. subangulata* subsp. *renigera* which had more condensed chromosomes than those of the presumed parental diploids (*M. dioica* and *M. cochinchinensis*). The average chromosome length and volume was minimum (0.93 lm and 0.48 lm3, respectively) in the tetraploid species *M. subangulata* subsp. *renigera* and maximum (2.62 lm) was found in *M. cymbalaria* (Bharathi *et al.*, 2011). One pair of satellite (SAT) chromosome was observed in *M. charantia* (Bhaduri and Bose, 1947; Bharathi *et al.*, 2011) and *M. balsamina* (Bharathi *et al.*, 2011). One pair of SAT chromosomes was observed in *M. dioica, M. sahyadrica, M. cochinchinensis* (Bharathi *et al.*, 2011) and two pairs in *M. subangulata* subsp. *renigera* (Sinha *et al.*, 1997; Bharathi *et al.*, 2011). The DNA content in *M. cochinchinensis* is 6.76 pg and *M. subangulata* subsp. *renigera* is 12.95 pg. No positive linear relation could be established between the amount of DNA and the total chromosome length (Sinha *et al.*, 1996, 1997).

M. charantia and *M. balsamina* have almost the same number of median and submedian chromosomes although the chromosomes of *M. balsamina* are slightly smaller (Trivedi and Roy, 1972). But the karyotype of *M. dioica* is more asymmetrical than that of *M. charantia* and *M. balsamina* and also differs from these species in chromosome number (Roy *et al.* 1966; Trivedi and Roy, 1972; Sinha *et al.*, 1997). Therefore, it is assumed that *M. dioica* may be the advanced taxon among these three species.

High numbers of submedian to nearly terminal chromosomes were observed in *M. subangulata* subsp. *renigera* which might be due to breakage and reunion of metacentric chromosomes and their duplication in the natural process of evolution (Bharathi *et al.*, 2011). Morphological analysis of *M. subangulata* subsp. *renigera* suggests an allopolyploid origin. Bharathi *et al.* (2010a) presented evidence elucidating the genomic relationships between *M. dioica, M. cochinchinensis* and *M. subangulata* subsp. *renigera* and reported a triploid *M. dioica* X *M. subangulata* subsp. *renigera* hybrid had an average of 12.76 bivalents, 13.84 univalents and 0.88 trivalents at metaphase I, while the *M. cochinchinensis* × *M. subangulata* subsp. *renigera* hybrid had an average of 13.08 bivalents, 12.96 univalents and 0.96 trivalents. F_1 hybrids of the two diploid species (*M. dioica* X *M. cochinchinensis*) had an average of 9.12 bivalents and 9.76 univalents, suggesting that the genomes of these species are only partially homologous. A higher number of bivalents in the triploid hybrids suggested that *M. subangulata* subsp. *renigera* is a segmental allopolyploid of *M. dioica* and *M. cochinchinensis* and that its genomes have diverged from the parental genomes.

Though cultivated bitter gourd is a diploid (2n = 22) species, natural triploid (2n = 33) has been discovered from Sikkim (Pandey and Saran, 1987). Kausar *et al.* (2015) reported tetraploid (2n = 44) and hexaploid (2n = 66) accessions in *M. charantia*. Hexaploid accessions had submetacentric and subtelocentric chromosomes. The chromosomes of hexaploid accessions were smaller than the diploid and tetraploid accessions, varying in length from 0.8 to 1.5μm. Synthetic tetraploid had also been induced by colchicine treatment of germinating seedling (Roy *et al.*, 1966) which are characterised by irregular meiosis and do not show much advantage over the diploid form in economic characters such as fruit size, number of fruits per plant, etc. During meiosis, the tetraploid species (*M. subangulata* subsp. *renigera*) showed the presence of uni-, bi-, tri- and quadrivalents with preponderance of bivalents and occasional presence of 28 bivalents in some pollen mother cells (PMC's) (Sinha *et al.*, 1997; Bharathi *et al.*, 2010a, 2011). The meiosis was highly irregular in a natural triploid *M. dioica* (Agarwal and Roy, 1976); in artificially induced colchiploid *M. charantia* (Roy *et al.*, 1966) and artificial triploids of *M. dioica* X *M. subangulata* subsp. *renigera* and *M. cochinchinensis* X *M. subangulata* subsp. *renigera* (Bharathi *et al.*, 2010b).

Prospects

Bitter gourd is an important vegetable crop and the fruit contain many bioactive compounds of medicinal importance. The other underutilized *Momordica* spp. may contribute in a significant way to the food and nutritional security and balanced diets.

Extensive collection, characterization, evaluation for resistance to biotic/abiotic stresses and conservation of wild *Momordica* genepool assumes great priority in *Momordica* improvement programmes. More effort needs to be made to understand the existing diversity pattern and phenetic relationships among the various *Momordica* spp. (both Asian and African spp.) and cultivated varieties included under Indian *Momordica* spp., using molecular markers. More molecular tools should be utilized to further elucidating the taxonomic relationship of *M. cymbalaria* with other *Momordica* species.

An understanding of complete crossability spectrum of Indian *Momordica* spp. with other Asian and African *Momordica* spp. will help in gene transfer in wide hybridization. Interspecific hybrids can be produced with introgression of genes for various traits using different species of *Momordica*. Further research on breaking the hybridization barrier and fertility restoration of inter-specific hybrid through hybridization and backcross using different *Momordica* spp. is required for genetic improvement of this underutilized genus.

Further confirmation of a segmental allopolyploid nature of *M. subangulata* subsp. *renigera* can be done by studying the inheritance pattern of various traits or by direct observation of the meiosis I (MI) pairing of individual chromosomes in the hybrids between *M. dioca* and *M. cochinchinensis*.

References

Agarwal, P.K. and Roy, R.P. 1976. Natural polyploids in Cucurbitaceae I. Cytogenetical studies in triploid *Momordica dioica* Roxb. *Caryologia*, 29:7–13.

Ames, O. 1939. Economic annuals and human cultures. Botanical Museum of Harvard University, Cambridge.

Backer, C.A. and Bakhuizen van den Brink, R.C. 1963 Flora of Java (Vol. 1). Rijks Herbarium, Leiden, 99 p.

Beevy, S.S. and Kuriachan, P. 1996. Chromosome numbers of South Indian Cucurbitaceae and a note on the cytological evolution of the family. *J. Cytol. Genet.*, 31:65–71.

Behera, T.K., Staub, J.E., Behera, S. and Simon, P.W. 2007 Bitter gourd and human health. *Medicinal and Aromatic Plant Science and Biotechnology*, 1: 224–226.

Behera, T.K., Behera, S., Bharathi, L.K., Joseph J.K., Simon, P.W. and Staub, J.E. 2010 Bitter gourd: Botany, horticulture, breeding. *Horticulture Reviews*, 37: 101–141.

Behera, T.K., Joseph J.K., Bharathi, L.K. and Karuppaiyan, R. 2011 *Momordica*. In: Kole, C. (Ed.), Wild Crop Relatives: Genomic and Breeding Resources–Vegetables. Springer, New York, USA, pp. 217–246.

Beloin, N., Gbeassor, M., Akpagana, K., Hudson, J., Soussa, K., Koumaglo, K. and Arnason, J.T. 2005 Ethnomedicinal uses of *Momordica charantia* (Cucurbitaceae) in Togo and relation to its phytochemistry and biological activity. *Journal of Ethnopharmacology*, 96: 49–55.

Bhaduri, P.N. and Bose, P.C. 1947 Cytogenetical investigation in some cucurbits with special reference to fragmentation of chromosomes as a physical basis of speciation. *J. Genet.*, 48:237–256.

Bharathi, L.K., Vishalnath, Naik, G., Joseph, J.K. and Anbu, S. 2009 Bitterless bitter gourds from nature's bounty. *Indian Hortic.*, 54:24–25.

Bharathi, L.K. 2010 Phylogenetic studies in Indian *Momordica* species. Dissertation, Indian Agricultural Research Institute, New Delhi.

Bharathi, L.K., Vinod, Munshi, A.D., Behera, T.K., Shanti, C., Kattukunnel, J.J., Das, A.B. and Vishal Nath 2010a Cytomorphological evidence for segmental allopolyploid origin of teasel gourd (*Momordica subangulata* subsp. *renigera*). *Euphytica*, 176:79–85.

Bharathi, L.K., Munshi, A.D., Behera, T.K. and Joseph, J.K. 2010b Relationship among Indian species of *Momordica* based on crossability studies. In: Abstracts of 4th Indian Horticulture Congress, IARI, New Delhi, 18–21 November 2010, pp. 271.

Bharathi, L.K., Munshi, A.D., Vinod, Shanti, C., Behera, T.K., Das, A.B., Joseph, J.K. and Vishal Nath 2011. Cyto-taxonomical analysis of *Momordica* L. (Cucurbitaceae) species of Indian occurrence. *J. Genet.*, 90:21–30.

Bharathi, L.K., Munshi, A.D., Behera, T.K., Vinod, Joseph, J.K., Bhat, K.V., Das, A.B. and Sidhu, A.S. 2012. Production and preliminary characterization of novel inter-specific hybrids derived from *Momordica* species. *Curr. Sci.*, 103:178–186.

Bharathi, L.K. and Joseph J.K. 2013. *Momordica* Genus in Asia: An Overview. Springer, New Delhi, 147 p.

Bharathi, L.K., Singh, H.S., Shivashankar, S. and Ganeshamurthy, A.N. 2014a. Characterization of a fertile backcross progeny derived from inter-specific hybrid of *Momordica dioica* and *M. subangulata* subsp. *renigera* and its implications on improvement of dioecious *Momordica* spp. *Scientia Horticulturae*, 172: 143–148.

Bharathi, L.K., Singh, H.S. and Joseph J.K. 2014b A novel synthetic species of *Momordica* (*M.* x *suboica* Bharathi) with potential as a new vegetable crop. *Genet. Resour. Crop Evol.*, 61: 875–878.

Blatter, E. 1919. Flora of Arabica, Rec Bot Surv India, Vol. viii, No. 1 (repr. 1978) Bishen Singh Mahendrapal Singh Publishers, Dehradun, 200 p.

Blume, C.L. 1826. Bijdragen tot de flora van Nederlandsch Indie. Ter Lands Drukkerij, Batavia, pp. 927–940.

Chakravarty, H.L. 1959. Monograph of Indian Cucurbitaceae. *Rec. Bot. Surv. India*, 17:81.

Chakravarty, H.L. 1982. Fascicles of flora of India II. Cucurbitaceae (Taxonomy and distribution). *Rec. Bot. Surv. India,* Howrah 17:1–234.

Chakravarty, H.L. 1990. Cucurbits of India and their role in the development of vegetable crops. In: Bates, D.M., Robinson, R.W. and Jeffrey, C. (Eds.), Biology and Utilization of Cucurbitaceae. Cornell Univ. Press, Ithaca, NY. pp. 325-334.

Clarke, C.B. 1879. Cucurbitaceae. In: Hooker, J.D. (Ed), Flora of British India. Reeve, London, pp. 604–635.

Coleman, J.R. 1982. Chromosome numbers of angiosperms collected in the state Sao Paulo. *Rev. Brasil Genet.*, 3:533–549.

Cooke, T. 1901–1908. The flora of presidency of Bombay, Vol. 1–3, London, Rep. ed. 1958, Kolkata.

De Candolle, A.P. 1828. Pandromus Systematis Naturalis Regni Vegetabilis.

Decker-Walters, D.S. 1998. Sanskrit, modern Indo-Aryan, and Dravidian names for cucurbits. Occasional Papers of the Cucurbit Network No. 1. The Cucurbit Network, Miami.

Decker-Walters, D.S. 1999. Cucurbits, Sanskrit, and the Indo-Aryas. *Econ. Bot.*, 53:98–112.

Degner 1947. Flora Hawaiiensis. Book 5. Privately Published, Honolulu.

De Wilde, W.J.J.O. and Duyfjes, B.E.E. 2002. Synopsis of *Momordica* (Cucurbitaceae) in SE-Asia and Malesia. *Bot. Z.*, 87:132–148.

Dey S.S., Singh, A.K., Chandel, D. and Behera, T.K. 2006. Genetic diversity of bitter gourd (*Momordica charantia* L.) genotypes revealed by RAPD markers and agronomic traits. *Scientia Horticulturae*, 109: 21–28.

Durry, H. 1864. Handbook of Indian Flora, Vol 1. Bishen Singh and Mahendrapal Singh Publishers, New Delhi.

Gaikwad, A.B., Behera, T.K., Singh, A.K., Chandel, D., Karihaloo J.L. and Staub, J.E. 2008. AFLP analysis provides strategies for improvement of bitter gourd (*Momordica charantia* L.). *HortScience*, 43: 127–133.

Gamble, J.S. and Fischer, C.E.C. 1919. Flora of the presidency of Madras (repr. edn. vol 1-3), Botanical Survey of India, Calcutta, 1957.

Genaust, H. 1996. Etymologisches Worterbuch der botanischen Pflanzennamen. Birkhauser Verlag.

Grover, J.K. and Yadav, S.P. 2004. Pharmacological actions and potential uses of *Momordica charantia*: A review. *Journal of Ethnopharmacology*, 93: 123–132.

Henderson, M.R. 1974. Malayan wild flowers, Dicotyledons. The Malayan Nature Society, Kualalampur, 156 p.

Hooker, J. 1871. *Momordica* L. In: Oliver, D. (Ed.) Flora of Tropical Africa, Vol. 2. Reeve & Co, London, pp. 534–540.

Jeffrey, C. 1980. A review of the Cucurbitaceae. *Botanical Journal of the Linnean Soc.*, 81:233-247.

Jeffrey, C. 1990. An outline classification of the Cucurbitaceae (Appendix). (In): Bates, D.M., Robinson, R.W. and Jeffrey, C. (Eds.), Biology and Utilization of the Cucurbitaceae. Cornell Univ. Press, Ithaca, New York. pp. 449-463.

Jha, U.C., Dutt, B. and Roy, R.P. 1989. Mitotic studies in *Momordica cochinchinensis* (Lour.) a new report. *Cell and Chromosome Research*, 12: 55–56.

Jha, U.C. and Trivedi, S.K. 1989. Meiotic analysis in three species of *Momordica* (Cucurbitaceae). In: Proceedings of the Indian Science Congress Association, Vol. 76, pp. 185.

Joseph, J.K. 2005. Studies on ecogeography and genetic diversity of the genus *Momordica* L. in India. Dissertation, Mahatma Gandhi University, Kottayam, Kerala.

Joseph, J.K. and Antony, V.T. 2007. *Momordica sahyadrica* sp.nov. (Cucurbitaceae), an endemic species of Western Ghats of India. *Nord. J. Bot.*, 24:539–542.

Joseph, J.K. and Antony, V.T. 2010. A taxonomic revision of the genus *Momordica* L. (Cucurbitaceae) in India. *Indian J. Plant Genet. Resour.*, 23:172–184.

Kajale, M.D. 1991. Current status of India palaeoethnobotany: Introduced and indigenous food plants with a discussion of the historical and evolutionary development of India agriculture and agricultural systems in general. (In): Renfrew, J.M. (Ed.), New Light on Early Farming. Recent Developments in Palaeoethnobotany. Edinburgh University Press, Edinburgh, UK. pp. 155–189.

Kanjilal, U.N., Kanjilal, P.C. and Das, A. 1938. Flora of Assam, Vol. 2. Government of Assam, Shillong, 409 p.

Kausar, N., Yousaf, Z., Younas, A., Sadia Ahmed, H., Rasheed, M., Arif, A. and Rehman, H.A. 2015. Karyological analysis of bitter gourd (*Momordica charantia* L., Cucurbitaceae) from Southeast Asian countries. *Plant Genetic Resources: Characterization and Utilization* 13(2): 180–182.

Keraudren, A.M. 1975. Flore du Combodge du Laos et du Viet-Nam. 15. Cucurbitaceae, pp. 36–44.

Li, H.L. 1970. The origin of cultivated plants in Southeast Asia. *Econ. Bot.*, 24:3–19.

Linnaeus, C. 1753. Species Plantarum (2 vols). Stockholm, a facsimile of the first ed published by Ray Society, London, in 1959.

Marr, K.L., Xia, Y.M. and Bhattarai, N.K. 2004. Allozyme, morphological and nutritional analysis bearing on the domestication of *Momordica charantia* L. (Cucurbitaceae). *Econ. Bot.*, 58:435-455.

Mathew, K.M. 1981. The flora of the Tamil Nadu Carnatic, vol 3, Rapinat herbarium. St. Joseph's College, Tiruchirapally, 648 p.

Mathew, K.M. 1983. Illustrations on the flora of Tamil Nadu Carnatic, vol 3, Rapinat herbarium. St. Joseph's College, Tiruchirapally, 648 p.

McKay, J.W. 1930. Chromosome numbers in the Cucurbitaceae. *Bot. Gaz.*, 89:416–417.

Mehetre, S.S. and Thombre, M.V. 1980. Meiotic studies in *Momordica cymbalaria* Fenzl. *Curr. Sci.*, 49:289.

Morton, J.F. 1967. The balsam pear: an edible, medicinal and toxic plant. *Econ. Bot.*, 21:57–68.

Okabe, H., Miyahara, Y. and Yamauchi, T. 1982. Studies on the constituents of *Momordica charantia* L. III. *Chem. Pharm. But.*, 30:3977-3986.

Pandey, V.S. and Saran, S. 1987. A natural triploid in *Momordica charantia* L. *Cell Chrom. Res.*, 11:33–35.

Raj, N.M., Prasanna, K.P. and Peter, K.V. 1993. Bitter gourd. (In): Kalloo, G. and Bergh, B.O. (Eds.), Genetic Improvement of Vegetable Plants. Pergamon Press, Oxford, pp. 239–246.

Raman, A. and Lau, C. 1996. Anti–diabetic properties and phytochemistry of *Momordica charantia* L. (Cucurbitaceae). *Phytomedicine*, 2: 349–362.

Reyes, M.E.C., Gildemacher, B.H. and Jansen, G.J. 1994. *Momordica* L. pp. 206-210. (In): Siemonsma, J.S. and Piluek, K. (Eds.), Plant resources of South-East Asia: Vegetables. Pudoc Scientific Publishers, Wageningen, the Netherlands.

Richharia, R.H. and Ghosh, P.N. 1953. Meiosis in *Momordica dioica* Roxb. *Curr. Sci.*, 22:17–18.

Robinson, R.W., and Decker-Walters, D.S. 1997. Cucurbits. CAB International, Wallingford, Oxford, UK.

Roy, R.P., Thakur, V. and Trivedi, R.N. 1966. Cytogenetical studies in *Momordica*. *J. Cytol. Genet.*, 1:30–40.

Roxburgh, W. 1832. Flora indica or descriptions of Indian plants. Today and Tomorrow Publishers (repr.edn.), New Delhi.

Saldanha, C.J. and Nicholson, D. 1976. Flora of Hassan District, Karnataka. Amarind Publishers, New Delhi.

Sands, W.N. 1928. The bitter cucumber of Paris. *Malayan Agric. J.*, 16:32–39.

Santapau, H. 1953. Flora of Khandala on the Western Ghats of India. *Rec. Bot. Surv. India*, 16:1–372.

Sarkar, D.D. and Majumdar, T. 1993. Cytological and palynological investigations in *Momordica subangulata* (Cucurbitaceae). *J. Econ. Tax. Bot.*, 17:151–153.

Schaefer, H. and Renner, S.S. 2010. A three-genome phylogeny of *Momordica* (Cucurbitaceae) suggests seven returns from dioecy to monoecy and recent long distance dispersal to Asia. *Mol. Phylogenet. Evol.*, 54:553–560.

Schaefer, H., Heibl, C. and Renner, S.S. 2009. Gourds afloat: a dated phylogeny reveals an Asian origin of the gourd family (Cucurbitaceae) and numerous oversea dispersal events. *Proc. R. Soc. B.*, 276:843–851.

Sen, R. and Dutta, K.B. 1975. Sexual dimorphism and polyploidy in *Momordica* L. (Cucurbitaceae). **(In)**: *Proceedings of the 62nd Indian Sci Cong,* Part III, pp. 127.

Seringe, N.C. 1828. Cucurbitaceae. (In): De Candolle, A.P. (Ed.) Prodromus Systematis Naturalis Regni Vegetabilis, Vol. 3. Parisiis: Treuttel & Wurtz, Paris, pp. 311–312.

Singh, A.K., Behera, T.K., Chandel, D., Sharma, P. and Singh. N.K. 2007. Assessing genetic relationships among bitter gourd (*Momordica charantia* L.) accessions using inter simple sequence repeat (ISSR) markers. *Journal of Horticultural Science and Biotechnology*, 82: 217–222.

Sinha, S., Debnath, B. and Sinha, R.K. 1996. Karyological studies in dioecious *Momordica cochinchinensis* with reference to average packing ratio. *Cytologia*, 61:297–300.

Sinha, S., Debnath, B. and Sinha, R.K. 1997. Differential condensation of chromosome complements of dioecious *Momordica dioica* Roxb. in relation to DNA content. *Indian J. Exp. Biol.*, 35:1246–1248.

Thwaites, G.H.K. 1864. Enumeratio Plantarum Zeylaniae (An enumeration of Ceylon Plants). Dulau, London, 126 p.

Tindall, H.D. 1983. Vegetables in the Tropics. Macmillan, London.

Trimen, H. 1893–1900. A hand book of flora of Ceylon, Vol. 1–5. Dulau & Co, London, pp. 248–250.

Trivedi, R.N. and Roy, R.P. 1972. Cytological studies in some species of *Momordica. Genetica*, 43:282–291.

Turner, R.L. 1966. A comparative dictionary of the Indo-Aryan languages. Oxford University Press, London.

Van Rheede, H.A. 1688. *Hortus Malabaricus*, Vol. 8. Joannis VS, Joannis DV (repr. ed. 2003), Amsterdam, pp. 17–36.

Varghese, B.N. 1973. Studies on cytology and evolution of South Indian Cucurbitaceae. Dissertation, Kerala University, Thiruvananthapuram.

Walters, T.W. and Decker-Walters, D.S. 1988. Balsam pear (*Momordica charantia*, Cucurbitaceae). *Econ. Bot.*, 42:286–288.

Whitaker, T.W. 1933. Cytological and phylogenetic studies in the Cucurbitaceae. *Bot, Gaz.*, 94:780–790.

Wight, R. and Arnott, G.A.W. 1841. Prodromus Florae Peninsulae India Orientalis. pp. 348.

Willdenow, C.L. 1805. Linnaei Species Plantarum, Vol 4. GC Nauk, Berolini, Vienna, pp. 601–605.

Yang, S.L. and Walters, T.W. 1992. Ethnobotany and the economic role of the Cucurbitaceae of China. *Econ. Bot.*, 46:349–367.

Yen, D.E. 1977. Hoabinhian horticulture? The evidence and the questions from Northwest Thailand. In: Allen, J., Golson, J. and Jones, R. (Eds.) Prehistoric studies in Southeast Asia, Melanesia and Australia. Academic Press, London, pp. 567–599.

Zeven, A.C. and de Wet, J.M.J. 1982. Dictionary of Cultivated Plants and their Regions of Diversity. Duckworth, London.

14

Onion

R P Gupta, R K Singh and V Mahajan

Onion is one of the most important vegetable crops spread worldwide, have both the food and medicinal values and grown in India in almost every state by big and marginal producer as well as also in kitchen garden. Because of its high export potential it comes under cash crop apart from vegetable (Gupta and Singh, 2010). It is used as spices, condiments and vegetables almost daily in every kitchen as a seasoning for wide varieties of dishes. Because of its special characteristics of pungency it has more value than other vegetables crops. It is one of the few versatile vegetable crops that can be kept for a fairly long period and can be safely withstand the hazards of rough handling including long distance transportation. This crop is known to the mankind since prehistoric times and found in the religious books like Bible, Koran and the inscriptions of ancient civilizations of Egypt, Greece and China. Greeks and Roman were using onion since 400-300 BC. In India, onion is grown since ancient times and it is known to have been cultivated since 600 BC. The Sanskrit name of onion is *Palandu*. The medicinal value of onion has been described in Charak Samhita, around 6th Century AD. Sanskrit language equivalent signifying Vedic period and Aryan usage is available for onion as "*Palandu*". Onion have biochemical components, *viz.*, thiosulfinates, thiosulfonates, allicin, aliin, ajoene and many others make them exclusive medicinal commodities too. The problems of heart diseases, rheumatism, cancer, digestive disorders, blood sugar and prolonged cough are known to be resolved by regular consumption of onion and garlic. The pesticidal and fungicidal properties of garlic are well studied and widely accepted.

It is widely cultivated for internal consumption as well as for the export. About 85.480 million tones of onions are produced in the world from 4.44 million ha area (FAO, 2015). Onion has primary centre of origin in Central Asia and secondary centre in the near East (McCollum, 1976).

General Characteristics of Genus *Allium*

Allium is a bulbous herb with characteristic odour. Leaves variable, fistular, flat, elliptic filiform or linear. Flowers white, yellow, red, rosy or purple, golden yellow, blue in capitates umbles, enclosed in 1-3 membranous spathes, stellate. Petals 6, free or fused into a corolla tube at the base. Stamens hypogynous or insereted on the perianth, filaments free or connate below, anther oblong. Ovary 5- fused carpels, 3- chambered each with many ovules. Style-1, filiform. Fruit a capsule, 3-valved, rarely a fleshy berry, small. Seeds few, compressed, black. It is very close to the family *Iridaceae*, distinguished by its 6 stamens. It is also distinguished from *Liliaceae* by a superior ovary. Flowers are surrounded in the bud by a papery spathe or flowers borne in branched clusters (Purseglove 1972, Polunin & Stainton 1984,).The genus *Allium* as a whole is assigned to the family "Liliacea" by some taxonomists (Melchior, 1964, Polunin, 1969) on account of the superior ovary, and to the Amryllidaceae by others (Traub, 1968) due to the umbellate inflorescence. Because of superior ovary (characteristic of family Liliaceae) and scapose umbellate inflorescence (flowers borne in a bracted umbel on top of a scape) with membranous bracts (characteristic of family Amaryllidaceae), Alliums have been placed in family Alliaceae by Purseglove (1972). Genus *Allium* L. includes more than 700 species, which wildly grow in the temperate, semi arid and arid regions of the northern hemisphere hence, results in a remarkable polymorphism and is one of the largest monocotyledonous genera (Hanelt *et al.*,1992 and Fritsch *et al.*, 2001). In the most recent and competent taxonomic classifications of the monocotyledons, *alliums* and its close relatives are recognised as a distinct family, the *Alliaceae*, close to *Amaryllidaceae*. The following classification of *Allium* has been adopted (Takhtajan, 1997).

Class:	*Liliopsida*
Sub class :	*Liliidae*
Order:	*Amaryllidales*
Super order:	*Liliianae*
Family:	*Alliaceae*
Sub family:	*Allioideae*
Tribe:	*Allieae*
Genus:	*Allium*
Species:	*cepa*

Cytogenetics Limitations

In the genus as a whole there are three basic chromosome number groups x=7, x=8, and x=9. The x=7 species are largely confined to the New World, while the majority of species (500) which are in the Old World are mainly x=8. There are only a few species with x=9. All of the cultivated edible forms have the same basic chromosome number of x=8 (Rabinowitch and Brewster, 1990). By way of general remarks it can be said at the outset that *Alliums* have large chromosomes and that their complements display a high degree of symmetry and uniformity. The onion nuclear genome is notable for its great size, 15 Gbp per 1C (Ricroch *et al.* 2005), one of the largest among cultivated plants. The 2C value for the species is 33.5 pg (Jones and Rees, 1963). Most of the chromosomes have median or sub median ceritromeres and there is a gradation in size from the longest down to the shortest member of the complement. Four of the seven species are polyploids, or have a polyploid series and there is much of interest to say about their chromosome behaviour at meiosis and about the use of species hybrids to mediate gene transfer for the purpose of crop improvement.

Allium cepa (2*n*=2x=16)

Alliums are bulbous biennial or perennial herbs which give off a distinctive and pungent odour when the tissues are crushed. *Allium cepa* is a diploid (2n=2x=16) with eight pairs of chromosomes (Stack and Coming, 1979). *A. cepa* is now a day considered as divisible into two large cultivar-groups: Common Onion Group, with large, normally solitary bulbs, reproduced from seed or from seed-grown bulblets ('sets'), and Aggregatum Group, with smaller, several bulbs forming an aggregated cluster, originating from a single mother bulb. A cultivar of *A. cepa* will be considered as a 'shallot' (belonging to Aggregatum Group) if more than 200 buds are present in one kg of mother bulbs, and if, under suitable climatic conditions, most of these buds give new bulbs (Messiaen and Rouamba, 2004). The species is an out breeder and current improvement programs utilize hybrid varieties as well as open pollinated selections. A critical review of the use of hybrid varieties is given in (Dowker and Gordon, 1983). Modern varieties sold by international seed companies, in particular F_1 hybrids which have a narrow genetic base, are replacing old varieties. Therefore, there is danger that old landraces with potentially valuable and adaptive genes, would be lost. The risk of genetic erosion due to the introduction of single new cultivars was considered especially high by Crisp and Astley (1985), for those vegetables that are built on a narrow genetic base. *A. cepa* is diploid and has a complement of eight pairs of metacentric or sub-metacentric chromosomes. As far as is known, all existing varieties and cultivars are diploid and it has evidently not yet been possible to exploit polyploidy in onion breeding. An early attempt at chromosome doubling with colchicine has been reported by (Toole, 1944), but it seems that chromosome pairing in the auto-tetraploids was highly irregular and the seed set poor.

Cultivated Forms of *Allium*

Species	Chromosome No.	Common name
Allium cepa L.	2n=2x=16	Onion
A. cepa var.*ascalonicum* Backer	2n=2x=16	Shallot
A. cepa var.*aggregatum* G. Don	2n=2x=16	Potato/multiplier onion
A. cepa var.*viviparum*(Metz) Mansf. / *proliferum*	2n=2x=16	Tree/Egyptian tree
A. sativum L.	2n=2x=16	Garlic
A. fistulosum L.	2n=2x=16	Welsh onion or Japanese bunching onion
A. porrum L.	2n=4x=32	Leek
A. kurrat Sfth. & Krause	2n=4x=32	Kurrat
A. ampeloprasum L.	2n=4x=32, 6x=48	Great headed garlic
A. schoenoprasum L.	2n=2x=3x=4x2	Chive
A. tuberosum Rottl. ex Spreng.	2n=4x=32	Chinese chives
A. chinense G. Don	2n=2x=16, 4x=32	Rakkyo

(Shinde and Sontakke, 1993)

Early History

This is especially true for long-living seed crops such as cereals, but much less so far for species the bulb onion, which have little chance of long-term preservation. Therefore, one has to rely mostly upon written records, carvings and paintings. Hence, the picture one obtains of the history of such species is fragmentary, at least for the earlier epochs. The conventional wisdom on the history of cultivation of the common onion has been summarized by Helm (1956), Jones and Mann (1963), Kazakova (1978) and Havey (1995) and was briefly discussed by Hanelt (1990).

Though several *Allium* species are found in Himalayas. In recent decades some of them appear to be rare, endemic or extinct. Since the work of Hooker (1892 & 1973), who compiled information on *Allium* species for the first time and reported 27 species of *Allium* in the *Flora of British India*, no other such work is available. At present information on some regional floras is available, but it also seems to be incomplete. Because onions have been cultivated for so long, and their bulb and inflorescence development must be closely adapted to the temperatures and photoperiods that prevail where they are grown, there exists a huge range of cultivars and landraces, developed over the centuries to fit the diverse climates and food preferences of the world (Brewster, 1994). Most of the 600 species of *Allium* are distributed in Afghanistan, Turkey, Iran and central Asia; Turkmen SSR, Uzbek SSR, Tadzhik SSR, Kirgiz SSR and Kazakh SSR; and Mongolia, the Tien-Shan Mountains and the Himalayas (Kotlinska *et al.*, 1990).

The domesticated forms which are grown for food are seven main species only in number. There are a number of minor edible species which are grown in some parts of the world; the several species which are eaten from the wild and many more are grown as ornamentals. *Alliums* have been under domestication for many centuries and they exhibit a wide diversity of forms which reflect numerous generations of selection in differing climates and cultural conditions. Only chives *(A. schoenoprasum)* and Chinese chive *(A. tuberosum)* resemble their wild forms and the origins of the others are largely unknown-except to say that their centres of origin are in the Near East and Far Eastern Asia.

Onion has been cultivated since more than 5000 years on the earth and thought to have been first domesticated in the mountainous region of *North West India, Afghanistan, Pakistan, Tajikistan* and *Uzbekistan. Western Asia and* the areas around the Mediterranean seas seem to be the secondary centres of origin (Castell and Portas, 1994). There exists a huge range of cultivars and landraces, developed over the centuries to fit the diverse climate and food preferences of the world (Brewster, 1994).

The closest among cultivated *Allium cepa* L. are *A. vavilovii* Popov. & Vved. from southern Turkmenistan and northern Iran and *A. asarense* R.M. Fritsch & Matin from Iran. *A. oschaninii* O.Fedtsch. (Uzbekistan and neighbouring countries), is considered to be the ancestor, though cannot be crossed successfully with *A. cepa* L. The *alliums* are distributed widely through temperate, warm temperate and boreal zone of the northern hemisphere. It is supposed that from Central Asia, the onion ancestor probably migrated first towards Mesopotamia, where onion is mentioned in Sumerian literature (2500 B.C.), then to Egypt (1600 B.C.), India and South East Asia. From Egypt *A. cepa* was introduced into Mediterranean area and from there to all Roman Empire. According to Vavilov, (1926) Southwest Asian gene centre is proposed as primary centre of domestication and variability of onion. Further, Vavilov and Burkinich (1929) confirmed that Afghanistan and adjacent countries is the genetic centre of origin of the cultivated forms of onion and garlic.

Adaptation of onion in India occurred from very early times before Christian era. Originally native of central Asia of temperate region with perennial/biennial habit and long day character, it has established well in India under tropical and short day (11-11.5 hrs) photoperiodic conditions (Seshadri and Chatterjee, 1996). Demand for highly pungent and pink skinned bulbs from Gulf countries made farmers of western India to select such types, which can produce seeds under same climatic conditions.

The current exploitation of *A. pskemense* can be used as an illustration or how early cultivation of the onion might have started. This species is consumed by inhabitants of the Pskem and Chatkal valleys, who frequently transplant it from the wild to their gardens, where it is cultivated and propagated (Levichev and Krassovskaja, 1981). Perhaps, thousands of years ago, over collecting made bulbs of the onions ancestor scarce, thus stimulating their transfer into gardens and so initiating domestication (Hanelt, I986a). Further human and natural selection probably favoured a change in allometric growth pattern towards bulbs and shortening the life cycle of the plants to biannual nature and adaptation to any environments (Hanelt, 1990).

In India, there are reports of onion in writings from the 6th century BC. In the Greek and Roman Empires, it was a common cultivated garden plant. Its medicinal properties and details on cultivation and recognition of different cultivars were described. It is thought that the Romans, who cultivated onions in special gardens (*cepinae*), took onions north of the Alps, as all the names for onion in West and Central European languages are derived from

Latin. Different cultivars of onion are listed in garden catalogues from the 9th century AD, but the onion became widespread as a crop in Europe only during the Middle Ages and was probably introduced into Russia in the 12th or 13th century.

The onion was among the first cultivated plant taken to the Americas from Europe, beginning with Columbus in the Caribbean. Later it was imported several times and established' in the early 17th century in what is now the Northern USA. Europeans took the species to East Asia during the 19th century. The indigenous cultivated species of this region, specially an *A. Fistulosum*, are still more widespread and popular for culinary uses there. This history of cultivation applies solely to the common onion group. The Aggregatum group is poorly documented in historical record. Most probably, the "Ascalonian onions" of the authors of antiquity were not shallots. The first reliable records are from the 12th and 13th centuries in France and 16th and 17th centuries in England and Germany. In the herbals of that time, there are good illustrations of this group (Helm, 1956).

Cultivated *Allium* Species and their Areas of Cultivation

Botanical names	Other names	Area of cultivation	English names
A. *altaicum* Pall.	A. *microbulbum* Prokh.	South Siberia	Altai onion
A. *ampeloprasum* L. Leek group	A. *porrum* L. A. *ampeloprasum* var. *porrum* (L.) J. Gay	Mainly Europe, North America	Leek
Kurral group	A. *kurrat* Schweinf. ex Krause	Egypt and adjacent areas	Kurrat. salad leek
Great-headed garlic group	A. *ampeloprasum* var. *holmense* (Mill.) Aschers. et Graebn.	Eastern Mediterranean. California	Great-headed garlic
Pearl-onion group	A. *ampeloprasum* var. *sectivum* Lued.	Atlantic and temperate Europe and Iran	Pearl onion
Taree group			Taree irani
A. canadense L.		Cuba	Canada onion
A. cepa L.			
Common onion group	*A. cepa* ssp. *Cepa/var. cepa,*	Worldwide	
Ever-ready onions	*A. cepa* ssp. *australe* Kazakova		Onion,
Aggregatum group	*A. cepa* var. *perutile* Steam		common onion
	A. ascalonicum auct. hort.,	Great Britain Nearly	Ever-ready *onion*
	A. cepa var. *aggregatum*	worldwide	Shallot.
	G. Don, var. *ascalonicum*		Potato onion.
	Backer,		multiplier onion
	ssp. *orientalis* Kazakova		
A. *consanguineum* Kunth		North-East India	
A. x *cornutum* Clem. ex Vis.*	*A. cepa* var. *viviparom* auct.*	Locally in South Asia, Europe, Canada, Antilles	Rakkyo, Japanese scallions
A. *chinense* G. Don	*A. bakeri* Regel	China, Korea, Japan, South-East Asia	Japanese bunching onion, Welsh onion
A. *fistulosum* L.		East Asia. temperate Europe and America	

Contd...

Botanical names	Other names	Area of cultivation	English names
A. hookeri Thw.		Bhutan, Yunnan, North-West Thailand Mexico	
		China, Korea, Japan	Chinese garlic. Japanese garlic Naples
A. *kunthii* G. Don	A. *longifolium* (Kunth) Humb.	Central Mexico	garlic
A. *macrostemon* Bunge	A. *uratense* Franch., *A. grayi*	West and South Siberia,	
A. *neapolitanum* Cyr.	Regel	Russia. Ukraine	
A. *nutans* L.	A. cowanii Lindl.		
A. *obliquum* L.		West Siberia. East Europe	Oblique onion
		France. Italy	French shallot*
A. *oschaninii* O. *Fedtsch.*			
A. x *proliferum* (Moench) Schrader			
East Asian group	A. *aobanum* Araki, *A. wakegi* Araki	China, Japan, South- East Asia	Wakegi onion
Eurasian group		North America, Europe,	Top onion, tree on-
	A. *cepa* var. *viviparum* (Metzg.) Alef., *A. cepa* var. *proliferum* (Moench) Alef.	North. East. Asia	ion, Egyptian onion, Catawissa onion
A. *pskemense* *B. Fedtsch.*		Uzbekistan, Kyrgyzstan, Kazakhstan	
ramsum L.	*A. odorum L., A. tuberosum Rottl. ex Sprengel*	China and Japan, world-wide now	Chinese chive, Chinese leek
A. rotundum L.	*A scorodoprasum ssp.rotundum* (L.) Stearn	Turkey	
A.sativum L.			Garlic
Common garlic group	*A. sativum var. sativum,*	Mediterranean area, also worldwide	
Longicuspis group	*A. sativum var. typicum Regel*		
	A. longicuspis Regel	Central to South and East Asia	
Ophioscorodon group	*A sativum var. ophioscorodon* (Link)Doll	Europe, also worldwide	
A. schoenoprasum L.	*A. sibiricum* L.	Worldwide in temperate	Chive
A. ursinum L.		Central and North Europe	Ramsons
A. victorialis L.	*A. microdictyon* Prokh., *A. ochotense* Prokh. *A. platyphyllum* Diels, *A. lancifolium* Steam	Caucasus, Japan, Korea, Eirope (formely) East Tibet	Long-root onion, Long-rooted garlic

(Fritsch and Friesen, 2002)

Recent History

In India systematic breeding programme was started as early as 1960 at Niphad, Nashik and later on ICAR-Indian Agricultural Research Institute (IARI), New Delhi and ICAR-Indian Institute of Horticultural Research (IIHR), Bangalore apart from these several agricultural universities initiated onion breeding programs. National Horticultural Research and Development Foundation (NHRDF), Nashik was established on November 3rd, 1977 under the Society Registration Act, 1860 at New Delhi for carrying out research and development activities on export oriented crops to begin with onion and garlic. NHRDF developed promising varieties and technology on several aspects. Development of multiplier onion varieties was done by Tamil Nadu Agricultural University (TNAU), Coimbatore. Prior to this, research on collection and maintenance of land races and standardization of agro-techniques was attempted by State Agricultural Departments. With the concept of coordinated projects, network project and Agricultural Universities, the work on onion research was strengthened, in terms of varietal development for different seasons and standardization of production techniques in early nineties. The Research and Development in onion got impetus with the establishment of National Research Center on Onion and Garlic at Nashik in 1994. This center was shifted to present location at Rajgurunagar in 1998 and upgraded to Directorate level with the addition of All India Network Research Project on Onion and Garlic in 2009.

Open Pollinated Varieties

Breeding programmes are used widely to improve onion quality and productivity. An important aspect of onion breeding programs involves the development of new varieties, with traits such as resistance to major pests and diseases and variations in bulb shape, colour and firmness. The breeding of new varieties containing higher concentrations of health-beneficial compounds is on the horizon with the development of analytical methodology and molecular genetics. A major contribution to onion breeding has been the development of new open pollinated varieties using a range of population improvement method. Open pollinated varieties are genetically variable populations, which are maintained and multiplied by mass pollination in isolation. Tremendous amount of variability is being utilized in onion using various breeding procedure such as mass selection, family selection, recurrent selection, reciprocal recurrent selection and pedigree selection.

The NHRDF initiated its research and development work with onion germplasm collections from different areas and locations. The germplasm collections at the institute were considerably augmented by introducing 862 accessions from exotic and indigenous sources in onion. The significant progress has taken place during the past three and half decades in development of nine high yielding varieties. During the course of evaluation/investigation/study of germplasm several genotypes selected for further recommendation for varieties development (Bhonde, *et al.*1992, Dubey *et al.*, 2004, Singh, *et al.* 2010, Singh and Gupta, 2010, Singh and Gupta, 2011, Singh and Dubey, 2011, Singh and Gupta, 2011, Singh and Gupta, 2014.)

The improvement of onion as like in other crops has not received much attraction of the breeders in India, perhaps, because of biennial habit of the crop requiring longer time for breeding and difficulties in attaining and maintaining genetic uniformity due to high nature cross pollination and rapid inbreeding depression. Besides, lack of facilities for storage of selected bulbs of breeding lines in controlled storage conditions is another factor for slow progress in onion breeding programme (Swarup, 1991). Organized breeding efforts are being made mainly in nine research centres of NARS *viz.*, ICAR-DOGR (Rajgurunagar), ICAR-IIHR (Bangalore), ICAR-IARI (New Delhi), NHRDF (Nashik), MPKV (Rahuri), PAU (Ludhiana) and TNAU (Coimbatore) in short day onion and at ICAR-CITH (Srinagar), ICAR-VPKAS (Almora) for long day onion. The onion breeding depends to some extent upon an understanding of the inheritance of desired traits, linkage relationship, and cytogenetics of interspecific hybridization. In onion need for breeding is mainly for (a) seasonal adoptability (b) processing quality (c) biotic and abiotic stress (d) export quality (e) short life cycle (f) green foliage (g) better keeping quality (h) ecological versatility and (i) high seed yields.

Open Pollinated Varieties Developed in India

In India till now 70 varieties of onion including 2 F1 hybrids and 6 varieties of multiplier onion have been developed and released for different colour of bulb and seasons by various organizations.

Agriculture Department Maharashtra: N-53, N-2-4-1 and N-257-9-1. MPKV, Rahuri: Baswant -780, Phule Safed, Phule Suvarna and Phule Samarth. ICAR-IARI, New Delhi: Pusa White Flat, Pusa White Round, Early Grano (Long Day), Brown Spanish (Long Day), Pusa Red, Pusa Ratnar, Pusa Madhavi, and Selection-126. ICAR-IIHR, Bangalore: Arka Pragati, Arka Niketan, Arka Kalyan, Arka Pitamber, Arka Bindu, Arka Ujjwal, (multiplier onion), Arka Swadista, Arka Vishwas (Rose onion), Arka Sona, Arka Bhim (tri-parental synthetic) and Arka Akshay (tri-parental synthetic), Arka Lalima (F_1), Arka Kirtiman (F_1). HAU, Hissar: Hissar-2 and HOS-1. NHRDF, Nashik: Agrifound Dark Red, Agrifound Light Red, NHRDF Red, NHRDF Red-2, NHRDF Red-3, NHRDF Red-4, Agrifound Rose, Agrifound Red (Multiplier) and Agrifound White. VPKAS, Almora: VL-67 (Long Day) and VL-3 (Long Day). RAU, Rajasthan: Udaipur 101, Udaipur 102 and Udaipur 103. PDKV, Akola: PKV White. GAU, Junagarh: Gujarat White Onion (GWO)-1. CSAUAT, Kanpur: Kalyanpur Red Round. PAU, Ludhiana: Punjab Selection, Punjab Red Round, Punjab-48, Punjab White and Punjab Naroya. TNAU, Coimbatore: Multiplier onion Co-1, Co-2, Co-3, Co-4, Co-5 and MDU-1. RARS, Durgapura: Rajasthan Onion-1 (RO-1), Arpita (RO-59) and RO 252. DOGR, Rajgurunagar: Bhima Super, Bhima Raj, Bhima Red, Bhima Shakti, Bhima Kiran, Bhima Shweta, Bhima Shubhra, Bhima Safed and Bhima Dark Red.

Worldwide, different groups of scientists are inventing the new techniques for population improvement in onion utilizing conventional and new techniques for developing varieties and hybrids suitable for different purposes have been summarized by Mahajan and Lawande (2011) in following table. It is expected that new open pollinated varieties will continue to be required in those countries where the growers are unable to pay for high cost of F1 hybrid seeds (Brewster, 2008).

List of different onion varieties released worldwide using population improvement methods

S. No.	Variety Released	Improvement Method Used	Scientists Involved/ References
1.	Improvement of ancient Russian variety 'Spasskii'	Mass selection and Intravarietal recurrent hybridization	Efimochkina, 1970
2.	Yalova 1, Yalova 3 and Yalova 12	Mass selection in Marmara population and Single plant selection in Thrace population.	Akgun, 1970
3.	N–53	Mass Selection (Collection from Nashik, Maharashtra)	MPKV, Rahuri, 1975
4.	Punjab Selection	Mass Selection in indigenous material (Collection from Punjab)	PAU, Ludhiana, 1975
5.	Pusa White Flat	Mass Selection (Local Collection)	IARI, New Delhi, 1975
6.	Pusa White Round	Mass Selection Local collection (106)	IARI, New Delhi, 1975
7.	Co 2	Mass Selection (Collection from Tamil Nadu)	TNAU, Coimbatore, 1978
8.	Punjab – 48	Mass Selection (Collection from Punjab)	PAU, Ludhiana, 1978
9.	Pusa Ratnar	Mass Selection (Selection from Red Granex from USA)	IARI, New Delhi, 1978
10.	Pusa Red	Mass Selection (Local Collection)	IARI, New Delhi, 1978
11.	Co 3	Mass Selection (Collection from Tamil Nadu)	TNAU, Coimbatore, 1982
12.	Kalyanpur Red Round	Mass Selection (Collection from U P)	CSUAT, Kanpur, 1983
13.	Arka Pragati	Mass Selection (Collection from Nashik, Maharashtra)	IIHR, Bangalore, 1984
14.	N –2–4–1	Mass Selection (Collection from Pune, Maharashtra)	MPKV, Rahuri, 1985
15.	Arka Niketan	Mass Selection (Mass selection from a local collection IIHR – 153)	IIHR, Bangalore, 1987
16.	Agrifound Dark Red	Mass Selection (Collection from Nashik, Maharashtra)	NHRDF, Nashik, 1987
17.	Pusa Madhavi	Mass Selection (Collection from Muzaffarnagar, U P)	IARI, New Delhi, 1987
18.	'Dorata di Parma' resistant for *F. oxysporum* f. sp. cepae Snyd. et Hans.	Mass and recurrent selection	Fantino, M.G. and Schiavi, M., 1987

Contd...

S. No.	Variety Released	Improvement Method Used	Scientists Involved/ References
19.	Arka Kalyan (Sel-14)	Mass Selection (Mass selection from a local collection IIHR – 145)	IIHR, Bangalore, 1987
20.	Baswant–780	Mass Selection (Collection from Pimpalgaon, Maharashtra)	MPKV, Rahuri, 1989
21.	'VL Piaz 3'	3 cycles of Mass selection after F_2 of cross 'In-13 x L-43'	Mani, Chauhan, Joshi, Tandon, 1999
22.	Screening and analysis of components of white shaft weight	Maternal pedigree selection in male sterile-plants and male-fertile plants (MPSMS and MPSMF)	Zhaoshui *et al.*, 1995
23.	'Composto IPA-6' and 'Belem IPA-9'	Breeding program tolerant to *C. gloeosporioides* and *T. tabaci*, good post-harvesting conservation qualities	De Franca and Candeia, 1997
24.	Cobriza INTA	Mass Selection from Valenciana type onions	Galmarini *et al.* 2001, Argentina
25.	Navideña INTA	Mass Selection from Torrentina local population	Galmarini *et al.* 2001
26.	Antártica INTA	Mass Selection from Valenciana type onions	Galmarini *et al.* 2001
27.	NuMex Chaco' Onion	Recurrent Selection	Cramer and Corgan, Las Cruces, 2001a
28.	NuMex Snowball' Onion	Recurrent Selection	Cramer and Corgan, Las Cruces, New Mexico, 2001b
29.	NuMex Arthur' Onion	Recurrent Selection	Wall and Corgan, Las Cruces, 2002
30.	Congregation of desirable genes in Gholy – Ghesseh Local Onion	Mass Selection	Javad *et. al.*, 2004,
31.	Purifying the popular land variety "Abu Ferewa"	Phenotypic recurrent mass selection and Inbreeding followed by bulking	Bakheet, 2008
32.	Genetic analysis in six generations (P1, P2, F1, F2, B1 and B2) of four onion crosses *viz.*, PBR 138 x AN 187, PBR 139 x AN 184, PBR 139 x AP 195 and PBR 140 x AN 184	Reciprocal recurrent selection: Exploit all gene actions simultaneously to develop a new resistant line/ variety, being best method to improve trait of resistance to purple blotch disease.	Evoor *et al.*, 2007,
33.	Arka Pitambhar	Pedigree selection from the cross U.D. 102 x IIHR-396	IIHR, Bangalore http://www.iihr.res.in/ frmVarities.aspx
34.	Bhima Super	Rigorous mass selection for single centeredness & bulb shape	ICAR-DOGR, Rajgurunagar, Pune 2007

Contd...

S. No.	Variety Released	Improvement Method Used	Scientists Involved/ References
35.	Bhima Red & Bhima Raj	Single bulb selection up to three generations followed by mass selection	DOGR, Rajgurunagar, Pune, 2009
36.	Bhima Shakti & Bhima Kiran	Mass selection for late *kharif* and *rabi* season with better storability & mass selection for rabi with good keeping quality	DOGR, Rajgurunagar, Pune, 2010a & b
37.	Bhima Sweta	Selection of elite lines from germplasm followed by random matting and mass selection for *rabi* season white onion	DOGR, Rajgurunagar, Pune, 2010
38.	Bhima Shubra	Selection of white segregating bulb from red germplasm followed by mass selection for white populations for kharif & late kharif season	DOGR, Rajgurunagar, Pune, 2010

Male Sterility

After the discovery of cytoplasmic genic male sterility by Jones and Clarke in 1925, the phenomenon of heterosis breeding in onion is being commercially exploited all over the world in long day as well as short day type's onions. Male sterility was first exploited by Jones and Clarke using a male-sterile genetic stock of cultivar 'Italian Red' found in breeding plots at Davis, California in 1925. Fortunately, when this plant was prevented from being cross-pollinated, bulbils were produced in the flower head and it could be propagated. Jones and Clarke (1943) published this classical work describing the genetics of male sterility and indicating how it could be used to produce hybrid cultivars. On the basis of these techniques, originally developed in onions, male sterility has been exploited in hybrid breeding in more than 150 crop species (Kale and Munjal, 2005). CGMS used presently world-wide in onion for commercial exploitation of heterosis was originally derived from the variety 'Italian Red 13-53'. The second source of CMS (*T*-cytoplasm) in onion was discovered in the French cultivar 'Jaune paille des venus'. This CMS (cytoplasmic male sterile) line was found to be different than that from 'Italian Red 13-53' as three independent segregating restorer loci were identified in this line, responsible for its complex inheritance. It has common occurrence of restorers, which makes this *T*-cytoplasm more difficult to use. Later, male-sterility has been observed in several other onion populations, mainly in long-day cultivars, e.g. Pukekohe Longkeeper, Red Wethersfield, Scott County Globe, Stuttgarter Riesen and Zittauer Glebe. The male sterility gene has been found to be widespread in several genotypes collected from many parts of the world (Little *et al.*, 1946; Davis, 1957). The male-sterile plants had anthers with a translucent, green appearance in contrast to normal, dark green anthers.

In India, male-sterility was identified in a local cultivar 'Nasik White Globe' at IIHR, Bangalore (Pathak *et al.*, 1980). Investigations on the causes of the cytoplasmic male sterility in onion indicated that tapetal abnormalities and histochemical changes were responsible for male sterility in onion and there was no role of meiotic abnormalities (Saraswathi and Veere Gowda, 2006).

Induction of Haploids

Haploids are required for further development of dihaploids for use as parents in development of F1 hybrids in order to exploit heterosis. *In vitro* haploid production through gynogenesis is generally used for one step inbred production. Induction of haploids is done through *in vitro* gynogenesis using unpollinated ovule culture, ovary and whole flower culture of long day cultivars (Campion *et al.*, 1992) and later got doubled haploids from gynogenic lines of onion through spontaneous and induced chromosome doubling in *A. cepa* using unfertilized ovary and flower culture by Campion *et al.* (1995).

Ionescu and Popandron (1995) studied the effect of genotype on gynogenesis with different explants like flower buds, ovary and ovules. The effect of immature flower buds and ovary on four Hungarian genotypes on BDS medium was studied by Gemesne and Martinovich (1995). Ovary culture gave the best result with 80% of the regenerants been haploid and 20%, doubled haploid. Michalik *et al.* (2000) reported that 3.5-4.5 mm long flower buds were the most responsive to gynogenesis. It was observed that the cultivars differed with regard to their demands for media composition and the yield of embryos strongly depended on the genotype. Different mitotic poisons like colchicines, oryzalin and APM have been evaluated and oryzalin and APM are routinely used for chromosome doubling in haploids (Jakse *et al.*, 2003).

Bohanec *et al.* (1995) were successful in inducing gynogenesis in four onion cultivars using ovule and ovary. Esterase isozyme analysis showed that 59% of regenerants were homozygous. The high genetic stability of onion homozygous lines passed through two cycles of gynogenesis *in vitro* using RAPD, has been reported by Javornik *et al.* (1998) revealed that the effect of genotype playing a major role in showing response for gynogenesis.

F_1 Hybrids Developed through Male Sterile Lines

Heterosis breeding provides an opportunity for improvement in productivity, earliness, uniformity and yield attributing characters. Till now, most of the hybrids were produced by private seed companies using male sterile lines. Various developed in the world by various public/private sector organizations are as follows:

Tamara: (Skof and Ugrinovic, 2004). Donglingbai: 244A × 244B, (Tong *et al.* 2005). Early Globe: W202A × S87-707, (Muro *et al.* 2006). Liaocong No.6: 2000Y24-3S98 × 244-152A, (Cui *et al.* 2007). Safrane, Hypark, Alonso: (Bimsteine *et al.* 2009). Musica, Vaquero, Manas, Sedona: (Popandron *et al.* 2009). Jinqiu: B2354A × 3104, (Cui *et al.* 2009). Jinxing: (Yan *et al.* 2009). Yeongpunghwang: MOS8 × Mokpo 11, (Lee *et al.* 2009). Jintianxing: HG02 A × K400C, (Ma *et al.* 2009). Wonye 30002: 402AC203 × M1, (Kim *et al.* 2010b). Wonye 30001: Ginque × YG-1-1, (Kim *et al.* 2010a). Quer-rich: NOR-1A × SRG-12, (Muro *et al.* 2010). Baifeng: (Zeng *et al.* 2011). Hybrid Nun-3001 and Hybrid Orient: (Gupta *et al.* (2011b). Sojiro: (Maekawa *et al.* 2012). 2572, 2573, 2578: (Faria *et al.* 2012). Hybrid Optima: (Boas *et al.* 2012). Daekwanhwang: Manchuhwang × NIHA-5001, (Kwon *et al.* 2013). Arka Kirtiman and Arka Lalima: (Veere Gowda *et al.* 1998).

Onion Tissue Culture

In vitro propagation of genetically uniform plants from numerous explants including shoot tips, basal plates, flower heads, leaf segment, roots and unripe aerial bulbils has been feasible. Large scale plant regeneration via direct organogenesis is suitable for cloning of desirable genotypes. This method is useful for multiplication of interspecific hybrids, polyploids and mutants of onion where high degree of sterility is a barrier to sexual reproduction. Protocols for plant regeneration from callus cultures are established for numerous *Allium* crops. In onion, callus cultures and plant regeneration have been established from different explants such as aerial bulbs, roots, basal plates, onion sets, flower heads, immature zygotic embryos and unfertilized ovules. Several authors (Hussey, 1978; Dunstan and Short, 1979). Kahane *et al*. (1992) succeeded in micropropagation of onion plants *in vitro* with a high potential for shoot regeneration from basal plate by destroying/injuring apex to obtain axillary bud multiplication. Mohamed *et al.* (1995) described procedure for bulb formation from explants using cut stem bases as explants. Adventitious shoots were induced firstly on twin scales cut from small bulbs and subsequently on split *in vitro* shoots used as secondary explants (Hussey and Falavigna, 1980). A novel method for direct organogenesis in onion (*A. cepa* L.) resulting in the formation of multiple shoot structures induced on mature flower buds or ovaries in a two-step culture procedure was described by Luthar and Bohanec (1999). Shoots have been produced in the explants from flower head receptacles in *A. porrum* (Novak and Havel, 1981). Different organogenic responses like induction of multiple shoots from shoot tip and root tip (Khar *et al.*, 2002b), seeds, flower bud (Asha Devi and Khar, 2000) have been studied in a few short day Indian onions. Somatic embryogenesis has been observed in callus cultures of *A. cepa, A. fistulosum* and in F1 hybrids between *A. cepa* and *A. fistulosum*. Seedling shoot meristem tip explants were used successfully by Phillips and Luteyn (1983) for obtaining high frequency somatic embryogenesis from two lines of onion, 'Yellow Grano' and 'Yellow Sweet Spanish'. *In vitro* vegetative propagation of male sterile lines can greatly simplify F1 hybrid seed production. Embryo rescue was used to overcome the problem of post zygotic incompatibility in interspecific hybrid of *Allium cepa* x *Allium fistulosum.*

Callus cultures generally grew better in suspension because *Allium* callus grows very slowly on semi solid medium. Although suspension cells capable of regeneration has been reported in *A. cepa, A. fistul*osum and interspecific hybrid *A. fistulosum* x *A. cepa,* the regeneration capacity is very low. These systems can generate new genetic variability via somaclonal variation which can be useful for improvement of onion. Callus initiation, multiplication and morphogenesis leading to plant regeneration and degree of genetic variability obtained with *in vitro* regenerated plants appears to depend on genotype, type and ontogenetical state of explants and the culture conditions used. In onion, in *vitro* screening was applied for *Fusarium* resistance, for pink rot disease resistance and more recently for resistance to purple blotch on media containing filtrates from the fungus.

Breeding and Genetic Transformation

Traditional breeding in onion particularly with reference to generating hybrids through the transfer of genes form wild species which have several desirable traits such as disease and pest resistance has proved to be difficult. There are very few reports of cross breeding such as those developed between *A. cepa* x *A. kermesinum* and between *A. roylei x A. cepa*. Overcoming the barrier between *A. cepa x A. fistulosum* has been especially difficult. Another approach has also been used to overcome the incompatibility barrier between *A. fistulosum x A. cepa* using *A. roylei* as a bridging species. The bridge cross populations [*A. cepa* x (*A. fistulosum x A. roylei*)] showed highly segregating population with many disease and pest resistance genes from *A. fistulosum* and *A. roylei*. Secondly, conventional breeding through wide crosses between species to improve cultivars is limited by open pollination, high degree of heterozygosity and poor fertility of F1 hybrids. The method which is being tried is that of interspecific hybridization followed by repeated backcrossing to the *A. cepa* parent. In this way it is hoped that recombination between the different chromosomes sets in the F_1 or backcross generations, combined with selection for resistant individuals, will allow for the possibility of recombination and the transfer of resistance genes from A. *fistulosum* into the *ascalonicum* chromosome set. The cytology of this hybrid has recently been studied by Peffley *et al.* (1984).

The production of onion inbreds required for hybrid cultivar development is a lengthy process because of the biennial and open-pollinated nature of the plant. Up to five generations of selfing are required to stabilize agronomically important traits. Onion is out breeder subject to severe inbreeding depressions with successive selfing for development of inbred, the lines become increasingly less vigorous and produce less seed. Inbred development can be accelerated by the generation of doubled haploid (DH) lines. These doubled haploid lines can also be more uniform and vigorous than inbred lines created by traditional means. The ability to produce inbred lines by the development of DH would therefore be of particular advantage for onions, considering the long development time of this crop species. This would have a major impact on development of improved open pollinated (OP) onion lines as well as hybrids.

Breeding for Disease and Pest Resistance

Though several reports of disease and insect resistance are available from India; none of the varieties have been found perfectly resistant. No accession of *A. cepa* screened so far in India has registered complete resistance to any of the above diseases. Interspecific hybridization with wild relatives is considered an alternative. Cultivated and wild Alliums possess many disease resistance traits, potentially useful in genetic improvement of bulb-onion. Fertile hybrids between *A. roylei* and *A. cepa* and successful transfer of downy mildew resistance were accomplished (Novak, 1986). Hybrids between *A. cepa* and *A. fistulosum* were known long ago but F_1 always showed extremely low fertility due to poor chromosomal pairing and the first attempt to introgress genes from *A. fistulosum* into *A. cepa* were reported by Emsweller and Jones, 1935 but these were not successful, and till lately all attempts to introgress genes from *A. fistulosum* to *A. cepa* have failed because of sterility in backcrossed generations. Low degree of fertility exhibited by hybrids between *A. cepa* and other *Allium* sp. restricts successful introgression of disease resistance genes. Ulloa *et al.* (1995), suggested that such sterility is due to an imbalance between nuclear and cytoplasmic genomes. Van der Meer and De Vries (1990) and McCollum (1982), showed that *A. roylei* ($2n=2x=16$) crosses readily with *A. cepa* and *A. fistulosum*, respectively. Hence, *A. roylei* can be used as a bridging species between *A. fistulosum* and *A. cepa*. By means of this bridge-cross not only genes from *A. fistulosum* can be introgressed into *A. cepa* but also simultaneously genes from *A. roylei*. In *A. fistulosum* resistance genes are present against *Botrytis squamosa* (Currah and Maude, 1984), *Pyrenochaeta terrestris* (Netzer *et al.*, 1985), *Colletotrichum gloeosporioides* (Galvan *et al.*, 1997), *Urocystis cepulea* and OYDV (Rabinowitch, 1997), and in *A. roylei* resistance is available against *Peronospora destructor* (Kofoet *et al.*, 1990) and *Botrytis squamosa* (De Vries *et al.*, 1992). Hence, via bridge-cross approach unique populations can be developed in which these resistance genes can be pooled. Introgression of *A. fistulosum* into the genome of *A. cepa* using *A. roylei* as a bridging species by means of genomic *in situ* hybridization was reported by Khrustaleva and Kik (2000) and is the first such successful effort.

The ICAR-IIHR, Bangalore, have done some work on purple blotch caused by *Alternaria porri* is most devastating disease of onion prevalent in different parts of the country. Lack of resistant variety in the gene pool is one of the major reasons for perpetuation of pathogen throughout the year and cause epidemic and farmers are forced to spray very high amount of pesticide on both onion bulb and seed crop. Development of hybrids through exploitation and introgression of purple blotch resistance available in the local germplasm provides a rational solution associated with onion production in India. Arka Kalyan is one of the improved varieties evolved for tolerance to purple blotch disease. Resistance source for purple blotch disease has been identified in breeding lines PBR-287 and PBR-296. CMS male sterile female lines MS 65, PBRMS 319 have also been developed and two F_1 hybrids resistant to purple blotch using MS lines as female parents (MS-65-268 x PBR-287 and MS-65-268 x PBR-296) have been developed.

Genetic Transformation in *Allium*

An efficient *in vitro* regeneration system is a prerequisite for genetic transformation, which provides opportunity to introduce genes across barriers of reproduction. Regeneration in different species of *Allium* has been reported through different methods which is prerequisite for transformation. Somatic embryogenesis, direct regeneration and through callusing has been achieved in onion (Eady *et al.*, 2000, Aswath *et al.*, 2006) and shallots (Zheng, 2000). For genetic transformation callus from immature embryo has been preferred in *A. cepa*. Other explants like callus from root and root tip (Zheng *et al.*, 2004), stem disc (Xu and Cui, 2007) and immature leaf tissue (Kenel *et al.*, 2010) have also proven to be useful for transformation. *Allium* species are recalcitrant to transformation. However, by the mid-nineties, both particle bombardment and *Agrobacterium* mediated DNA delivery systems were used successfully in onion for transformation with the *uid a* (β -glucuronidase) reporter gene. Microprojectile bombardment for transformation has been employed for onion by Xu and Cui (2007). Degradation of naked DNA in the tissue affects the transformation efficiency and treating the tissues with nuclease inhibitor (aurintricarboxylic acid (2 mM)) improved transformation efficacy (Barandiaran *et al.*, 1998). But the *Agrobacterium* mediated transformation is frequently used method and has been proven useful in different species of *Allium*.

A successful transformation of an onion cultivar mediated by *A. tumefaciens* was reported by Eady *et al.* (2000) using immature embryos as inoculated explants. Zheng *et al.* (2001) developed a reproducible *A. tumefaciens* mediated transformation system both for onion and shallot with young callus derived from mature embryos with two different *Agrobacterium* strains. Use of immature embryos is a common feature among transformation in onion due to their excellent morphogenetic competence (Eady *et al.* 2000). Zheng *et al.* (2001) used mature embryos, which is tedious to remove and requires stereomicroscope to identify the shoot apex portion. Klein *et al.* (1987) and Scott *et al.* (1999) used epidermal tissues with high velocity microprojectiles.

Molecular Markers in Onion

Allium is a large genus of approximately 600 species and classification of such a large genus has proved difficult and many ambiguities still remain. Havey (1991) suggested that there could be a role for genetic markers in systematic studies of *Allium*. Bark *et al.* (1994) studied introgression of *A. fistulosum* genes into *A. cepa* background using restriction fragment length polymorphism (RFLP) analysis. Van Heusden *et al.* (2000) presented a genetic map based on amplified fragment length polymorphism (AFLP) in an interspecific cross of *A. roylei* and *A. cepa* and reported one of the allinase genes (a key enzyme in sulphur metabolism) and a Sequence Characterised Amplified Region (SCAR) marker linked to the disease resistance gene for downy mildew on the map. Gokce *et al.* (2002) sequenced the genomic region corresponding to the cDNA revealing the closest RFLP to Male sterility (Ms) gene in their efforts on molecular tagging of the Ms locus in onion. Mapping studies in onion have thus far been scarce. King *et al.* (1998) presented a low-density genetic map of restriction fragment length polymorphism (RFLP) based on an interspecific cross showing, that, genomic organization of onion was complex and involved duplicated loci. Reasons for delay in molecular marker studies in onion are: biennial nature of onion, it's severe inbreeding depression and its huge genome size. While RAPDs have been used successfully for genetic studies in *Allium*, the size of the genome may cause many problems, such as rather poor reproducibility and high backgrounds. Simple Sequence Repeats (SSRs), also called microsatellite markers, are codominantly inherited and reveal high levels of polymorphism. Fischer and Bachmann (2000) identified a set of informative STMS (Sequence Tagged Microsatellite Sites) markers for which can be used for distinguishing accessions and for studying interspecific taxonomic analysis using close relatives of *A. cepa*. Molecular characterization has been carried out in 20 accessions belonging to 14 *Allium* species using 5 isozyme and 15 RAPD markers by Lakhanpaul *et al.*, 1996, indicated the presence of narrow intra-specific and high inter-specific variations in *Allium* species. The RAPD markers were useful for assessing relatedness and genetic diversity in onion cultivars. Mahajan *et al.* (2009) studied 14 short day and 2 long day cultivars of onion at 24 microsatellite loci. Twenty-one primer pairs were polymorphic. Nashik Red and Poona Red cultivars showed 100% similarity and were indistinguishable. Cultivars N-53 and Bombay Red showed 95% similar fragments. It was followed by Arka Niketan and Poona Red (91.3% similar fragments) and Arka Niketan and Nashik Red (90.5% similar fragments pair), which were quite close. Alisa Craig showed maximum diversity with about 70% dissimilar fragments pairs compared to other cultivars and showed 25.7% similar fragment pairs with Brigham Yellow Globe. Some fragments were unique and were only present or absent in a cultivar at particular loci can be used as cultivar specific markers for rapid discrimination. Based on Jaccard's coefficient for SSRs, cultivars were grouped into 5 main groups. Exotic cultivars Alisa Craig and Brigham Yellow Globe were different compared to the Indian cultivars. Nashik Red and Pune Red were indistinguishable and similarly N-53 and Bombay Red were quite close. It will also be helpful in selecting the diverse parents (inbreds) for the development of suitable hybrids of onion, for the DNA fingerprinting of the cultivars and promising germplasm for their protection under Plant Cultivar Protection act.

Prospects

The genus *Allium* offers excellent materials for cytogenetical studies and for the application of cytogenetics to crop improvement. Among the cultivated edible forms all have large chromosome which are few in number and which can be readily and easily observed. Modern classifications accept more than 750 species and about 60 taxonomic groups at sub-generic, sectional and sub-sectional ranks. Further progress in compiling a phylo-genetically based natural *Allium*, classification will mainly depend on the accessibility of living material from the hitherto under-investigated end areas of South-West, southern Central and western East Asia. Common onion is species of worldwide economic importance and they consist of several infra-specific groups. Their cultivation traces back to very ancient times, and thus their direct wild ancestors and places of domestication remain unknown. Other *Allium* species of minor economic importance, such as leek, chives, etc., as well as about two dozen species and hybrids grown sporadically or in restricted regions only, have been mostly taken into cultivation in the historical period.

The yield level in developed varieties has reached to a plateau. Variability in the germplasm is also exhausted. There are two alternatives to create variability one by mutation breeding and second by hybridization/heterosis breeding. Absence of established male sterile lines along with maintainer in short day onion always remained a bottleneck in heterosis breeding programme in India. Besides, narrow base of inbred lines involved, fail to exhibit high heterosis for yield. F_1 hybrids will play an increasing part in future breeding because of the advantages to the breeder of exclusive ownership of the parent lines. In recent findings indicated that the hybrids are high yielder and they are most needed. Uniform maturity, size, shape and colour of bulb can be monitored effectively with the help of F_1 hybrids besides added advantage of high yield. Further, there can be good control over seed production and distribution as hybrids involve three parents. Maintenance and up gradation of parents and F_1 seed production is skilful job. Mutation breeding though creates variability but results are not predictable. The alternate way is to make crosses between long day or intermediate day type exotic onion in temperate conditions i.e. in phytotron or in temperate northern hills. Thus direct hybrids after testing can be exploited or further selection can be made for desirable characters and inbred lines can be developed for the development of stable hybrids.

Further biometrical work is needed to predict the vigour of recombinant inbred derivable from appropriate crosses; in parallel, more rapid ways of obtaining homozygous inbred lines, e.g., by anther-culture, are required. There are obvious dangers in relying on a male sterility system based on a single sterile cytoplasm type, especially one which shows some evidence of instability. Work is needed at the molecular level to determine the biochemical basis for male sterility and hence possible ways of manipulating and controlling.

References

Akgun, H. 1970. Selection in native onion populations as a means of producing white-fleshed, yellow skinned keeping varieties of onion suitable for export. Yalova Bahce Kultuleri Arastirma ve *Egitim Merkezi Dergisi,* 3(4): 20-32.

Asha Devi, A. and Khar, A. 2000. Preliminary studies on *in vitro* regeneration from flower bud explants of onion (*Allium cepa* L.). In: *Proceedings of National Symposium on Onion-Garlic Production and Post Harvest Management Challenges and Strategies,* Maharashtra, pp.194.

Aswath, C., Mo, S., Kim, D. and Park, S. W. 2006. Agrobacterium and biolistic transformation of onion using non-antibiotic selection marker phosphomannose isomerase, *Plant Cell Reports* 25(2), 92-99.

Bakheet, K. A., 2008. Assessing relative efficiency of two breeding methods for the improvement of yield and quality of the local sudanese onion variety (*Allium cepa* L.) abu ferewa. Thesis (M.Sc. in Horticulture), Sudan Academy of Sciences, Khartoum (Sudan) pp. 49.

Barandiaran, X., Di Pietro, A. and Martín, J. 1998. Biolistic transfer and expression of a *uid*A reporter gene in different tissues of *Allium sativum* L. *Plant Cell Reports.* 17, 737–741.

Bark, O.H., Havey, M. J. and Corgan, J.N. 1994. Restriction fragment length polymorphism (RFLP) analysis of progeny from an *Allium fistulosum x A. cepa* hybrid. *J. Amer. Soc. Hort. Sci.,* 119:1046-1049

Bhonde, S. R., Srivashtava, K. J., and Singh, K. N. 1992. Evaluation of varieties for late *Kharif* (Rangda) crop of onion in Nasik areas. *AADF, News Letter* 12 (1):1-2

Bimsteine, G., Lepse, L. and Bankina, B. 2009. Possibilities of integrated management of onion downy mildew. *Sodininkysteir Darzininkyste* 28(3): 11-17.

Boas, R.C.V., Pereira, G.M., Souza, R.J., Geisenhoff, L.O. and Lima, J. A. 2012. Development and production of two onion cultivars irrigated by drip system. *Revista Brasileira de Engenharia Agricola e Ambieental.* 16 (7): 706-713.

Bohanec, B., Jakse, M., Ihan, A. and Javornik, B. 1995. Studies of gynogenesis in onion (*Allium cepa* L.): Induction procedures and genetic analysis of regenerants. *Plant Science (Limerick)* 104(2): 215- 224.

Brewster, J. L. 2008. Crop Production Science in Horticulture 15. *Onion and Other vegetable Alliums* 2nd Edition. CAB International.

Brewster, J. L. 1994 *Onions and Other Vegetable Alliums.*CAB International, Wallingford, UK.pp. 236.

Dubey, B. K., Bhonde, S. R., and Singh, Lallan. 2004. Studies on the performance of red onion advance lines under Nasik region of Maharashtra. *NHRDF, News Letter* 24 (2):1-3.

Campion, B., Azzimonti, M.T., Vicini, E., Schiavi, M. and Falavigna, A. 1992. Advances in haploid plant induction in onion (*Allium cepa* L.) through *in vitro* gynogenesis. *Plant Science (Limerick)*. 86(1): 97-104.

Campion, B., Bohanec, B. and Javornik, B. 1995. Gynogenic lines of onion (*Allium cepa* L.): evidence of their homozygosity. *Theoretical and Applied Genetics*. 91(4): 598-602.

Castell, V. R. and Portas, C. M. 1994. *Alliaceae* production systems in the Iberian Peninsula: facts and figures of potential interest for a worldwide R&D network. *Acta. Hort.*, 358, 43-47.

Cramer, C. S. and Cogan, J. N. 2001a. "NuMex Chao" Onion. *Hort. Science*, 36 (7):1337-1338.

Cramer, C. S. and Cogan, J. N. 2001b. "NuMex Snowbal" Onion. *Hort. Science*, 36 (7):1339-1340.

Crisp, P. and Astley, D. 1985. Genetic resources in vegetables. (In:) G.E. Russel (Ed.). *Progress in Plant Breeding –1.* Butterworths, London, pp. 281-310.

Cui, C. R., Jia, T. J., Ma, Y. H., Xu, Q. J. 2009. A new long-day onion F_1 hybrid 'Jinqiu'. *China Vegetables* 22: 80-81.

Cui, L.W., Tong, C. F. and Du, X. J. 2007. A new welsh onion variety -'Liaocong No.6'. *China Vegetables* 10: 39-40.

Currah L and Maude R B. 1984. Laboratory tests for leaf resistance to *Botrytis squamosa* in onions. *Annals of Applied Biology* 105: 277-83.

Davis, G.H. 1957. The distribution of male sterility gene in onion. *Proceedings of American Society for Hotricultural Science*. 70. pp. 316-318.

De Franca, J.G.E and Candeia, J.A. 1997. Development of short-day yellow onion for tropical environment of the Brazilian northeast. ISHS Acta Horticulturae 433: I International Symposium on Edible Alliaceae. http://www.actahort.org/books/433/433_29.htm

De Vries J. N, Wietsma W A and Jongerius M C. 1992. Introgression of leaf blight resistance from *A. roylei* Stearn into onion (*A. cepa* L.) *Euphytica* 62: 127-33.

Dowker, B. D. and Gordon, G. H. 1983. Heterosis and hybrid cultivars in onions, in *Monogr. on Theor. and Appl. Genet.*, Vol. 6, Frankel, R., Ed., Springer-Verlag. Heidelberg, 220

Dunstan, D.I. and Short, K.C. 1979 Shoot production from the flower head of *Allium cepa* L. *Scientia Horticulturae* 10, 345-356.

Eady, C.C., Weld, R.J. and Lister, C.E. 2000 Agrobacterium tumefaciens-mediated transformation and transgenic-plant regeneration of onion (*Allium cepa* L.). *Plant Cell Reports* 19, 376–381.

Efimochkina, O. N. 1970. The Spasskii onion. Tr. Po selektsii semenovodstvu ovoshch. Kul'tur. Grivov. Ovoshch. Selects. St. I: 68-78.

Emsweller S. L. and Jones H. A. 1935. An interspecific hybrid in *Allium*. *Hilgardia* **9**: 265-73.

Evoor, S., Veere Gowda R. and Ganeshan G. 2007 Genetics of Resistance to Purple Blotch Disease in Onion (*Allium cepa* L.) Karnataka *J. Agric. Sci.*, 20 (4): 810-812

Fantino M. G. and Schiavi, M., 1987. Onion breeding for tolerance to Fusarium oxysporum f. sp. cepae, in Italy. Phytopathologia Mediterranea [Phytopathol. Mediterr.] 26(2): 108-112

Faria, M. V., Morales, R. G. F., Resende, J. T. V. de., Zanin, D. S., Menezes, C. B. de., Kobori, R. F. 2012. Agronomic performance and heterosis of onion genotypes. *Horticultura Brasileira*. 30(2): 220-225.

FAO, 2015. FAO, website.

Fischer, D. and Bachmann, K. 2000. Onion microsatellites for germplasm analysis and their use in assessing intra- and interspecific relatedness within the subgenus *Rhizirideum*. *Theor Appl Genet.*, 101:153-164.

Fritsch, R. M and Friesen, N. 2002. Evolution, domestication and taxonomy. CAB International. 2002. *Allium Crop Science: Recent Advances* (eds: H. D. Rabinowitch and L. Currah). 5-30

Fritsch., R. M., Farideh M. and Manfred K. 2001. Allium vavilovii M. Popov et Vved. and a new Iranian species are the closest among the known relatives of the common onion *A. cepa* L. (Alliaceae). *Genetic Resources and Crop Evolution* 48, 401–408.

Galmarini, C.R., Della Gaspera, P. and Fuligna, H. 2001, New argentine onion cultivars. ISHS Acta Horticulturae 555: II International Symposium on Edible Alliaceae. http://www.actahort.org/books/555/555_39.htm

Galvan G. A., Wietsma W. A., Putrasemedja S., Permadi A. H. and Kik C. 1997. Screening for resistance to anthracnose (*Colletotrichum gloeosporioides* Penz.) in *Allium cepa* and wild relatives. *Euphytica* 95: 173-78.

Gemesne, J. A. and Martinovich, L. 1995. *In vitro* gynogenesis induction in Hungarian lines of onion (*Allium cepa* L.). ZKI Bulletin, Kecskemet. 27: 37-43.

Gokce, A.F., McCallum, J., Sato, Y. and Havey, M.J. 2002. Molecular tagging of the Ms locus in onion. *J. Amer. Soc. Hort. Sci.*, 127:576-582

Gupta, R. P and Singh, R. K. 2010. Onion Production in India. Published by Director, National Horticultural Research and Development Foundation Chitegaon Phata, Post-Darna Sangavi, Taluqa-Niphad, Dist- Nashik, Maharashtra. Malhotra Publishing house, B-6, DSIDC Complex, Kirti Nagar, New Delhi. 1-88 pp.

Hanelt P., Schultze-Motel J, Fritsch R, Kruse J, Maas H.I, Ohle H and Pistrick K 1992. Infrageneric Grouping of Allium – the Gatersleben Approach.In : Hanelt, P., Hammer, K, and Knupffer, H (Ed.), *The Genus Allium–Taxonomic Problems and Genetic Resources,* Gatersleben, IPK, pp. 107-123.

Hanelt, P. 1990. Texonomy, evolution and history. In: Rabinowitch, H. D. and Wrewster, J. L. (eds.) *Onion and allied crops*, Vol. I *Botany, Physiology and Genetics*. CRC, Press, Boca Raton. Florida pp. 1-26.

Havey, M. J. 1995. Identification of cytoplasms using the polymerase chain reaction to aid in the extraction of maintainer lines from open-pollinated populations of onion. *Theoretical and Applied Genetics*, 90: 263-268.

Havey, M.J. 1991. Phylogenetic relationships among cultivated *Allium* species from restriction enzyme analysis of the chloroplast genome. *Theor Appl Genet.*, 81:752-757

Helm, J. 1956. Die zn Wurz-und Speisezwecken kultivierten Arten der Gattung *Allium* L. *Kulturpflanze* 4, 130-180.

Henelt, P. 1986a. Formal and Informal classification of the infra-specific variability of cultivated plants-advantages and limitation, (In:) Styles, B. T (Eds.) *Infraspecific Classification of Wild and Cultivated Plants.* Clarendone Press, Oxford. pp. 139-156.

Hooker JD 1892; rev. 1973. Liliaceae: *Allium* L. *Flora of British India, vol VI*, repr. edn. Bishen Singh and Mahindra Pal Singh, Dehradun, pp. 337–345.

Hussey, G. 1978. *In vitro* propagation of the onion *Allium cepa* by axillary and adventitious shoot proliferation. *Scientia Horticulturae* 9, 227-236.

Hussey, G. and Falavigna, A. 1980. Origin and production of *in vitro* adventitious shoots in the onion, *Allium cepa* L. *Journal of Experimental Botany* 31, 1675-1686.

Ionescu, A. and Popandron, N. 1995. Research on the *in vitro* culture of ovules, ovaries and flower buds of some *Allium* genotypes. *Anale Institutul de Cercetari Pentru Legumicultura Si Floricultura Vidra.* 13:17-22.

Jakse, M., Havey, M. J. and Bohanec, B. 2003. Chromosome doubling procedures of onion (*Allium cepa* L.) gynogenic embryos. *Plant Cell Rep* 21:905–910.

Javad, L. H., Behrouz, M. and Seyyad Hamid, M. 2004. Investigation and mass selection on Gholy-Ghesseh Local onion. Agricultural and Natural Resources Research Center of Zanjan provinceZanjan (Iran) pp. 31.

Javornik, B., Bohanec, B. and Campion, B. 1998. Second cycle gynogenesis in onion, *Allium cepa* L. and genetic analysis of the plants. *Plant Breeding*, 117(3): 275-278.

Jones H A and Rees H. 1963. Nuclear DNA variation in Allium. *Heridity* 23: 591.

Jones, H. A. and Clarke, A. E. 1943. Inheritance of male sterility in the onion and the production of hybrid seed. *Proceedings of the American Society for Horticultural Science*, 43:189-194.

Jones, H. A. and Mann. L.K. 1963. Onion and their allies: *Botany Cultivation and Utilization.* Leonard Hill. London and Interspecific .New York. 285 pp.

Kahane, R., Teyssendier de la Serve, B. and Rancillac, M. 1992 Bulbing in long day onion (*Allium cepa* L.) cultured *in vitro*: comparison between sugar feeding and light induction. *Annals of Botany* 69, 551-555.

Kale, A. A. and Munjal, S. V. 2005. Molecular analysis of mitochondrial DNA of lines representing a specific CMS-fertility-restorer system of pearl millet [*Pennisetum glaucum* (L.) R. Br.] by RAPD markers, *Indian J. Genet.*, 65(1), pp.1-4.

Kazakova, A. A. 1978. *Luc.* Kulturnaja Flors SSSR, X. Kolos, Leningrade, USSR, 264pp.

Kenel, F., Eady, C. and Brinch, S. 2010. Efficient Agrobacterium tumefaciens-mediated transformation and regeneration of garlic (*Allium sativum*) immature leaf tissue. *Plant Cell Reports* 29, 223–230.

Khar, A., Bhutani, R.D., Chowdhury, V. K., Lawande, K.E. and Asha Devi, A. 2002b. Effects of genotype and media on direct and indirect organogenesis in onion (*Allium cepa* L.). In: *Proceedings of International Conference on Vegetables* Bangalore, Karnataka, pp.144.

Khrustaleva L. I. and Kik C. 2000. Introgression of *Allium fistulosum* into *A. cepa* mediated by *A. roylei*. *Theor Appl Genet* 100: 17-26.

Kim, C. W., Lee E. T., Choi, I. H., Jang, Y. S. and Suh, S. J. 2010b. Early maturing male sterile line of onion (*Allium cepa* L.) 'Wonye 30002'. *Korean J. Breeding Science*. 42(3): 298-301.

Kim, C. W., Lee E.T., Choi, I., Jang, Y. S. and Suh, S. J. 2010a. Mid-late maturing male sterile line of onion (*Allium cepa* L.) 'Wonye 30001'. *Korean J. Breeding Science*. 42(3): 294-297.

King, J.J., Brandeen, J., Bark, O., McCallum, J. and Havey, M.J. 1998. A low density genetic map of onion reveals a role for tandem duplication in the evolution of an extremely large diploid genome. *Theor. Appl. Genet.*, 96:52-56

Klein T.M., Wolf E.D., Wu R. and Sanford J.C. 1987. High-velocity microprojectiles for delivering nucleic acids into living cells. *Nature* 327, 70-73.

Kofoet A., Kik C., Wietsma W. A. and de Vries. 1990. Inheritance of resistance to downey mildew (*Pernospora destructor* (Berk.) Casp.)from *Allium roylei* Stearn in the back-cross *Allium cepa* L. × (*A. roylei* × *A. cepa*). *Plant Breeding* 105: 144-49.

Kotlinska, T. P., Havranek, M., Navratill, L., Gerasimova, A., Pimakov and Neikov, S. 1990. collecting onion, garlic and wild species of *Allium* in central Asia, USSR. *Plant Genetic Resources Newsletter*, 31-32.

Kwon, Y. S., Cho, K. S., Lee, E. H., Jang, S. W., Suh, J. T., Kim J. S., Kim, W. B., Im, J. S. and Lee, J. N. 2013. High-quality and long-storability of F_1 hybrid long-day onion. *'Daekwanhwang' Korean Journal of Breeding Science*. 45(1): 61-65.

Lakhanpaul, S., Bhat, K.V. and Chandel, K.P.S. 1996. Identification of onion (*Allium cepa* L.) cultivars using AP-PCR. (In:) fourth International conference on DNA fingerprinting. 3-7 Dec., Melbourne, Australia.

Lee, E. T., Kim, C. W., Choi, I. H., Jang, Y. S., Bang, J. K., Bae, S. G., Hyun, D. Y., Jung, J. M., Ha, I. J. and Kim, S. B. 2009. New mid-late maturing F1 hybrid onion cultivar, "Yeongpunghwang. *Korean Journal of Breeding Science.* 41(4): 587-590.

Levichev, I. G. and Krassovskaja, L. S. 1981. The Pskemski onion *Allium pskemense* B. Fedtsch, in the Sothern part of its range. *Bjulletin Moskovskogo Obshchestva Ispytatelej Prirody, Otdel Biologycheskij* 86, 105-112 (In Russian).

Little, J.M., Jones, H.A. and Clarke, A.E. 1946. The distribution of male sterile gene in varieties of onion. *Hereditas*, 111, 310-312.

Luthar, Z. and Bohanec, B. 1999 Induction of direct somatic organogenesis in onion (*Allium cepa* L.) using a two step flower or ovary culture. *Plant Cell Reports* 18, 797-802.

Ma, Y. H., Liu. W. and Cui, C. R. 2009. A new onion cultivar 'Jintianxing'. *Acta Horticulturae Sinica.* 36 (9): 1401.

Maekawa, K., Tanaka, S., Yanagida, D., Koyano, S. Noda, T., Akiba, M., Majima, T., Kondo, K., Hagihara, T. and Mori, N. 2012. 'Sojiro':extremely early maturity spring-sowing onion hybrid with resistance to Fusarium basal rot. *Acta Horticulturae.* 969: 103-106.

Mahajan V and Lawande K E. 2011. Genetic diversity and crop improvement in onion and garlic, pp. 19-40. (Souvenir). *(In): Exploiting Spices Production Potential of the Deccan Region*, SYMSAC-VI, Indian Society for Spices held at 8th-10th December 2011 at Dharwad, Karnataka, India.

Mahajan V., Jakse J, Havey M.J. and Lawande K.E. 2009. Genetic fingerprinting of onion cultivars using SSR markers., *Indian J. Hort.* 66(1): 62-68.

Mani, V. P, Chauhan, V. S., Joshi, H. S. and Tandon, J. P. 1999. Exploiting gene effects for improving bulb yield in onion. *The Ind. J. Gen. Pl. Breed.* 59 (4): 511-514.

McCollum G. D. 1982. Experimental hybrids between *Allium fistulosum* and *A. roylei. Botanical Gazat* 143: 238-42.

McCollum, G.D. 1976. Onion and Allies In: Simmonds N. W. (ed.).Evolution of Crop Plants. London and New York, Longman, pp.186-190.

Melchoior, H. 1964. Reihe *Liliflorae* (In.) Melchior, H (Edn.) A. Englers syllabus derpflanzenfamilien.12. Auglge. Gebruder Borntraeger, Berlin-Nikolassee. pp. 513-543.

Messiaen C M and Rouamba A. 2004. *Allium ampeloprasum* L. Internet Record from PROTA4U. PROTA (Plant Resources of Tropical Africa/Resources vegetables de l'Afrique tropicale), Wageningen, Netherlands. (http://www.prota4u.org/search. asp).

Michalik, B., Adamus, A. and Nowak, E. 2000. Gynogenesis in Polish onion cultivars. *Journal of Plant Physiology.* 156 (2): 211-216.

Mohamed Yasseen, Y., Barringer, S.A. and Splittstoesser, W.E. 1995. *In vitro* bulb production from *Allium* spp. *In vitro Cell Developmental Biology* 31, 51-52.

Muro, T., Ito, K., Sato, Y., Abe, H., Turui, H., Asari, H., Minagawa, Y. and Harada, H. 2006. Breeding of early-maturing hybrid onion variety 'Early Globe' and its characteristics. *Research Bulletin of the National Agricultural Research Center for Hokkaido Region.* 184: 45-55.

Muro, T., Noguchi, Y., Morishita, M., Ito, K., Sugiyama, K., Kondo, T., Kurenuma, Y. and Ono, T., 2010. 'Quer-rich', a new variety of hybrid red onion with high quercetin content . *Research Bulletin of the National Agricultural Research Center for Hokkaido Region.* 192: 25-32.

Netzer D., Rabinowitch H. D. and Weintal C. H. 1985. Greenhouse technique to evaluate pink root disease caused by *Pyrenochaeta terrestris. Euphytica* 34: 385-91.

Novak F. J., Havel L. and Dolezel J. 1986. *Allium*, p 419. *(In): Handbook of Plant Cell Culture, Techniques and Applications*, Vol. 4. Evans D, Sharp W and Ammirato P. (Eds) MacMillan, New York,USA.

Novak, F.J. and Havel, L. 1981. Shoot production from *in vitro* cultured flower heads of *Allium porrum* L. *Biologia Plantarum* 23, 266-269.

Pathak, C. S., Singh, D. P. and Deshpande, A. A. 1980. Annual Report of Indian Institute of Horticultural Research. pp. 34-36.

Peffley, E. B., Corgan, J. N., Horak, K. E. and Tanksley, S. D., 1984. Electrophoratic analysis of *Allium* alien addition lines, *Theor. Appl. Genet.*71, 176.

Phillips, G.C. and Luteyn, K.J. 1983. Effects of picloram and other auxins on onion tissue cultures. *Journal of American Society of Horticultural Science* 108(6), 948-953.

Polunin, O. 1969. *Flowers of Europe: A Field Guide.* Oxford University Press, London, pp. 682.

Polunin, O. and Stainton, A. 1984. *Flowers of the Himalaya.* Oxford University Press, New Delhi, pp. 413–416.

Popandron, N., Basturea, M. and Tudora, M. 2009. Researches regarding the behaviour of some onion cultivars in the vegetable agro-system in southern Romania. *Lucrari Stiintifice, Universitatea de Stiinte Agricole Si Medicina Veterinara "Ion Ionescu de la Brad" Iasi, Seria Horticultura.* 52: 477-480.

Purseglove, J.W. (1972) *Alliaceae. In: Tropical Crops (monocotyledons).*Longman Group Ltd., pp. 37–57.

Rabinowitch H. D. 1997. Breeding alliaceous crops for pest resistance. *Acta. Horticulture* 433: 223-46.

Rabinowitch, H. D and Brewster, J. L. 1990. Onion and Allied Crops. Volume-I, Botany, Physiology and Genetics. *CRC* Press, INC. Boca Raton, Florida pp. 200.

Ricroch A, Yockteng R, Brown S C and Nadot S. 2005. Evolution of genome size across some cultivated *Allium* species. *Genome* 48: 511-20.

Saraswathi, K.M. and Veere Gowda, R. 2006. Studies on causes of male sterility in short day onion (*Allium cepa*, L). In Recent advances in Allium Research (Eds.) Singh, U. P., Singh D. P. and Sharma, B. K., Published by Association of Allium workers in India, Varanasi. pp. 107-127.

Scott, A., Wyatt, S., Tsou, P.L., Robertson, D. and Allen, N.S. 1999. Modelsystem for plant cell biology: GFP imaging in living onion epidermal cells. *Biotechniques* 26 (6), 1125, 1128-1132.

Seshadri, V. S. and Chatterjee, S. S. 1996. The history and adaptation of some introduced vegetable crops in India. *Vegetable Science* 23 (2), 114-140.

Shinde, N.N. and Sontakke, M.B. 1993. Bulbb crops: Onion. (In:) Bose, T.K., Som, M.G. and Kabir, J. (Ed.), Vegetable crops, Naya Prokash, Calcutta: pp. 641 – 685.

Singh, R. K., Dubey, B. K. Bhonde, S. R. and Gupta, R. P. 2010. Estimates of genetic variability, heritability and correlation in red onion (*Allium cepa* L.) advance lines. *Indian Journal of Agriculture Science* 80 (2): 160-163.

Singh, R. K., Dubey, B. K. Bhonde, S. R. and Gupta, R. P. 2010. Variability studies for some quantitative characters in white onion (*Allium cepa* L.) advance lines. *Vegetable Science*, 37(1):105-107.

Singh, R. K., Bhonde, S. R. and Gupta, R. P. 2011. Performance studies on onion promising lines for yield and quality. *Green Farming. An International Journal of Agriculture, Horticulture and Applied Science*, 2 (2):170-172.

Singh, R. K. and Dubey, B. K. 2011. Studies on genetic divergence in onion advance lines. *Ind. J. Hort.* 68 (1):123-127.

Singh, R. K., Bhonde, S. R. and Gupta, R. P. 2011. Studies on genetic variability in late *kharif* (Ragada) onion (*A. cepa* L.) *Journal of Applied Horticulture*, 13:79-82.

Singh, R. K., Gupta, P. K. and Gupta, R. P. 2014. Studies on comparison of (*Allium cepa* L.) genotypes for quantitative traits. *International Journal of Innovative Horticulture* 2(2):153-158

Skof, M. and Ugrinovic, K. 2004. Storage capability of different onion cultivars. *Novi izzivi-v-poljedelstvu 2004 Zbornik simpozija, Ljubljana, Slovenia* 13-14-Decembra 2004. 269-275.

Stack S and Coming D. 1979. The chromosomes and DNA of Allium cepa, *Chromosoma* 70: 161.

Swarup, V. 1991. Breeding Procedures for Crops Pollinated Vegetable Crops. ICAR, New Delhi, pp.118.

Takhtajan. A. 1997. *Diversity and Classification of Flowering Plants.* Columbia University Press, New York, 643 pp.

Tong, C. F., Tang, C. Y., Cui, L. W., Wang, G. Z. and Du, D. J., 2005. Selection and utilization of the male-sterile line "244A" in Welsh onion. *Acta. Horticulture.* 688: 151-157.

Toole, M. G. 1944. Chromosome behaviour and fertility of colchicine-induced tetraploids in *Allium cepa* and *A. fistulosum*, *Herbertia* 11, 295.

Traub H.P. 1968. The subgenera, sections and subsections of Allium L. *Plant Life* 24, 147–163.

Ulloa M, Corgan J N and Dunford M. 1995. Evidence for nuclear-cytoplasmic incompatibility between *Allium fistulosum* and *A. cepa*. *Theor Appl Genet* 90: 746-54.

Van der Meer, Q. P. and De Vries J. N. 1990. An interspecific cross between *Allium roylei* Stearn and *Allium cepa* L. and its backcross to *A. cepa*. *Euphytica* 47: 29-31.

Van Heusden, A.W., Van Ooijen, J.W., Vrielink van Ginkel, R, Verbeek, W.H.J., Wietsma W.A and Kik, C. 2000. A genetic map of an interspecific cross in *Allium* based on amplified fragment length polymorphism (AFLPTM) markers. *Theor. Appl. Genet.*,100:118-126

Vavilov, N. I. 1926. *Origin and geography of cultivated plants*. English translation by D Love (1992) Cambridge Univ. Press, Cambridge, U.K.

Vavilov, N. I. and Burkinich, D. D. 1929. Zemeledeleheskii Afganistan. Tr. Po. Prikl. Botanike, Genetike I selekcii V/R. T.Z. pp. 156-158.

Veere Gowda, R., Pathak, C. S., Singh, D. P. and Deshpande, A. A. 1998. Onion hybrids: Arka Kirthiman and Arka Lalima, *Indian Horticulture* 43(3):20.

Wall, M. and Corgan, J. N. 2002. "NuMex Arthur" Onion. *Hort. Science*, 37(4):707-708.

Xu, Q.J. and Cui, C.R. 2007. Genetic transformation of OSISAP1 gene to onion (*Allium cepa* L.) mediated by a microprojectile bombardment. *Zhi Wu Sheng Li Yu Fen Zi Sheng Wu Xue Xue Bao*. 33, 188-196.

Yam, J. Y., Xu, H. L., Zhuang, Y., Gao, B., Hong, J. Z. and Li, B. 2009. A new onion F_1 hybrid-'Jinxing'. *China Vegetables*. 16: 71-73.

Zeng, A. S., Yan, J. Y., Yun, J. S., Song, L. X. and Gao. B. 2011. A new onion F_1 hybrid for dehydration -'Baifeng'. *China Vegetables*. 12: 103-105.

Zhaoshui, L., Qipei, Z. and Dehang, C. 1995. Effect of two methods of maternal pedigree selection on components of white shaft weight in green onion. *Acta. Horticulture* 402: International Symposium on Cultivar Improvements of Horticultural Crops. Part I: Vegetable Crops.

Zheng, S. J. 2000. Towards onions and shallots (*Allium cepa* L.) resistant to beet armyworm (*Spodoptera exigua* Hübner) by transgenesis and conventional breeding. Thesis. Wageningen, Netherlands, pp. 146.

Zheng, S.J., Henken, B., Ahn, Y.K., Krens, F.A. and Kik, C. 2004. The development of a reproducible *Agrobacterium tumefaciens* transformation system for garlic (*Allium sativum L.*) and the production of transgenic resistant to beet armyworm (*Spodoptera exigua* Hübner). *Molecular Breeding*14, 293-307.

Zheng, S.J., Khrustaleva, L., Henken, B., Sofiari, E., Jacobsen, E., Kik, C. and Krens, F.A 2001.

15

Potato onion

Vijay Mahajan

Onions including potato onion or shallot are one of the oldest cultivated vegetables since more than 4000 years. It appears as carving on pyramid walls and in tombs from the third and fourth dynasty. The medicinal value of onion has been described by Charak in *Charak Sanhita*, a famous early medical treaty in India. Adoption of onion in India is from very early times before Christian era. In India, there are reports of onion in writing from 6th century BC. It is probably originated from Central Asia (between Turkmenistan and Afghanistan) where some of its relatives still grow in the wild. The closest among them are *Allium vavilovii* Popov & Vved. from southern Turkmenistan and northern Iran, with which it gives 100% fertile hybrids, and *Allium asarense* R.M. Fritsch & Matin from Iran. *Allium oschaninii* O.Fedtsch. (Uzbekistan and neighbouring countries), which used to be considered the ancestor of *Allium cepa* cannot be crossed successfully with the cultivated onion, but its domestication seems to be the origin of some European 'shallots'. From Central Asia the supposed onion ancestor probably migrated first towards Mesopotamia, where onion is mentioned in Sumerian literature (2500 BC), then to Egypt (1600 BC), India and South-East Asia. From Egypt, *Allium cepa* was introduced into the Mediterranean area and from there to all the Roman empire. According to Vavilov (1926) the Southwest Asian gene centre is proposed as the primary centre of domestication and variability of onion. Further, Vavilov and Bnkinich (1929), on the basis of ecotypes and wildforms, confirmed that Afghanistan and adjacent countries are the genetic centre of origin of the cultivated forms of onion and garlic. In addition, the secondary centre of origin in the Mediterranean gene centre represents the area from which onions with large bulbs were selected (Castell and Portas, 1994).

The Mediterranean countries and these regions are most important sources of genetic diversity for onions showing particular diversity (Astley *et al.*, 1982). Shallots probably originated in Central or South-East Asia, traveling from there to India and the eastern Mediterranean. The name "shallot" comes from Ashkelon, an Israeli city, where people in classical Greek times believed shallots originated (Anonymous 2010). Most of the 600 species of *Allium* are distributed in Afghanistan, Turkey, Iran and central Asia; Tnrkmen SSR, Uzbek SSR, Tadzhik SSR, Kirgiz SSR and Kazakh SSR; and Mongolia, the Tien-Shan Mountains and the Himalayas (Kotlinska *et al.* 1990). In Africa traditional tropical cultivars may have been introduced either from southern Egypt, or from India via Sudan to Central and West Africa (Messiaen & Rouamba, 2004). In Africa after selections from genetically heterogeneous seed or bulb lots, by local farmers adapted for seed-propagated onions, or selected to become shallots. *Allium cepa* as bulb onion and/or shallot is probably cultivated in all countries of tropical Africa. Important production areas for bulb onion are Senegal, Mali, Burkina Faso, Ghana, Niger, Nigeria, Chad, Sudan, Ethiopia, Kenya, Tanzania, Uganda, Zambia and Zimbabwe. In the lowlands between 10°N and 10°S shallots replace onions because the temperature is too high for vernalization and seed production, and the climate too humid. The short vegetative cycle of shallots (60–75 days) gives the possibility of two crops a year, especially in the four season climate along the Gulf of Guinea. Yellow or red/purple shallots are grown in Guinea, Côte d'Ivoire, Ghana, Benin, Nigeria, Sudan, Ethiopia, Uganda, Kenya, Tanzania, and on both banks of the Congo River near Brazzaville (Congo) and Kinshasa (DR Congo). The spicy taste and high dry matter content (15–18%) of shallots have made them attractive for growers farther from the equator, in many areas where common onions are also produced, e.g. by the Dogon in Mali, or in Cape Verde. It grows wild across the Zagros Mountains in different provinces of Iran (Ebrahimia *et al.*, 2009).

Potato Onions were once quite common, being mentioned in old gardening books of the 1800s. By 1940 they had fallen out of favour, and became quite rare and once a mainstay in many homestead gardens, they were gradually replaced by today's common onion, *Allium cepa*, as modern transportation for shipping and grocery stores replaced the need for people to grow their own onions (Winterton, 2011). The Green Revolution of the 1950s also dealt a blow to the popularity of the Potato Onion because modern agriculture and farming methods focused on the larger and more uniform size of common onions. Today, Potato Onions are grown only by a small percentage of gardeners who are interested in preserving this heirloom crop, or gardeners. The origins of Potato Onions previous to 1800 are mostly unknown. Old books presume their origin to be the northern areas of Europe. William (1997), reported that the Potato Onion was introduced into England in the 1790s under the name Egyptian onion, a name also applied to tree onions in common parlance. Johnson 1844, stated that the onion had been first introduced at Edinburgh by a certain Captain Burns, and for this reason it was sometimes called the Burns onion. There were two varieties, one that set bulbs on top like a tree onion, the other never sending up flowers. Pliny the Elder, in his Natural History, mentioned the Ascalon onion (the shallot,

Allium ascalonicum) as one of six types of onions known to the Greeks (Fenwick and Hanley 1985). It came from Ascalon in Syria, and Joseph Michaud's history of the Crusades affirmed this origin. Shallots were known in Spain, Italy, France, and Germany by 1554, had entered England from France by 1633, and were grown in American gardens by 1806 (Hedrick,1972).

Allium cepa is now divided into two large cultivar-groups: Common Onion Group, with large, normally solitary bulbs, reproduced from seed or from seed-grown bulblets ('sets'), and Aggregatum Group, with smaller, several to many bulbs forming an aggregated cluster, originating from a single mother bulb. A cultivar of *Allium cepa* will be considered as a 'shallot' (belonging to Aggregatum Group) if more than 200 buds are present in one kg of mother bulbs, and if, under suitable climatic conditions, most of these buds give new bulbs (Messiaen & Rouamba, 2004). The aggregatum group is usually vegetatively propagated but, recently, improved seed reproduced varieties of shallots have been bred and are being widely grown in Europe, Israel and North America. Hybrid cultivars have been developed in Israel and Holland using cytoplasmic male sterility (Rabinowitch and Kamenetsky, 2002). There is some dispute whether such seed-reproduced cultivars truly constitute shallots from the producers of traditional, vegetatively propagated varieties in France. Even if flowers, they are interfertile with the Common Onion group, and therefore they are the same species and thus the specific name A. ascalonicum, used for shallots in the past, is not justified. The shallot, is also called as "multiplier onion" *Allium cepa* L. var. *aggregatum* which was formerly classified as a species A. ascalonicum, which is a synonym (Anonymous 2010). The term "shallot" is also used for the French gray challot or griselle, *Allium oschaninii*, which has been considered to be the "true shallot" by many. It is a species that grows wild from Central to Southwest Asia.

The aggregatum group are not so important commercially as the Common Onion group, and many are grown as home garden crops; however, large-scale cultivation of shallots takes place in Europe, North America, Argentina and in some tropical regions. These are are traditionally grown in Finland and northern Russia (Aura, 1963). Shallots and multiplier onion are also suitable for such high-latitude regions with a short growing season besides in the humid tropics – particularly in lowland coastal regions (Currah, 2002). Shallots are of much less economic importance than their larger-bulbed relatives. They are mainly produced by small-scale and home-garden growers, but they are particularly important in the humid tropics since the local strains have the pest and disease resistance necessary to grow in that environment (Currah and Proctor, 1990). In France, they are an important commercial crop with about 2400 ha under cultivation, giving an annual production of about 50,000 tons (FAO, 2007).

Bulbs of aggregatum group are smaller and rapidly divide as compared with common onion and thus form laterals resulting into clusters (Jones and Mann (1963) of two types called multiplier onion or potato onion and shallots. Multiplier onions are wider and long encased with a outer skin whereas shallots form cluster of narrow, separate bulbs with leaves and flowers usually smaller than common onion. Mostly aggregatum group are vegetatively propagated. Some even flowers but are mostly infertile except few. Hence, some have openion that name *A. ascalonicum* used for shallots in the past is not justified. In general aggragatum group is not as much popular as common onion and restricted mostly to kitchen garden except in some areas. Its cultivation on large scale is taken in France and some tropical regions like West Africa, Caribbean area due to its bolting resistance and somewhat resistant to local diseases (Rabinowitch & Brewster, 1990) and southern parts if India. Shallot and multiplier onion can complete a cycle of foliage growth and bulbing in 60-75 days where common onion requires longer season of foliage growth and hence are vulnerable to pest and diseases as compared with aggragatum group (Currah and Proctor, 1990). Like garlic, shallots are formed in clusters of offsets with a head composed of multiple cloves. Their skin color can vary from golden brown to gray to rose red, and their off-white flesh is usually tinged with green or magenta. Shallots are much favoured by chefs because of their firm texture and sweet, aromatic, yet pungent, flavor. Shallot is sweeter with mild flavour compared with common onion and have better keeping quality of about six months. *A. cepa* var. *ascalonicum* (Shallot) is perennial onion rarely produces seeds, propagated through bulbs and produces bulbs in cluster on the surface of soil, have on an average two eyes per asymmetrical bulbs. Whereas, *A. cepa* var. *aggregatum* (Multiplier onion or potato onion) produces small bulbs born in cluster in the soil, have usually only one eye regardless of size of bulb (Vishnu Swarup, 2006, Paramguru *et al.*, 2015).

Adoption of onion in India is carried through from very early times before Christian era. Originally, native of Central Asia of temperate region with perennial/ biennial habit and long day bulbing characters, it has established well in India under tropical and short day (11-11.5 hrs.) photoperiodic conditions. During acclimatization, farmers applied selection pressure involuntarily to meet the market preferences. Among the cultivated species of *Allium*, onion (*A. cepa* L.), leek (*A. porrum* L.), shallot (*A. ascalonicum* L.) and chives (*A. schoenoprasum* L.) are well known vegetable crops grown in different part of India (Chadha 1985; Pandey *et al.*., 2005). The tropicalization progresses further southwards towards Bellary region in Northern Karnataka and finally a vegetatively propagated multiplier onion (aggregatum) type got established in Tamil Nadu (6 – 8 ^{0}N latitude). The adoption to hardy conditions of high rainfall, high temperature and short day photoperiod typical of rainy (*kharif*) season of Western India has not been chronologically documented (Sheshadri & Chaterjee, 1996). Aggregatum grown in 26,491 hectares area with 2.56 lakh tonnes production (Paramguru *et al.*, 2015) and it is gaining momentum among the growers. Ceratin types of aggregatum onion grown particularly near seashores of east cost of Tamil Nadu exhibit seed setting ability. It is also grown in kitchen gardens or in small areas in North Eastern Hill region and some parts of Orissa, where germplasm from these palces have been collected.

Shallots and multiplier onion (*Allium cepa* var. *aggregatum*, *A. ascalonicum*) are known by various names in different languages (Slobodan 2009), viz. English: potato onion, multiplier onion, Arabic: bassal el shallut, Chinese: fen nie yang cong, Danish: skalotteløg, kartoffeløg, Dutch: sjalot, Filipino: sibuyas talalog, French: échalote, oignon-paatate, German: Eschlauch, Hindi: kanda or gandana, Italian: scalogno, Japanese: sharotto, Malay: daun bawang, bawang merah, bawang kecil, Portuguese: chalota, Russian: luk-salot, Spanish: chalote, escaluña, ascalonia, Thai: homon deng, Vietnamese: hành huong, Australia; scallions.

Taxonomy

The taxonomic position of Alliums is still a matter of controversy. The genus Allium as a whole is assigned to the family "Liliacea" by some taxonomists (Polunin, 1969) to which it belongs on account of the superior ovary and to the Amryllidaceae by others (Traub, 1968) because of the umbellate inflorescence. Genus *Allium* characterized by superior ovary (characteristic of family Liliaceae) and scapose umbellate inflorescence (flowers borne in a bracted umbel on top of a scape) with membranous bracts (characteristic of family Amaryllidaceae) has been placed in family Alliaceae (Purseglove, 1972). Others have opted for a subfamily - Allioideae-of the Liliaceae (McCollum, 1976). Fay and Chase (1996), Friesen *et al.* (2000) and Chase *et al.* (2009), considered *Allium* (including Caloscordum Herb., Milula Prain and Nectaroscordum Lindl.) is the only genus in tribe Allieae. A separate family altogether, Alliacea- within which can be placed all those other genera, e.g., *Agapanthus, Brodiaea, Triteleia, Tulbaghia* which have superior ovaries and umbellate inflorescence.The following hierarchy regarding its classification has been adopted (Rabinowitch and Brewster, 1990).

Kingdom:	Plantae
Class:	Monocotyledones
Super Order:	Liliiflorae
Order:	Asparagales
Family:	Alliaceae
Tribe:	Allieae
Genus:	*Allium*
Species:	*cepa*
Variety:	*aggregatum and ascalonicum*

Cytology

Allium cepa var. *ascalonicum* (Shallot) have chromosome 2n=2x=16. Some genotype of shallot produce fertile flowers and even set seed. Its flowers are similar to those of common onion and hybrids between these two shows regular meiosis and are fully fertile with no loss of seedling viability in F2 or further generations thus shallot is given a varietal status (Jones and Mann 1963). Mitotic chromosomes of shallot are indistinguishable from common onion. There is close relationship in terms of morphology and c-band size on the basis of karyotype analysis where Vosa (1976) recorded widespread polymorphism for the terminal heterochromatic c-band regions. Battagalia (1957) and El-Gadi and Elkington (1975) recorded identical nature of the two karyotypes . Rajender *et al.* (1978) found G+C base pair percentage of 33.6% for the onion and 30.6% for the shallot whereas in terms of repetitions of DNA content it was 50% for *A. cepa* and 44% for the shallot. Meiosis is similar to *A. cepa* except chiasmata which tend to be locally mainly in distal and interstitial locations within the bivalents at metaphase (Cochran, 1953 and Jones and Kehr, 1957).

Chromosome number and chromosomes of *A. cepa* var. *aggregatum* (potato or multiplier onion) is similar to *A. cepa*. Only difference is that they form perennial clumps of bulbs underground and rarely produce any seed (Helm, 1956).

Shinde & Sontakke (1993) described multiplier onion and shallot as follows:

1. *A.cepa* var.*ascalonicum*	2n=2x=16	Commonly known as shallot, perpetuated by bulbs which form in clusters on soil surface.
2. *A.cepa* var.*aggregatum*	2n=2x=16	Commonly known as potato onion/underground onion/multiplier onion/Egyptian ground onion. It grows as closely packed clusters of bulbs underground rather than on surface like shallot.

Nutritive Value and Its Use

They are employed as a seasoning in stews and soups but can also be used in the raw state, diced in salads, or sprinkled over steaks and chops. Shallots also make excellent pickles (Hedrick 1972). The nutritional content of shallots depends on variety, ecological conditions, and climate. According to the Nutrition Coordinating Center (1994), 100 g (3.53 ounces or 0.44 cup) of shallots yields 32 kilocalories of energy, 1.83 g of protein, 0.19 g of fat, and 7.34 g of carbohydrate. One-half cup of fresh shallots provides a male between 18 and 25 years of age approximately 9 percent of the RDA of calcium (72 mg), 14.8 percent of iron (1.48 mg), 31.3 percent of vitamin C (18.8 mg), and 32 percent of folacin (64 mg) (National Research Council 1989). Shallots are high in potassium (276 mg) and low in sodium (16 mg). They contain small amounts of copper, magnesium, phosphorus, selenium, and zinc, and also have 38.42 mcg RE of vitamin A (230.54 mcg of beta-carotene) and 0.46 mg of vitamin E (alpha-tocopherol 0.37 mg, beta-tocopherol 0.17 mg, gamma-tocopherol 0.17 mg, and delta-tocopherol 0.09 mg), as well as small amounts of thiamine, riboflavin, niacin, pantothenic acid, and vitamin B6. Shallots have the same flavor components as onions but generally contain more methyl, propyl, and (1-propenyl) di- and trisulfides (Wu *et al.* 1982). A study of the volatile oils from raw, baked, and deep-fried shallots identified sulfides, disulfides, trisulfides, thiophene derivatives, and oxygenated compounds. The oils from baked or fried shallots contain decreased amounts of alkyl propenyl disulfides and increased amounts of dimethyl thiophenes (Carson, 1987). Similar to onions, raw shallots release chemicals that irritate the eye when sliced, resulting in tears. Shallots appear to contain more flavonoids and phenols than other members of the onion family (Yang *et al.*, 2004).

Shallots are extensively cultivated for fresh use in cooking, in addition to pickles. Finely sliced deep-fried shallots are used as a condiment in Asian cuisine (Anonymous, 2010). Raw shallot can be pickled along with cucumbers in mild vinegar solution. It is chopped finely, fried until golden brown for preparing crispy shallot chips called 'bawang goreng' (fried onions) in Indonesian language, sold in supermarkets. It enhances flavor of many South East Asian dishes including fried rice variants. Crispy shallot chips are also used in Southern Chinese cuisine. In Indonesia, it is also used as pickle in variable kinds of traditional food. It is widely used in the southern part of India. In Kannada it is known as 'Chikk-Eerulli' used extensively in snacks, salads, curries and rice varieties. In Malayalam it is called as 'Chuvannulli' used in Sambar (a tamarind flavoured lentil soup) and different types of curry.

Breeding and Varieties

Varieties of Multiplier onion in India (*Allium cepa* var. *aggregatum*)

Work on potato onion is restricted in India mainly to Tamil Nadu. Some of the local cultivars grown in Tamil Nadu are Cuddalore, Podusu, Dindigul Red, Mutlore Natu and Pallunvengayam (Paramguru *et al.*, 2015). Besides, some varieties were developed by TNAU and NHRDF. Details of varieties of Multiplier onion (*Allium cepa* var. *aggregatum*) released from various research institutes/ universities in India are as follows (Singh *et al.*, 2009, Mahajan 2011, Paramguru *et al.*, 2015).

CO-1: It is a clonal selection from a germplasm CS 450 and evolved at TNAU, Coimbatore. Leaves are light green. Composite bulb weighing 55-60g contains 7-8 red coloured bulblets. Bulblets are fairly pungent with TSS 8-9%. It produces bulb yield of 100 q/ha in 90 days cop duration. The variety is grown throughout Tamil Nadu.

CO-2: It is a clonal selection from a germplasm CS 911 and evolved at TNAU, Coimbatore. Leaves are cylindrical and green. Composite bulb weighing 60-65g contains 7-9 crimson coloured bulblets. Bulblets are fairly pungent with 12% TSS. It produces bulb yield of 120 q/ha in 65 days crop duration. The variety is grown throughout Tamil Nadu.

CO-3: It is a clonal selection from open pollinated progenies of CS 450 and developed at TNAU, Coimbatore. Leaves are erect, light green and cylindrical. Composite bulb weighing 75g contains 8-10 pink coloured bulblets. Bulblets are fairly pungent with 13% TSS. It produces bulb yield of 158 q/ha in 65 days crop duration. The variety is grown throughout Tamil Nadu. The bulbs can be stored for about 4 months.

CO-4: This variety was also developed at TNAU, Coimbatore. Plants are comparatively taller and have cylindrical green leaves. Composite bulb weighing 90g contains 8-10 light brown coloured bulblets. It produces bulb yield of 180 q/ha in 60- 65 days crop duration. The variety is grown throughout Tamil Nadu. The bulbs can be stored for about 5 months.

MDU-1: The variety was developed at the TNAU, Coimbatore, Madurai campus. Plants have medium height. Leaves are cylindrical, light to dark green in colour. Uniform big sized, composite bulbs weighing 75 g and contains 9-11 bulblets of bright red colour. It is adopted in southern district of Tamil Nadu. It produces bulb yield of 150 q/ha in 60-75 days crop duration. Bulbs can be stored for longer period.

Agrifound Red: It is developed at the National Horticultural Research and Development Foundation, Dindigul center. Composite bulb weighing 65g contains 6 light red coloured bulblets. Bulblets are fairly pungent with 15-16% TSS. It produces bulb yield of 180-200q/ha in 65 days crop duration. The variety is grown throughout Tamil Nadu.

CO-5: The variety has been developed by TNAU, Coimbatore. It is high yielding variety with attractive pink colour and bold bulblets. It is free flowering type with seed setting ability and can be propagated through seeds as well as bulblets. It gives the bulb yield of 190 q/ha. Seed crop takes 95-100 days and produce 250-300 kg seed/ha.

Improvement in Multiplier Onion/Shallot

In France, clonal selection combined with virus eradication by meristem tip-culture started in 1970. After 1984, hybridization was used and selection in the progenies obtained by intraspecific hybridization in the Jersey shallot type let to the production of new cvs recently released. In order to widen the cultivar types, interspecific hybridization, including the Jersey type, grey shallot, *Allium roylei*, and *Allium oschaninii*, was used. In this case *in vitro* embryo rescue was necessary to obtain plants which are now under study. Haplodiploidization was also used in order to obtain homozygous lines which could be used in the production of F1 hybrid varieties Cohat *et al.* (2001).

Fifteen accessions of shallot and multiplier onion, which are subgroups of aggregatum, common bulbing onion group, and inter-varietal hybrids of A. cepa, were characterized by Sumanaratne *et al.* (2005), based on morphological traits, isoenzyme and random amplified polymorphic DNA markers with confirmation of hybridity. The selected crosses harboured a wide genetic base, as the parents were genetically very distant. Therefore, the vegetative propagation of these hybrids or advanced generations is useful for developing new cultivars.

Flowering in multiplier were studied by Bunao, 1989 using different levels of phosphorus and gibberellic acid concentrations. The levels of phosphorus were: 0, 50, 100, and 150 kg P_2O_5/ha and the gibberellic acid concentrations were: 0, 75, 150, 300, 600, and 1200 ppm. Where no GA_3 was applied, application of phosphorus from 0 to 100 kg/ha increased the height of plants at maturity. The effect of phosphorus was not observed when GA_3 (gibberellic acid) was applied regardless of rates. The weight of seeds per umbel significantly increased with the application of higher levels of phophorus. At 100 and 150 kg P_2O_5/ha, weight of seeds per umbel increased from 377.11 to 426.11 and 466.05 mg. respectively. Days to bolting, number of plants bolted, number of days to seed maturity and seed per plot were not affected by the different treatments and similarly flowering characteristics, seed yield components, germination percentage, and seed vigor index were not affected by phosphorus and gibberellic acid applications.

In shallots, experiments to induce flowering in traditional vegetatively propagated types have shown that some flower readily whereas others do not flower (Krontal *et al.*, 2000; Esnault *et al.*, 2005). The optimal temperature for floral induction is 5–10°C; inflorescences can be induced in growing plants. Inflorescence initials formes after the production of only six leaves, including primordia, following a mid-October sowing in Israel (Krontal *et al.*, 1998). Bolting-susceptible cvs of onion become sensitive to vernalizing temperatures after initiation of seven leaves, but at least one and usually two or three further leaf primordia are initiated before the shoot apex becomes floral, thereby giving a minimum of about nine leaf initials before the floral apex. In case of tropical shallot juvenility ends at an earlier physiological age than in any bulb onion so far investigated. Tabor *et al.* (2005) studied of two tropical and one temperate shallot cvs grown from seed and found that the tropical cvs needed, respectively, 3.50 or 4.75 visible leaves which is to be 50% responsive to vernalizing temperatures whereas, temperate cv needed seven visible leaves. Esnault *et al.* (2005) investigated the conditions needed to induce flowering in the traditional vegetatively propogated shallots, in France and found range in the flowering responses of shallots. Some tropical types flower easily when exposed to cool temperatures, while some temperate types can be induced to flower given sufficient time under cool conditions when they are large enough. However, some strains fail to flower when exposed to long periods at temperatures normally favourable for flower initiation.

In India most of the improvement work is concentrated in southern parts of ther country. Major breeding work done at TNAU and some work at regional centers of NHRDF in south. Twelve seed setting types were collected at TNAU and evaluated with CO-5 by Kumaravelu, 2009, where cultivar Aca12, Aca3 and Aca 7 were found promising. Work at DOGR has been also initiated and germplasm of multiplier onions were collected from different parts of the country. Further evaluation is in progress in red and white multiplier onion and there is wide scope for its popularization in other parts of the country and in non-traditional areas.

References

Anonymous 2010. *Allium ascalonicum* information". *Germplasm Resources Information Network*. USDA. Retrieved 2010-08-20. http://wikipedia.unicefuganda.org/latest/A/Shallot.html

Astley, D., Innes, N.L. and Van Der Meer Q. P. 1982. Genetic Resources of Allium Species–A Global Report. International Board for Plant Genetic Resources, Rome, Italy. *Report No: FAO-AGPG-IBPGR/81/77*, pp. 41.

Aura, K. 1963. Studies on the vegetatively propagated onions cultivated in Finland, with special reference to flowering and storage. *Annales Agriculturae Fenniae* 2 (Suppl. 5).

Battagalia, E. 1957. *Allium ascalonicum* L., *A. fistulosum* L., *A. cepa* L. analisi cariotipica. *Caryologia*, 10, 1.

Bunao, C.B. 1989. Influence of gibberellic acid and phosphorus on seed production of multiplier onion.. Thesis (M.S. in Horticulture), Munoz, Nueva Ecija (Philippines). Mar 1988. 88 leaves.

Carson, J.F. 1987. Chenistry and biological properties of onion and garlic. *Food Rev. Internat.* 3 (1 & 2): 71-103.

Castell V.R. and Portas C.M. 1994. Alliacea production systems in the Iberian Peninsula: facts and figures of potential interest for a worldwide R&D network. *Acta Horticulture* 358, 43-47.

Chadha, Y.R. 1985. The Wealth of India - Raw Materials Series. NISCAIR, CSIR.

Chase M.W., Reveal J.L and Fay M.F. 2009. A subfamilial classification for the expanded asparagalean families Amaryllidaceae, Asparagaceae and Xanthorrhoeaceae. *Botanical Journal of the Linnean Society* 161, issue 2, 132–136.

Cochran, F.D. 1953. Cytogenetic studies of the species hybrid *Allium fistulosum* x *Allium ascalonicum* and its backcross progenies. In La state University Study Biological Science, 2nd series, 1.

Cohat, J., Chauvin, J.E. and Le Nard, M. 2001. Shallot (*Allium cepa* var. *aggregatum*) production and breeding in France ISHS *Acta Horticulturae* 555: II International Symposium on Edible Alliaceae, 10.17660/ActaHortic.2001.555.32.

Currah, L. 2002. Onions in the tropics: cultivars and country reports. (In:) Rabinowitch, H.D. and Currah, L. (Eds) *Allium Crop Science: Recent Advances.* CAB International, Wallingford, UK, pp. 379–407.

Currah, L. and Proctor, F.S. 1990. Onion in Tropical Regions. NRI, UK. Bulletin No. 35.

Cuthbert W. *Johnson*, C. W. 1844. *The Farmer's Encyclopedia*, p. 861.

Ebrahimia, R., Zamani, Z. and Kash, A. 2009. "Genetic diversity evaluation of wild Persian shallot (*Allium hirtifolium* Boiss.) using morphological and RAPD markers". *Scientia Horticulturae* 119 (4): 345–351.

El-Gadi, A. and Elkington, T.T. 1975. Comparision of the Giemsa C-band karyotypes and the relationships of *Allium cepa, A. fistulosum* and *A. galanthum. Chromosoma*, 51, 19.

Esnault, F., Tromeur, C., Kermarrec, M.P. and Pouliquen, R. 2005. Effect of temperature on growth and flowering ability of three Jersey shallot (Allium cepa L. group Aggregatum) cultivars. *Journal of Horticultural Science and Biotechnology.* 80, 413–420.

FAO 2007. *Agriculture Production and Trade Statistics.* Food and Agriculture. Organization of the United Nations, Rome (available at: http://faostat.fao.org).

Fay M.F. and Chase M.W. 1996. Resurrection of Themidaceae for the Brodiaea alliance, and recircumscription of Alliaceae, Amaryllidaceae and Agapanthoideae. *Taxon* 45, 441–451.

Fenwick, G. R. and Hanley, A. B. 1985. The genus Allium. Part-I. *CRC Crit Rev. Food Sci. Nutr.* 22, 273-340.

Friesen N., Fritsch R.M, Pollner S and Blattner F.R. 2000. Molecular and morphological evidence for an origin of the aberrant genus Milula within Himalayan species of *Allium* (Alliaceae). *Molecular Phylogenetics and Evolution* 17, 209–218.

Hedrick. U.P. 1972. Sturtevant's Edible Plants of the World. Dover Publications. ISBN 0– 486 – 20459 - 6

Helm, J. 1956. Die zu wurz- und speisezwecken kultivierten arten gattung Allium L. Die Kulturpflanze, 4, 130.

Jones, H.A. and Mann, L.K. 1963. *Onions and their Allies.* Leonard Hill, London.

Jones, S.T. and Kehr, A.E. 1957. The cytology and plant characteristics of amphidiploids derived from *Allium ascalonicum* x *A. fistulosum. Am. J. Bot.*, 44, 523.

Kotlinska T., P. Havranek, M. Navratill, L. Gerasimova, A. Pimakhov and S. Neikov. 1990. Collecting Onion, Garlic and Wild Species of Allium in Central Asia, USSR. *FAO/ IBPGR Plant Genetic Resources Newsletter*, 83184, 31-32.

Krontal, Y., Kamenetsky, R. and Rabinowitch, H.D. 1998. Lateral development and florogenesis of a tropical shallot: a comparison with bulb onion. *International Journal of Plant Science.* 159, 57–64.

Krontal, Y., Kamenetsky, R. and Rabinowitch, H.D. 2000. Flowering physiology and some vegetative traits of short-day shallot: a comparison with bulb onion. *Journal of Horticultural Science and Biotechnology.* 75, 35–41.

Kumaravelu, K.V. 2009. Evaluation of seed propagated cultivars of multiplier onion (*Allium cepa.* Var. *aggregatum* Don.). M.Sc. (Hort.) Thesis, Dept. of Vegetable Crops, HC & RI, TNAU, Coimbatore, Tamil Nadu, India.

Mahajan Vijay and Lawande K.E. 2011. Genetic diversity and crop improvement in onion and garlic. *Souvenir: Exploiting spices production potential of the Deccan region,* SYMSAC-VI, Indian Society for Spices, 8-10 Dec., 2011, Dharwad, 19-40.

McCollum, G.D. 1976. Onion and Allies In: Simmonds N. W. (Ed.).Evolution of Crop Plants. London and New York, Longman, pp.186-190.

Messiaen, C.M. and Rouamba, A. 2004. *Allium ampeloprasum* L. Internet Record from PROTA4U. PROTA (Plant Resources of Tropical Africa / Ressources végétales de l'Afrique tropicale), Wageningen, Netherlands. (<http://www.prota4u.org/search.asp).

National Research Council 1989. Recommended Dietary Allowances. 10th Edition, Food and Nutrition Board. Commission on Life Sciences. National Academy Press, Washington, D.C.

Nutrition Coordinating Center 1994. Nutrition data. System Version 2.6/8A/23, St. Paul, Univ. of Minnesota, Minnesota.

Pandey U.B., Kumar A., Pandey R. and Venkateswaran, K. 2005. Bulbous crops – cultivated Alliums. In: Dhillon BS, Tyagi RK, Saxena S, Randhawa GJ (Eds) Plant genetic resources of vegetable crops. Narosa Publishing House Pvt Ltd., New Delhi, pp. 108–120.

Paramaguru, P., Velmurugan, M. and Pugalendhi, L. 2015. Aggregatum onions and Shallots. In The Onion. (Ed.) Krishna Kumar, N.K., Jai Gopal and Parthasarthy, V.A. Pub. ICAR-DKMA, ICAR, KAB, N.Delhi.

Polunin, O. 1969. *Flowers of Europe: A Field Guide.* Oxford University Press, London, pp.682.

Purseglove, J.W. 1972. *Alliaceae.* (In:) *Tropical Crops (monocotyledons).* Longman Group Ltd., pp. 37–57.

Rabinowitch H.D, and Brewster, J.L. 1990. Onion and Allied Crops, Vol.I, Botany, Physiology and Genetics, CRC Press, Inc. Florida, pp. 273.

Rabinowitch, H.D. and Kamenetsky, R. 2002. Shallot (*Allium cepa,* Aggregatum group). (In:) Rabinowitch, H.D. and Currah, L. (Eds) *Allium Crop Science: Recent Advances.* CAB International, Wallingford, UK, pp. 187–232.

Rajender, P.K., Pallotta, D. and Lafontaine, J.G. 1978. Analysis of plant genomes. V. Comparative study of molecular properties of DNAs of seven *Allium* species. *Biochem. Genet.,* 16, 957.

Sheshadri, V. S. and Chaterjee, S. S. 1996. The history and adaptability of some introduced vegetable crops in India. *Vegetable Science* 23 (2), 114 – 140.

Shinde, N.N. and Sontakke, M.B. 1993. Bulbb crops: Onion. (In:) Bose, T.K., Som, M.G. and Kabir, J. (Ed.), *Vegetable Crops,* Naya Prokash, Calcutta, pp. 641 – 685.

Singh, H.P., Rai, M., Pandey, S. and Kumar, S. 2009. Vegetable varieties of India. Studium Press (India) Pvt. Ltd. pp. 99-108.

Slobodan V. 2009. Asian vegetables glossary. Published by Department of Environment and Primary Industries, 1 Spring Street, Melbourne, Victoria. http://agriculture.vic.gov.au/agriculture/horticulture/vegetables/vegetables-a-z/asian-vegetables/asian-vegetables-glossary.

Sumanaratne, J. P., Samarasinghe, W. L. G., Wanigadeva, S. M. S. W., Gunasekara, I. A., Ranjanee, S. and Kumary, L. G. S. 2005. Characterization of onions (Allium cepa L.) by morphological traits, isozymes and randomly amplified polymorphic DNA markers. Annals-of-the-Sri-Lanka-Department-of-Agriculture. 2005; 7: 253-270.

Tabor, G., Stuetzel, H. and Zelleke, A. 2005. Juvenility and bolting in shallot (*Allium cepa* L. var. *Ascalonicum* Backer). *Journal of Horticultural Science and Biotechnology.* 80, 751–759.

Traub H.P. 1968. The subgenera, sections and subsections of *Allium* L. *Plant Life* 24, 147–163.

Vavilov N.I. 1926. *Origin and Geography of Cultivated Plants.* English translation by D Love (1992). Cambridge Univ Press, Cambridge, U.K.

Vavilov, N.I. and Bukinich, D.D. 1929. Zemeledelcheskii Afganistan.Tr. po. prikl. Botanike, genetike i selekcii VIR. T.Z. pp.156-158.

Vishnu Shwarup 2006. Vegetable Science and Technology in India. Kalyani publishers, New Delhi, India.

Vosa, C.G. 1976. Heterochrometic patterns in Alliums. I. The relationships between the species of the *Cepa* group and its allies. *Heredity,* 36, 383.

William W. W. 1997. Heirloom Vegetable Gardening: A Master Gardener's Guide to Planting, Seed Saving, and Cultural History. Pub. Henry Holt & Co; 1st edition.

Winterton K. 2011. Potato Onions in past history. Booklet: Resurrecting the Potato Onion. https://docs.google.com/document/d/1jnqst7- 9YfWFovhqjARtcZZVJC0TPzKsow_5mdAwnyA/edit?pref=2&pli=1

Wu, I. L., Chou, C.C., Chen, M.H. and Wu, C.M. 1982. Volatile flavor component from Shallots. *J. Food Sci.* 47: 606-608.

Yang, J., Meyers, K.J., van der Heide, J. and Liu, R.H. 2004. "Varietal differences in phenolic content, and antioxidant and antiproliferative activities of onions". *J. Agric. Food Chem* 52 (21): 6787–6793.

16

Pineapple

P P Joy and Anjane Sumathi

Christopher Columbus, an Italian explorer, navigator and colonizer who discovered pineapple described it as "Fruits like artichoke, four times as tall, fruit in the shape of a pine cone, twice as big, fruit is excellent and can be cut with a knife like a turnip and it seems to be wholesome." Pineapple (*Ananas comosus* (L.) Merr.), is found in almost all the tropical and subtropical areas of the world, and it ranks third in production of tropical fruits, behind bananas and citrus (Paull and Duarte, 2011). According to the Food and Agriculture Organization (FAO) statistics (http://apps.fao.org), world pineapple production increased from 3,833,137 tons in 1961 to 15,287,413 tons in 2004. Five countries, namely Thailand (1,7000,000 t), the Philippines (1,650,000 t), Brazil (1,435,600 t), China (1,475,000), and India (1,300,000) contributed with about half of the world production in 2004. Pineapple is also a source of bromelain, used as a meat-tenderizingenzyme, and high quality fibre (Coppensd'Eeckenbrugge *et al.*, 2011).It contains considerable calcium, potassium, fibre, and vitamin C. It is low in fat and cholesterol. It is also a good source of vitamin B_1, vitamin B_6, copper and dietary fibre. Pineapple is a digestive aid and a natural Anti-Inflammatory fruit (Joy, 2010).

Fig. 1: Mauritius pineapple (field in Kerala, India)

Fig. 2: Christopher Columbus and native Americans (Image from the Florida Memory Project.)

Discovery

Pineapple had been an integral part of diet in America years before it was discovered by the Italian explorer, Christopher Columbus on 4th November 1493. The natives of the Guadaloupe Island served him and his colleagues this delicious fruit, pineapple. He carried the fruit to the Europe. The European travelers were greatly impressed by this fruit that they often mentioned about them in their chronicles (Morrison, 1963; Collins, 1960). The people of South America called it, 'nanas' or 'ananas' (in Guarani language 'ananas' meant 'excellent fruit'). The people around there literally cultivated it with a definite protocol and even they were so keen to select the superior types to obtain higher fruit yield and quality. They developed an in depth knowledge about the crop agronomy and its production. They also practised some of the processing standards continuing throughout the world till now. The native Americans prepared pineapple wine, extracted fibre and met some pharmacological needs like emmenagogue, abortifacient, antiamoebic, vermifuge, stomach disorders, poisoning arrow heads (Leal and Coppens d'Eeckenbrugge, 1996). Leal and Antoni in 1980 proposed the center of origin of pineapple, an areafurther north, between 10° N and S latitudes and 55–75° W longitude covering the areas of north-western and eastern Brazil, Columbia, Guyana and Venezuela. The varieties grown during the time of Columbus arrival were all seedless types and many typical wild cultivars were missing (Collins, 1948). As part of colonization the Spanish and Portuguese navigators played a pivotal role in distribution and spread of pineapple to all other parts of the world (www.daf.qld.gov.au).

Domestication

All crops cultivated now were once originated at one place and were transported to other for better utilization and undergone lots of changes gradually. Same goes to pineapple as well. It also went through this evolutionary process in which new types were constantly being selected to meet new demands by the cultivator. At the same time they were moved further away morphologically, physiologically and genetically from their wild progenitors (Ladizinsky, 2012).

Two types of domestication can be distinguished conceptually: landscape domestication and plant population domestication. Both are interrelatedbecause domesticated populations require some kind of landscape management, especially cultivation.Plant population domestication is a co-evolutionary process by which human selection on the phenotypes of promoted, managed or cultivated individual plants results in changes in the descendent population's phenotypes and genotypes that make them more useful to humans and better adapted to human management of the landscape (Clement, 1999). A domesticated population has been further selected for adaptation to human-modified landscapes, especially cultivated gardens and fields, and has lost its original ecological adaptations for survival without humans, especially its original dispersal mechanisms and survival capabilities (Clement *et al.*, 2010).

Domestication of Cultivars

'Smooth Cayenne' and 'Queen' were the two early cultivars distributed from Europe to all tropical and subtropical countries (Collins, 1951). The Spaniards and Portuguese dispersed other varieties, including 'Singapore Spanish', to Africa and Asia during the great voyages of the 16th and 17th centuries. Both Smooth Cayenne and Singapore Spanish can be called true cultivars. The pineapples were taken up by the Spanish and Portuguese explorers to the tropical world. It was introduced into Africa at an early date and reached southern India in 1548. Before the end of the 16th century, it had become established in China, Java and the Philippines (Collins, 1949). A kind of cloth was being made from pineapple leaf fibres (piña cloth) in the Philippines in the 1500's. They were first grown to fruiting inHolland about 1690. The first successful greenhouse cultivation was by Le Cour, or La Court, at the end of the 17th century near Leyden. He published a treatise on pineapple horticulture, including 'forcing' the plants to flower. Pineapple plants were distributed from The Netherlands to English gardeners in 1719 and to France in 1730 (Gibault, 1912). As pineapple cultivation in European greenhouses expanded during the 18th and 19th centuries, many varieties were imported, mostly from the Antilles. The now famous variety Cayenne Lisse ('Smooth Cayenne') was introduced from French Guiana by Perrotet in 1819 (Perrotet, 1825). The first pineapples (rough-leafed) are thought to have been introduced to Australia from India in 1838 by a German missionary, although some records indicate that pineapples were grown near Sydney as early as 1824 (www.daf.qld.gov.au., 2015).The Cayenne variety of pineapple was first mentioned in an English horticultural journal in 1841. Evidence is presented to show that it came from French Guiana in 1820. It is presumed to have been grown by the Maipure Indians in the upper Orinoco River valley long before it reached French Guiana. The manner and time of its origin are obscure. The Cayenne variety reached Jamaica in 1870, although it came to Jamaica via Florida. Australia attributed a lot during the 19th century for the pineapple canneries in Hawaii. European propagation was accomplished in France. Hawaii has been a major source of distribution during the first half of the twentieth century. It was imported from Ceylon into South Africa. In Australia its development was fairly rapid and from 1890 to 1895 that country was able to furnish many slips and suckers for expanding the pineapple industry in the Hawaiian Islands, although the first Cayennes came to Hawaii from Florida in 1885 and Jamaica in 1886. The decade from 1885 to 1895 was a period of accumulation of the Cayenne variety in the Hawaiian Islands. Introductions of pineapple plants were made from 11 different tropical countries (Florida, England, Jamaica, Bahamas, Trinidad, Puerto Rico, Mexico, Australia, Singapore, Samoa and Algeria) four of which were known to have included the Cayenne variety. Hawaii has been a major source of distribution during the first half of the twentieth century (Collins, 1934).

Domestication of Varieties

The genus Ananas is ideal for domestication studies, with multiple processes in time and space, and specialization related to the major uses as a food or as a source of fibers. Selection for fruit characteristics took place where the diversity and quality of spontaneous materials allowed it. The fruit quality induced the crop‘s dispersal, which in turn induced further diversification and environmental specialization (Clement *et al.*, 2010). Two hot spots for cultivated *A. comosus var. comosus* diversity were found. The first one lies in the eastern Guiana Shield and hosts a wide nuclear and cytoplasmic diversity along with a number of intermediate forms between *A. comosus var. comosus* and the wild *A. comosus var. ananassoides* that is commonly observed in the forest. These intermediate forms are noticeable by their variation in fruit size. These data point out this region as a likely primary center of domestication for the fruit. The second hot spot liesin the upper Amazon. No wild or intermediate forms have been found in this region, which appears as an important center of diversification of agriculture (Schultes, 1984; Clement, 1989) and could be a center of diversification for the domesticated pineapple. The plant would have been brought there by humans, which allowed for completion of the domestication process while in the absence of counteracting gene flow from wild forms.

'Curagua' (*A. comosus* var. *erectifolius* [L.B. Smith] Coppens and Leal) developed as a fiber crop *via* selection from *A. comosus var. ananassoides*. It was commonly cultivated north of the Amazon and Solimões rivers, as well as in the Antilles in pre-Columbian times. Its characteristic dense, erect and smooth foliage are the likely result of selection for an abundance of long easily-extractable fibers. Genetic affinity of the 'Curagua' with different lineages of var. ananassoides indicates multiple and independent domestication events (Duval *et al.*, 2001; 2003). Their antiquity is probably variable, as some clones have reduced fruit production, while others are remarkably fertile.

The domestication process for *A. comosus* var. *bracteatus*, also cultivated for its fiber in Paraguay (Bertoni, 1919), may have simply consisted of the direct vegetative propagation of rare interspecific hybrids, as this botanical variety has very limited variability. Furthermore, the chloroplast haplotype of the rarest form is very similar to that of *A. macrodontes* (Duval *et al.*, 2003).

A. bracteatus sensu is limited to the southeast of theSubcontinent where it is grown as a fence. This form is very homogenous, displays the most common cytoplasmic haplotype shared with other cultivated forms and the variety ananassoides, and shares nuclear markers specific to the southern group constituted by *A. macrodontes* and *A. fritzmuelleri* Camargo. These data point out this form as a hybrid between representatives of these two groups (Coppens d'Eeckenbrugge*et al.*, 2011).

Glottochronology

The glottochronology (The study of the historical relationships between languages) of pineapple in Ancient Mesoamerica suggests that the crop was significant by 2,500 years ago. Thus, domesticated pineapple was traded and adopted as an important fruit crop on a continental scale more than 3,000 years ago. Given the rarity of sexual reproduction in *A. comosus* var. *comosus,* the development of tradable cultivars was necessarily a long and slow process, certainly counted in millennia. Thus, a likely time frame for the divergence between wild and cultivated pineapple lies between 6,000 and 10,000 years ago (Clement *et al.*, 2010).

Phylogeography

Origin, dispersal and genetic diversity in living populations of native Amazonian crops are studied *via* phylogeography (Avise, 2000). It is the analysis of the geographic distribution of genetic variants, especially lineages of genes, which is generally due to dispersal of organisms (seed dispersal in plants) and thus provides insight into the history of a species. The same information permits inferences about the domestication process (Emshwiller, 2006; Zeder, 2006; Pickersgill, 2007). This contribution identifies emergent patterns that can be used to interpret crop domestication and dispersal before conquest.

The studies using enzymatic systems concluded a clear separation of two groups with *A. comosus* var. *bracteatus sensu* Smith & Downs and *A. macrodontes* (Coppensd' Eeckenbrugge, *et al.*, 2011). A low cytoplasmic diversity was reported with only one polymorphic probe-endonuclease combination (Noyer, 1991). Later nuclear *r*DNA studies revealed a group of six in the whole Ananas genus. The first and largest group include all the clones of the variety comosus except one, parguazensis, and ananassoides from Venezuela. The second group consists of *A. comosus* var. *bracteatu* saccessions (Noyer *et al.,* 1998). A French–Brazilian pineapple prospecting expeditions jointly conducted a nuclear DNA RFLP analysis of 301 samples of S. America (Duval *et al.,* 2001b). A large distribution range High levels of variation was found within *A. macrodontes* and the wild forms *A. comosus* var. *ananassoides* and *A. comosus* var. *parguazensis.* Genetic diversity varied within cultivated forms, ranging from very low (*A. bracteatussensu* Smith & Downs), to very high (*A comosus* var. *erectifolius). A. macrodontes* separated well butshared 58.7% of the markers with Ananas and wasvery close to the diploid *A. fritzmuelleri* Camargo. Within Ananas, only *A. comosus* var. *parguazensis* accessions form a consistent cluster. The scattering of botanical varieties and the occurrence of intermediate forms indicates a very probable gene flow, which is consistent with the lack of reproductive barriers between them. Chloroplast restriction site variation was then usedto study a subsample of 97 accessions of Ananaschosen for their genetic diversity and 14 accessions from other genera of the Bromeliaceae for phylogenetic purposes (Duval *et al.,* 2003). No sister group was evidenced among these bromeliads.

A. macrodontes and *A. comosus* varieties were represented by 11 haplotypes and formed a monophyletic assemblage withthree strongly supported groups. Two of these groupsare consistent with the nuclear data analysis and with geographical data.**The first group** includes the *tetraploid A. macrodontes,*represented by only one haplotype and the diploid *A. fritzmuelleri Camargo,* both from the southof the subcontinent and adapted to low light conditions.The contrast in *A. macrodontes* exhibiting high nuclear but low cytoplasmic diversity favor's the hypothesis of a recent speciation process by autopolyploidization. The nature of its parental relationship with *A. fritzmuelleri Camargo* is difficult to evaluate because of the extreme rarity of the latter(no accession could be recovered during the 1990s prospecting expeditions).**The second group** includes the majority of *A. comosus* var. *parguazensis* accessions, all from the RioNegro region. **The third** and largest group includes cultivated forms, *A. comosus* var. *comosus* and *A. comosus* var. *erectifolius,* as well as wild forms, *A. comosus*var. *ananassoides,* and the remaining accessions of *A. comosus*var. *parguazensis,* from thewhole Ananas distribution range.The comparison of molecular data obtained using uniparentally and biparentally inherited markers indicate hybridization between these groups in the RioNegro region, as well as the hybrid status of *A. bracteatussensu* Smith & Downs from the south(Coppensd'Eeckenbrugge, *et al.,* 2011).

Pineapple Culture in Societal Settings-Some Historical Artifacts

The European gardeners took a greater effort to design a hot house for the growth of pineapple. There were records of the King Ferdinand of Spain eating pineapple in 1530. Later, by the end of 16th century, Portuguese and Spanish explorers introduced pineapples into many of their Asian, African and South Pacific colonies like Philippines, Hawaii and Guam. Portuguese traders brought seeds of pineapple from Moluccas to India in 1548. During 1500's clothes were made from pineapple fibres in Philippines and called as pina cloth. They also introduced the crop to the east and west coasts of Africa. By 1594, it was seen growing in China too. No one is certain of when pineapples were first grown in Hawaii, but historians believe that a Spanish shipwreck in 1527 on the South Kona coast on the Big Island of Hawaii brought tools, stores, garments and plants, including pineapples, from Mexico to Hawaii.

In 1655, pineapple was known to be grown in South Africa. In 1675, the King Charles II of England posed for an official portrait by Hendrick Danckurts which showed him receiving a pineapple as a gift from his royal gardener, John Rose. It was then considered as a symbol of royal privilege to receive such a rare gift. During 1690, it was first cultivated in Holland and after 30 years moved on to England (Collins, 1950). During 17th century in the Americas it was regarded as the symbol of hospitality. Also the fruit was hanged before the houses of shipmen as a custom for their safe return after long voyages.The pineapple is described in seventeenth- century books as having been brought to China from Brazil via the East Indies.

The cultivation of pineapple was unsuccessful in England till 1712. Later by 1700's greenhouse culture flourished in England. Le Cour made the first greenhouse cultivation of pineapple near Leyden. His treatise on pineapple horticulture which included 'forcing' the plants to flower. This could be the first attempt of artificial flower induction in pineapple. In 1719, the pineapple plants were brought to English gardeners from Netherland. Also the plants were distributed to France from Netherland in 1730 (Gibault, 1912).

In 1777, Captain Cook planted pineapple in South Pacific and some nearby islands. In later years, more Spanish explorers arrived in Hawaii, planting pineapples among other fruits. Francisco de Paula Marin, a Spanish adventurer who arrived in Hawaii in 1794 and became a trusted friend and advisor to King Kamehameha I the Great, experimented with raising pineapples in the early 1800s.He was a big experimenter of Hawaiian herbs and the ancient folklore that surrounded their healing properties. He was also well known for his unique gardens that derived from Spanish-born seeds.

The first written record of pineapple in Hawaii was a diary note of Don Francisco de Paula Marın on 21stJanuary 1813 saying "This day I planted pineapples and an orange tree" (Collins, 1960).

The "Wild Kailua" pineapple was found growing in the Kona area as early as 1816. In 1838, Lutheran missionaries in Brisbane, Australia, imported plants from India. In 1840 the first commercial cultivation was introduced in Nundah (a part of Brsibane) (Lewcock, 1939).Smooth Cayenne was brought to Australia from Kew Gardens in the year 1858 (Collins, 1960). Now pineapples are commercially grown in Queensland. There is a Big Pineapple in Queensland which was a tourism icon when it was first opened in 1971. In 2015, it is opened for the tourists and has a clear depiction of iconic history of pineapple farming (Fig.3).

'Smooth cayenne' was first identified from French Guiana by Perrotet in 1819 (Perrotet, 1825). 'Smooth cayenne' and 'Queen' cultivars were distributed from Europe to tropical and sub-tropical regions. Now a days they were imported from West Indies to the European countries.

The first reference in the literature to the Cayenne variety appears to be the short notice carried in the Gardeners' Chronicle (England) of March 6, 1841, under the column heading of "Foreign Correspondence"- Only four kinds are considered desirable for general cultivation; of these, however, more than 1000 plants are annually fruited, namely 700 Queens and 300 Cayennes, Endville, and Providence.

Jamesm Dole who pioneered the industry and became popularly known as the "Pineapple King" founded a company in Hawaii in 1851, the company built its reputation on its commitment to "quality, and quality, and quality." These were the words of James Drummond Dole's "Statement of Principles," upon which he founded and operated the Company. James Dole, formed the nucleus of what would eventually become the largest pineapple industry in the world (Auchter, 1951; Larsen and Marks, 2010). He quickly established relationships with prominent citizens in Hawaii, including Governor Sanford B. Dole, his second cousin. These

Fig.3: Big Pineapple in Australia (Queensland)

relationships helped to assure that his venture into pineapple growing and canning was not starved for capital (Marks, 2010).The pineapple fruit was first canned in Baltimore in 1865. There the pineapples were imported from the Bahamas and later also from Cuba (Hawkins, 1995).Pineapples were first canned in Malaya by a retired sailor, in 1888 and exporting from Singapore soon followed. In 1860, fields were established on Plantation Key and Merritt's Island. It reached Jamaica in 1870, although it came to Jamaica via Florida.And in 1876 planting material from the Keys was set out all along the central Florida east coast. Shipping to the North began in 1879.

The basis for the modern Hawaii industry was begun when John Kidwell, a trained horticulturist, arrived in Honolulu from San Francisco in 1882 and established a nursery in Manoa Valley. 'Smooth Cayenne' was first introduced to Hawaii in 1886 from Florida. Captain John Kidwell is credited with founding Hawaii's pineapple industry. In the 1880's he imported and tested a number of varieties and selected Smooth Cayenne for its cylindrical form and uniform texture. Kidwell was encouraged by Charles Henson, a local horticulturist and fruit broker, to grow pineapples because he liked to include a few fresh pineapples in his banana shipments to the U.S. mainland (Auchter, 1951). The commercial Hawaiian pineapple canning industry began in 1889 when Kidwell's business associate, John Emmeluth, a Honolulu hardware merchant and plumber, produced commercial quantities of canned pineapple. The decade from 1885 to 1895 was a period of accumulation of the Cayenne variety in the Hawaiian Islands.

James Dole established the Hawaiian Pineapple Company (HPC) in 1901 and is "usually considered to have produced the first commercial pack of 1,893 cases of canned pineapple in 1903" (Auchter, 1951). By 1900, shipments reached a half million cases. One early planter on Eden Island moved his farm to the mainland because bears ate the ripe fruits. With the coming of the railroad in 1894, pineapple growing expanded. The 1908-09 crop was 1,110,547 crates. Then Cuban competition for U.S. markets caused prices to fall and many Florida growers gave up. It was then believed in those days that the pineapple benefitted by closeness to salt water. World War I brought on a shortage of fertilizer, then several freezes in 1917 and 1918 devastated the industry. A commercial industry took form in 1924. In the early 1930's, the United Fruit Company supplied slips for a new field at White City but the pressure of coastal development soon reduced this to a small patch. Shortly after World War II, a plantation of 'Natal Queen' and 'Eleuthera' was established in North Miami but, after a few years, the operation was shifted inland to Sebring, in Highlands County, Central Florida, where it still produces on a small scale.

The first plantings in Israel were made in 1938 when 200 plants were brought from South Africa. In 1939, 1350 plants were imported from the East Indies and Australia but the climate was not a favourable one for this crop. Over the past 100 years, the pineapple has become one of the leading commercial fruit crops of the tropics.

A modern canning plant was erected in about 1946. South Africa produces 2.7 million cartons of canned pineapple yearly and exports 2.4 million Cartons. In addition, 31,000 tons of fresh pineapple were sold on the domestic market and 500,000 cartons exported yearly. As in many areas, pineapple culture existed on a small scale on the Ivory Coast until post World War II when cultural efforts were stepped up. By 1950, annual production amounted to 1800 tons.In 1952-53, world production was close to 1,500,000 tons and reportedly nearly doubled during the next decade.

The industry alternately grew and declined, and then ceased entirely for three and a half years during World War II. The Malaysian Pineapple Industry Board was established in 1959. Thereafter there has been steady progress. The pineapple, was a very minor crop in Thailand until 1966 when the first large cannery was built. Others followed. Since then, processing and exporting have risen rapidly. By 1968, the total crop had risen to 3,600,000 tons, of which only 100,000 tons were shipped fresh (mainly from Mexico, Brazil and Puerto Rico) and 925,000 tons were processed. In the period 1961-66, imports of fresh pineapples into Europe rose by 70%. Soon many new markets were opening.

As of 1971, the main leading exporters of fresh pineapples were (in descending order): Taiwan (39,621 tons), Puerto Rico, Hawaii, Ivory Coast, and Brazil. The main leading exporters of processed pineapples were (in descending order): Hawaii, Philippines, Taiwan, South Africaand Malaysia (Singapore).By 1972, it had risen to 200,000 tons for shipment, fresh or canned, to Western Europe. Cameroun's annual production was about 6,000 tons.

In 1973, the total crop was estimated at 4,000,000 tons with 2.2 million tons processed. The cannery (Hawaii) was closed in October 1973 because of high shipment charges compared to Thailand and Philippines.

In Puerto Rico, the pineapple was the leading fruit crop, 95% produced, processed and marketed by the Puerto Rico Land Authority. In 1980,the crop was 42,493 tons having a farm value of 6.8 million dollars.The increased worldwide demand for canned fruit has greatly stimulated plantings in Africa and Latin America.

For years, Hawaii supplied 70% of the world's canned pineapple and 85% of canned pineapple juice, but labor costs have shifted a large segment of the industry from Hawaii to the Philippines. Because production costs in Hawaii (which were 50% labor) have increased 25%or more, Dole has transferred 75% of its operation to the Philippines, in 1983. By1992, DoleThai was operating the third largestpineapple cannery in the world with PPCranked first and DoleFil (Dole Philippines) second (Larsen andMarks, 2010).

In the Azores, pineapples have been grown in green-houses for many years for export mainly to Portugal and Madeira. They are of luxury quality, carefully tended and blemish free, graded for uniform size and well-padded in each box for shipment.The technique of forced induction of flowering of pineapple using smoke was accidentally done in the latter part of the 19thcentury in the Azores Islands where pineapples were being grown in greenhouses (Collins, 1960). Rodriquez (1932) found that the active ingredient in smoke was ethylene and a technique was developed for its application to pineapple (Kerns and Collins, 1937).

'MD-2' pineapples, trademarked Del Monte Gold, were officially introduced to U.S. and European markets in 1996. In 2010, the 'MD-2' pineapple was named the American Society for Horticultural Science's 2010 Outstanding Fruit Cultivar (Anonymous, 2010).

Till 19th century pineapple has been used as a fruit itself. By 20th century pineapple has been used in the canning industry. While the pineapple is considered a tropical fruit it has been grown commercially from latitude 27°N (Okinawa & Florida) to latitude 34°S (South Africa), with the great mass of production within the tropics only a few degrees north and south of the equator.

Social History

During colonial period, in America, pineapples were recognized as the symbol of hospitality. The sailors placed pineapples outside their homes as an announcement of their safe return and an invitation for their friends to visit. Many pictures and statues were built around homes. They made pineapple center pieces for their dining tables and decors. The pineapple was an epitome of higher status in the society. The decors included carvings, still life paintings, wallpaper and sculptures. Also many notable buildings were constructed in the shape of pineapple (Fig.4).

Fig. 4: (Top left to right) pineapple photo frame, Gate stump, floor décor

They made public buildings notable with pineapple sculptures. They also carved pineapple into door lintels; stenciled pineapples on walls and canvas mats; wove pineapples into tablecloths, napkins, carpets and draperies; and cast pineapples into metal hot plates. There were whole pineapples carved of wood; pineapples executed in the finest china kilns; pineapples painted onto the backs of chairs and tops of chests.

"The Big Pineapple" located in Bathurst, South Africa is an exact copy of the Big Pineapple in Queensland, Australia. It is around 16.7 m tall (where as Queensland Big Pineapple is only 16m tall). It is three storied building focused on research, history and sale of pineapple goods. The Bathurst pineapple building was made larger so that Bathurst could claim that they have "the world's biggest pineapple". Ancient Romans created depictions of pineapple.

Now a day's pineapple related many festivals are organized to boost up its popularity. (Fig. 5-9).

Fig. 5: In 1596: Jan Huygen van Linschoten: Travel account of the voyage of the sailor Jan Huyghen van Linschoten to the Portuguese East India, (r., engraving by Johannis Baptista van Doetecum)

Fig. 6: In 1666: Sir Christopher Wren: design of St. Paul's Cathedral with elongated pineapple (left, All Souls' College, Oxford) & pineapple on the towers of present St. Paul's Cathedral, London (middle and right.)

Fig. 7: An old picture having pineapple - oil on canvas, Rijks museum (Anonymous, 1666)

Fig. 8: In 1781: after Jean Michel Moreau le Jeune, engraving by Isidore Stanislas Helman: 'Le souper fin' (left, Musée Carnavalet, Paris) 1783: Charles Bretherton: 'Mrs. Weltje in her shop in Pall Mall' (right, engraving)

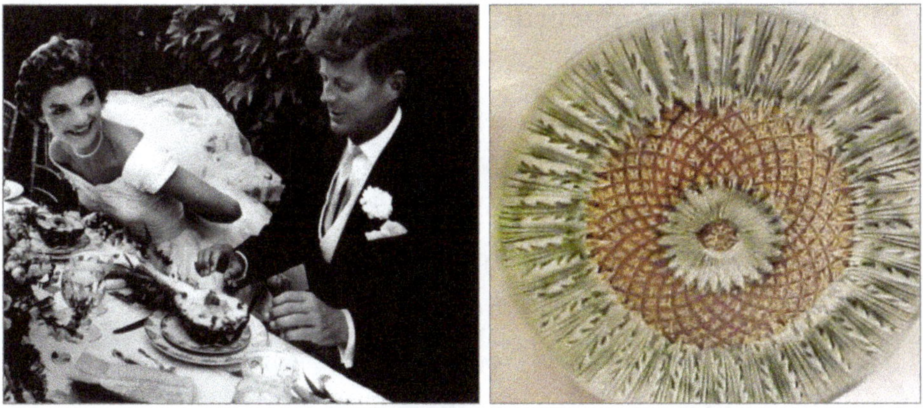

Fig. 9: John F. Kennedy wedding, 1953. Note the pineapple used as a distinctive feature in the fruit course b: Pineapple themed plate (Berg, 2010)

Taxonomy

Taxonomic History

The first botanical description of cultivated pineapple was by Charles Plumier at the end of the 17th century when he created the genusBromelia for the plants called karatas, in honour of the Swedish physician Olaf Bromel andalso described Ananas as "*Ananas aculeatus fructuovato, carne albida*".

In 1753, Linnaeus in his Species Plantarum designated the pineapple as *Bromelia ananas* and *Bromelia comosa*. However, Miller (1754, 1768) maintained the name Ananas, with all six cultivated varieties.In 18th and 19th century, pineapple classification resulted in a number of different names (Leal*et al.*, 1996). To simplify classification, Mez (1892) recognized in the Flora Brasiliensis only one species, *Ananas sativus*, with five botanical varieties. In 1917, Merrill established the binomial *Ananas comosus*. In 1919, Hassler divided the genus Ananas into two sections Euananas and Pseudananas. Pseudananas was raised to genus by Harms in 1930. In 1934, L. B. Smith and F. Camargo divided the genus Ananas and renamed and multiplied species. This resulted in 2 genera and nine species recognized in 1979 (Rohrbach *et al.*,2003). This classification has been criticized on the basis of practicality and inconsistency with available data on reproductive behavior and morphological, biochemical and molecular diversity (Leal, 1990; Loison-Cabot, 1992; Leal and Coppensd'Eeckenbrugge, 1996; Coppensd'Eeckenbrugge *et al.*, 1997; Leal *et al*, 1998) and therefore a much simpler and consistent classification has been prepared taking the above information into consideration (Leal and Coppensd'Eeckenbrugge, 1996; Leal *et al*, 1998).The **present classification is** as follows (Coppens d'Eeckenbrugge and Leal, 2003) **on thebasis of new data on reproduction** (Coppens d'Eeckenbrugge *et al.*, 1993), **morphological** (Duval and Coppensd'Eeckenbrugge, 1993), **biochemical** (Garcı´a, 1988, Aradhya *et al.*, 1994), and **molecular diversity**.

Classification by Coppensd'Eeckenbrugge and Leal (2003)

Ananas comosus (L.) Merril

A. comosus var. *ananassoides* (Baker) Coppens & Leal: (formerly two species: *A. ananassoides* and *A. nanus*)

A. comosus var. *erectifolius* (L.B. Smith) Coppens& Leal: formerly *A. lucidus*

A. comosus var. *parguazensis* (Camargo & L.B. Smith) Coppens & Leal

A. comosus var. *comosus*

A. comosus var. *bracteatus* (Lindl.) Coppens & Leal: (formerly two species: *A. bracteatus* and *A. fritzmuelleri*)

Ananas macrodontes Morren: (formerly *Pseudananas sagenarius*)

According to Luther and Sieff (1998), this is the largest family whose natural distribution is restricted to the New World, with the exception of *Pitcairnia feliciana* (Aug. Chev.) Harms & Mildbr., which is native to Guinea. Their unified geographical distribution and their strong adaptation towards an epiphytic mode of life indicate that this is quite a young family.

Systematic Position

Kingdom:	Plantae – plantes, Planta, Vegetal, plants
Subkingdom:	Viridiplantae
Infrakingdom:	Streptophyta – land plants
Superdivision:	Embryophyta
Division:	Tracheophyta – vascular plants, tracheophytes
Subdivision:	Spermatophytina – spermatophytes, seed plants, phanérogames
Class:	Magnoliopsida
Superorder:	Lilianae – monocots, monocotyledons, monocotylédones
Order:	Poales
Family:	Bromeliaceae
Sub family:	Bromelioideae
Genus:	Ananas Mill. – pineapple
Species:	*Ananas comosus* (L.) Merr. – Piña, pineapple

USDA Plants (2007-10)

Botanical Varieties and Characteristics

Pineapple is a perennial monocot belonging to the order Bromeliales, family Bromeliaceae and subfamily Bromelioideae. The Bromeliaceae comprises 56 genera with 2,921 species (Luther 2002), classified into three subfamilies: Pitcarnioideae, Tillandsioideae and Bromelioideae. This last subfamily shows a tendency towards the fusion of floral parts, a trait most developed in Ananas, the only genus whose flowers and bracts are completely merged into a single sorose-type parthenocarpic fruit formed by 50 to 200 coalescent berries (Rohrbach *et al.*,2003).

A. macrodontes is a vigorous self-fertile tetraploid (2n = 4x = 100), with spiny leaves, 2-3 m long and 7 cm wide, propagating by elongate basal stolons. The syncarp lacks the leafy crown. It naturally inhabits humid forest areas, under semi-dense shade, in coastal and southern Brazil and in the drainage of the Paraguay and Parana′ rivers(Coppensd'Eeckenbrugge *et al.*,1997). The species even tolerates short periods of flooding (Bertoni, 1919). The selfed, clones of *A. macrodontes* produce uniform progenies (Collins, 1960). It has been traditionally used as a source of long and strong fibers.

A. comosus* var. *comosus is the most widely cultivated pineapple and the basis of the world trade in fresh and processed fruit. It is generally diploid (2n = 50), self-incompatible, and propagates vegetatively by suckers (borne on the stem), slips (borne on the peduncle), and the fruit crown(s).Its leaves are relatively wide (more than 5 cm), spiny(antrorse spines), partially spiny or smooth, and its strong peduncle bears a fruit whose size may reach several kilograms. The fruit has many fruitlets ("eyes"). Seeds are rare in the fruits, because of reduced fertility, conjugated with stronger self-incompatibility and monoclonal cultivation. Its wider and longer stem allows a larger starch storage capacity.It was planted throughout tropical America at the time of the Conquest. Its fruit was widely consumed and particularly appreciated in the form of fermented drinks (Patin˜o, 2002). Rotted pineapple was used on arrows and spear heads for poisoning (Leal and Coppensd'Eeckenbrugge, 1996).

A. comosus* var. *ananassoides (Baker) and *A. nanus* (L.B. Smith) L.B. Smith) corresponds to the most common wild form, with thinner spiny leaves (up to 2 m long and lessthan 4 cm wide) and a much smaller fruit on a long, slender scape. It is the mostlikely ancestor of the cultivated pineapple. It is found in most tropical regions of South America east of theAndes, generally in savannahs or clear forest, growing on soils with limited water-holding capacity (sand,rocks) and forming populations of variable densities. They are monoclonal, but some are polyclonal, with variation of probably recent sexual origin (Duval *et al.*,1997).The fruit peduncle is elongate (most often more than 40 cm), slender (usually less than 15 mm wide). Its inflorescence is small to medium insize, globose to cylindrical, and it shows little growth after anthesis, so it has little flesh. The pulp is white oryellow, firm and fibrous, and palatable, with a high sugar and acidity content, with numerous seeds. The crown resumes fast growth after fruit maturation. They are sometimes cultivated or tolerated in gardens. Such pineapples may have served as a basis for domestication. At the other end, some dwarf types have recently been cultivated as ornamentals for thecut flower market, at both national and international levels.

A. comosus* var. *parguazensis (Camargo & L.B. Smith) is another wild form with wider leaves, constricted at their base, antrorse and retrorse spines, and a globose fruit. The two remaining botanical varieties are cultivated. It is very similar to *A. comosus* var. *ananassoides,* from which it differs by wider leaves, slightly constricted at their base, and larger spines. Its distribution mostly corresponds to the basins of the Orinoco and upper Rio Negro, with a fewobservations in eastern Colombia and in the northeastern

Amazon (Coppensd'Eeckenbrugge *et al.*,1997). It grows in lowland forests, under canopies of variable densities, from clearings or river banks to dense forest. It seems restricted to shadier environments, because oflower water use efficiency (Leal and Medina, 1995).

A. comosus* var. *erectifolius (L.B. Smith) is very similar to *A. comosus* var. *ananassoides*, except for its smooth fibrous leaves, which are used by Amerindians to make hammocks, fishing lines and nets. Plants are medium sized, with abundant shoots, frequent crownlets at the base of the main crown, numerous erectfibrous -a trait which is under monogenic control (Collins, 1960) spineless leaves making it a potential fibre crop (Ray, 2002). It has recently found a new economic use in the production of cut flowers.

A. comosus* var. *bracteatus (Lindl.) is a very vigorous and spiny plant, producing a medium-sized fruit with long bracts. It is cultivated as a living fence. Its fruit was also collected for juice and it is still found as a sub-spontaneous plant in ancient settlements of southern South America. A variegated variant has become a common ornamental of tropical gardens (Carlier, *et al.*, 2006). It was cultivated as a living hedge and harvested for fiberand fruit juice, or for traditional medicine, in southernBrazil and Paraguay (Bertoni, 1919). Indeed, its dense,long, and wide leaves are strongly armed by large antrorse spines, forming impenetrable barriers. It isvery robust and still thrives in abandoned plantations,but it seems unable to colonize new habitats. The syncarp is of intermediate size (0.5–1 kg), borne by a strong scape. The inflorescence has as pectacular appearance.

A. fritzmuelleri shares an additional trait with *A. macrodontes*, as it exhibits retrorse spines at theleaf base. According to Camargo (1943) and Smithand Downs (1979), it was also used in making fences. It is a very rare form, whose diversity has not been documented, only one clone being conserved in Brazil,by EMBRAPA (The Brazilian Agricultural Research Corporation) and the botanical garden of Rio de Janeiro. Nuclear and chloroplast DNA data confirm its closer proximity with *A. macrodontes*. The chromosome number is $2n = 2x = 50$ (Camargo, 1943).

Pineapple plants from most cultivars can survive when their cultivation is abandoned, resisting competition in sufficiently openvegetation and even dry edaphic or climatic conditions; however, they do not propagate efficiently to form sub-spontaneous feral populations.

Cultivars

Pineapple isozyme variation indicates five genetically diverse groups (Loison-Cabot, 1992; Aradhya *et al.*, 1994). A brief discussion of the five horticultural groups follows.

Cayenne group: 'Smooth Cayenne' is the standard for processing and for the fresh fruit trade because of its cylindrical shape, shallow eyes, yellow flesh colour, mild acid taste and high yields. In most areas, 'Smooth Cayenne 'constitutes a mixture of clones due to new introductions from mutations, lack of roguing and other various sources. Local selections are mostly known by their areas of origin, such as 'Sarawak' in Malaysia. 'Champaka' is a selection of 'Smooth Cayenne' originating in India and widely grown in Hawaii. The group is susceptible to mealy bug wilt and nematodes.

Queen group: This group generally produces smaller plants and fruit with spiny, shorter leaves than the 'Cayenne' group. 'Queen' is grown in South Africa, Australia and India for the fresh fruit market. 'Z-Queen' or 'James Queen' is reported to be a mutant of 'Natal Queen' and is a natural tetraploid.

Spanish group: The plants are generally small to medium, spiny-leaved, vigorous and resistant to mealy bug wilt, but susceptible to gummosis caused by the larvae of the Batrachedra moth. It is acceptable for the fresh fruit market but not favoured for canning, due to deep eyes and poor flesh colour. 'Red Spanish' or 'Espanola roja' is the major cultivar in the Caribbean region. 'Singapore Spanish', or 'Singapore Canning' and 'Nanas Merah', are the principal canning pineapple in West Malaysia because of their adaptability to peat soil. The flesh has a bright yellow colour. Other Malaysian cultivars are 'Masmerah', a spineless type with large fruit, and 'Nanas Jabor', a Cayenne–

Spanish hybrid that is susceptible to fruit marbling and cork spot. 'Cabezona', a natural triploid, is an exception, having large plants and fruit weighing 4.5–6.5 kg. It is grown primarily in the Tabasco State of Mexico and a small area of Puerto Rico where local consumers prefer the larger fruit. The PuertoRico clone PR 1-67 is suspected to be a hybrid between 'Red Spanish' and 'Smooth Cayenne', as these were the only clones grown in adjacent fields. The fruit has light yellow flesh with adequate sugar and resistance to gummosis, is fairly tolerant to mealy bug wilt, and has good slip production and good shipping qualities.

Abacaxi group: This group is grown mostly in Latin America and inthe Caribbean region. Py *et al.* (1987) called this the Pernambuco group. The fruit is not considered suitable for canning or for fresh fruit export, but the juicy, sweet flavour of the fruit is favored in the local markets. 'Perola',' Pernambuco', 'Eleuthera' and 'Abacaxi' are the principal clones in Brazil, along the eastern Espirito Santo in the south through Bahia and Pernambucoto Paraibo.

Maipure group: This group is cultivated in Central and South America as fresh fruit for the local markets. These clones may be of interest to breeders in the western hemisphere as they constitute a gene pool of adapted forms almost unused in breeding programmes.

The 'Smooth Cayenne' cultivar dominates commercial production for canning and is also one of the major fresh fruit varieties. However, 'Smooth Cayenne' has objectionably high acidity during the winter months, so newer hybrids such as 73-114 (MD-2, MG-3), which have comparable yield and a better sugar to acid balance during the winter months, have rapidly expanded in importance as fresh fruit varieties and now dominate international trade. Other varieties of some importance commercially include 'Queen' and 'Spanish', both of which are primarily consumed fresh.

Morphology and Reproduction

Morphology

The bromeliads show some significant differences from other monocots like their inhabitancy in varying ecological conditions. Most of the species are epiphytic or saxicolous or terrestrial. They are having a short stem, rosette of stiff leaves with terminal inflorescence.

The flowers are actinomorphic, trimerous with differentiated floral parts. They produce a coenocarpium (a multiple fruit derived from ovaries, floral parts and receptacles of many coalesced flowers) (Paull and Duarte, 2011). Each fruitlet has small naked, winged or plumose seeds. Their endosperm is reduced and with a small embryo. Cayenne cultivars produce seedless fruits because of self-incompatibility. Even then they produce germ cells. The germ cells of 'Queen' and 'Cayenne' cultivars are mutually compatible and gametes are produced.

Reproduction

Pineapple is a self-incompatible fruit crop which naturally prefers vegetative propagation. This kind of preference may be a result of domestication and selection process they undergone for seedless fruits (Coppens d'Eeckenbrugge *et al.,* 1993). The self-sterility is mainly due to the inhibition of pollen tube growth in the upper third of the style (Kerns, 1932; Majumder, *et al.,* 1964; Brewbaker and Gorrez, 1967). They propagate through crowns, slips, suckers and butts. Each of the new clone will be a genetically identical copy of their mother making it a true to type.

Certain varieties of the Ananas genus shows natural cross pollination like *A. macrodontes.* This is also called as pseudo-self-compatibility (Coppens d'Eeckenbrugge *et al.,* 1993; Muller, 1994). They are self-fertile and the off springs are very homogenous (Collins, 1960) and are autogamous. The pollinating agents are mostly humming birds while bees and ants are secondary agents. *A. macrodontes*also are stoloniferous in nature (Coppens d'Eekenbrugge *et al.*, 1997).

The chromosome number of pineapple is 50 and are diploid in nature. Rarely triploids (75 chromosomes) also been seen in case of *A. comosus* var. *comosus* and *A. comosus* var. *ananassoides. A. macrodontes* Morren is a tetraploid (Collins, 1960; Lin *et al.,* 1987). The viability of pollens are greatly variable between varieties, cultivars, and even between clones from the same cultivar (Coppens d'Eeckenbrugge *et al.,* 1993; Muller 1994).

For hybridization, breeders prefer artificial pollination methods. Hand pollination is also possible in the absence of natural pollination (Kumar, 2006). Generally, seeded pineapples are not preferred in the market or processing arena. Hybrid clones like "Golds" are kept in laboratory for better development as seedlings rarely survive under natural conditions (Coppens d'Eekenbrugge *et al.,*1997).

Evolution of Breeding

Pineapple breeding throughout the world mainly focused on the following characters and most of the time one or two among them had been their objectives.

Objectives of breeding

- To develop high yielding cultivars.
- To develop cultivars suitable for table purposes-medium sized, 1-2 kg, cylindrical, sweeter in taste.

- To develop cultivars suitable for canning purposes- bigger sized, ≥2 kg, cylindrical, sweeter in taste, higher juice content.
- Resistant to pests and diseases especially, heart rot and root rot (Kumar, 2006).

Breeding

Valentine Holt in 1914 started to breed pineapple at the Federal Experiment Station in Honolulu, **Hawaii.**The pineapple growers imported a wide range of world pineapple varieties. In 1916, R. E. Doty germinated the 'Smooth Cayenne' seeds from the canneries in Honolulu, also the crosses made between 'Cayenne' and 'Queen' to get the vigor of 'Queen' and fruit qualities of 'Cayenne'. He found out that the progenies from 'Cayenne' x 'Queen' showed promising. By 1923, back cross has been made using the best of the hybrids and the results of screening is published.

W.A. Wendt in 1922, noted that promising seedling lots included 'Cayenne' x 'Smooth Gautemala', x 'Taboga' and x 'Ruby' and that back crossing 'cayenne' x 'Queen', selections gave vigorous plants and by the third propagation cycle after selection enough plants were available to test fruit cannability and plant disease resistance. In 1927 *A. ananassoides* (with the habit of ground suckering and vigour) was crossed onto Cayenne and several thousand hybrids were obtained. Testing of these hybrids showed them to have resistance to heart rot, better tolerance to nematodes and higher brix value. In 1929, J.C. Collins joined the breeding programme as geneticist and identified the resistant clones for plantation and parental use. Those plants were experimented on rot affected areas to study their performance. Highly replicated randomized small plot designs with susceptible and resistant checks added to the rating efficiency of those Phytophthora trials. Screening for wilt resistance/tolerance using inoculation with mealy bugs planting in known wilt problem areas were initially promising. The resistant varieties were selected and was used in the breeding programmes. *A. comosus* var. *comosus* x *A. comosus* var. *ananassoides* hybrid progeny called "Lot 520" was considered as highly tolerant of the nematode, Meloidogyne sp.

The screening of other significant characteristics were also used in the breeding trials like yield, vitamin C, fibrosity, acidity were also carried out. At some point sensory evaluation of the fruits were carried out for its acceptance with meal.

In 1930, Collins obtained hybrids of 'Cayenne' x 'Wild Brazil'. 53-116 was an excellent product of the breeding programme with excellent appearance and crop yield but unsuitable for canning. 52-323 and 59-656 were found to be resistant to Phytophthora wilt. Tests of various hybrids developed by the PRI indicated a level of resistance particularly in hybrids involving *A. comosus* var. *bracteatus* and *A. comosus* var. *ananassoides*. In effect, the result of all those breeding trials were not promising and had to suspend the programmes because of some unavoidable poor traits were incorporated. Only seven varieties and two species appear in the parentage of retained varieties (Williams and Fleisch, 1992). Resistant varieties include "PRI-10388" syn. 'Spanish Jewel', 'PRI-59-656', 'PRI-52-323', and 'PRI-61- 2223' (Smith 1965; Rohrbach and Johnson 2003).Two of these, 'PRI-59-656' and 'PRI-52-323', were grown commercially on a small scale in Hawaii before improved chemical control methods and high yielding 'Smooth Cayenne' clones became available (Williams and Fleisch, 1993).The variety 'PRI-59-656' is claimed to possess good resistance to both the pathogens (Smith, 1965).

'MD-2' Pineapple

In Hawaii, when the cannery business was becoming less profitable, there was a gradual shift to fresh fruit production in 1960 until 2007.In the mid-1980s DM(Del Monte) shipped plants of PRI hybrid 73-114 (named 'MD-2' after manager Frank Dillard's wife Millie) from Hawaii to Costa Rica (Bartholomew, 2009). 'MD-2' pineapples, trademarked Del Monte Gold, were officially introduced to U.S. andEuropean markets in 1996. Del Monte was purchased by R.J. Reynolds in 1979 and the fresh fruit business was sold in 1989 by what was then RJR Nabisco and renamed Fresh DelMonte Produce (FDMP).In 2010, the 'MD-2' pineapple was named the American Society for Horticultural Science's 2010 **Outstanding Fruit Cultivar** (Anonymous, 2010). 'MD-2' was one of only two commercially successful pineapple cultivars produced by the world's largest pineapple breeding program. The other cultivar, PRI hybrid 73-50, named 'MD-1' by DM, was later patented by DM as 'CO-2' (Bartholomew *et al.*, 2010). Both 'MD-2' and 'CO-2' have slightly higher brix than 'Smooth Cayenne' but their superiority as fresh fruits was the result of significantly lower acidity, especially during winter, and greatly improved storability.As a result of the market shift from 'Smooth Cayenne' to 'MD-2', independent growers in Ghana and the Ivory Coast quite rapidly abandoned 'Smooth Cayenne' and rapidly expanded the area planted to 'MD-2' as did independent growers in Costa Rica, Ecuador, and the Philippines. As a result of this rapid expansion in the supply of 'MD-2' fruits, fresh pineapple has become cheaper, resulting in "heterogeneous quality" (Loeillet*et al.*, 2011).The pineapple business of MPC (Maui Pineapple Company) closed in 2009 and a new company, Maui Gold Pineapple Co., began growing Maui Gold('CO-2') in 2010.

During 1921, 'Smooth Cayenne' was introduced to the **Philippines**. They crossed 'Red Spanish', 'Smooth Cayenne' and 'Buitenzorg' (Queen). The resulted progenies were of three groups like the one with qualities of 'Red Spanish', the one like 'Smooth Cayenne' parent with more propagation capacity and the one with 'Queen' characters and yielding larger fruit. The studies were not further remarkable after that (Mendiola*et al.*, 1951). Further in 1990's in the Institute of Plant Breeding they were being tried to develop a variety, spineless 'Queen'. They could produce a hybrid with such traits. Later they also tried to develop a variety which could be a good source of fruit and fibre. Hybrids produced between crossing 'Singapore Spanish', 'Smooth Cayenne' and 'Queen' Cultivars are being evaluated (Villegas *et al.*, 1996).

In 1924, at Kagi Experimental Station, **Taiwan**, a breeding programme was conducted between Smooth cayenne from Hawaii with local cultivars ('Ohi', 'Uhi', 'Anpi' and 'Seihi'). The final outcome was unknown. Later, 'Smooth Cayenne' and 'Queen' were crossed to yield 'Tainung' 1 to 8. Of these 'Tainung 4' (Easy Peeler) was selected which resembled its parent Queen (Fitchet, 1989).

From 1962s to 1970s **Malaysia** focused on clonal selection. Malaysian Pineapple Industry Board (MPIB) in 1965 concentrated in hybridization 'Sarawak' and 'Singapore Spanish' (Wee, 1974). In 1974 Malaysian Agriculture Research Divisional Institute (MARDI) took up MPIB. Evaluation and vigorous selection of F_1 hybrids resulted in the development of two promising lines-SS SC-1-AB and SRK SS-3 having qualities from both parents. In

1984, a systematic diallel cross started and the progenies were called MARDI hybrid 1 and MARDI hybrid 3 respectively (**Chan, 1992**). Among them MARDI hybrid 1 was accepted and named as 'Nanas Johor', a new canning variety (Chan and Lee, 1985; Chan, 1986). Nanas Johor was a high yielding, widely adaptable cultivar, has better canning properties, replacing the older 'Singapore Spanish' (Hybrid between Kew and 'Ripley Queen').The variety is no longer in use because of Marbling and cork spotdisease susceptibility (Leal and Geo Coppensd'Eeckenbrugge, 1996). A kind of transgressive segregations were observed when a diallel cross between 'Moris' ('Queen'), 'Masmerah' ('Singapore Spanish') 'Sarawak' ('Cayenne') and 'Johor' (Chan, 1991, 1993). Around 303 clones were selected having the features like fruit size, square shouldered fruit shape, flesh colour core diameter, absence of spines and total soluble solids (TSS). Finally A20-3, D4-37 and A25-34 were selected and recommended for field trial (Chan, 1995).

In **Brazil**, 'Perola' and 'Smooth Cayenne' were the cultivars grown. They were susceptible to fusarosis and breeding studies were conducted for resistant varieties. 'Perolera', 'Primavera' and 'Sao Bento' were the hybrids with Fusarium resistance. Resistant cultivars like Perolera and Primavera were crossed with 'Perola' and 'Smooth Cayenne'. The greatest number of seeds were obtained from the 'Perolera' x 'Smooth Cayenne' and 'Perolera' x 'Perola' crosses where 'Perolera' was the female parent with an average of 2,187 and 1,247 seeds/fruit, respectively. The promising genotypes were recommended for planting as fusariosis resistant cultivars if they could retain their desirable characters during clonal evaluation. Maximum genotypes selected were of the cross 'Perolera' x 'S. Cayenne' (Cabral *et al.*). Almost all the cultivated varieties were susceptible to the disease and hence the Brazilian work was a land mark in the history of pineapple breeding (Leal and Coppensd' Eeckenbrugge, 1996).

In 1978, fruit department in **France** "Centre de cooperation international en recherché agronomique pour le development' (CIRAD-FLHOR) started a breeding programme for both fresh and processed fruits. The crossing was done between 'Smooth Cayenne' and 'Perolera'. The selections were based on a multitrait phenotypic index. 'Smooth Cayenne'- 409 and 'Perolera'- 101 were selected.

In **Australia**, crossing was done between 'Smooth Cayenne', Hawaiian hybrids and 'Queen' for resistance of Phytophthora disease. Some of the hybrids showed fruits similar to 'Queen' but were on smooth leaves (Winks *et al.*, 1985).

In **India**, hybridization work was taken up at Pineapple Research Centre, Kerala Agricultural University, Vellanikkara, and developed a variety called 'Amritha' (Hybrid of Kew and Ripley Queen)- yield of 85 tonnes per hectare. Each fruit weighs more than 2 kg, has single, small crown, golden yellow colour with desirable cylindrical shape. It has the added advantage of flesh colour, pleasant aroma, high TSS and total sugars and low acidity.

Interspecific Hybridization:

A. comosus x *A. ananassoides* (high sugar and acid, small core, resistant to nematode, wilt, heart rot and root rot)

A. comosus x *A. bracteotus* (bigger fruit size than *A. ananassoides*, small core, resistant to wilt, heart rot and root rot)

A. comosus x *A. segenarius* (immune to heart rot root rot and wilt)

F_1 hybrids are the ones with desirable traits from these related species (Kumar, 2006).

In Kerala, India, irradiation of the plants of the cultivars Kew and Mauritius led to growth retardation and in one plant to retardation to premature sucker (Anon., 1964). Marz (1964) reported induction of self-fertile mutants by X irradiation of pollen during meiosis. Technique for applying certain chemical mutagens like ethylene imine (EI), N-nitroso-N methyl-urethane (NMU) and diethyl sulphate (DES) applied on to detached slips of 1-1.5 months old (Singh and Iyer, 1974). As a result several spineless plants were produced from 'Queen' and was economically significant (Broerties and Harten).

Hybridization programmes were also reported in Puerto Rico, Cuba and Japan (Leal and Geo Coppensd'Eeckenbrugge, 1996).

Breeding for Garden Varieties

The Bromeliaceae family is well diverse to have many ornamental varieties in it. Lately,the thought of exploiting its colorful appearance in breeding was introduced.

Small but increasing quantities of Ananas plants and blooms are now being marketed in various countries for their ornamental appeal, usually *A. comosus* var. *bracteatus* "Tricolor" and *A. comosus* var. *erectifolious* "Selvagem 6". Both these varieties, while currently commercially exploited, have limitations and do not incorporate the breadth of ornamental potential within the Ananas gene pool. Breeding programs for ornamental pineapple are reported for Brazil, Australia, France, and Malaysia (Duval *et al.*, 2001a; Chan 2006; Souza et al. 2006, 2009; Sanewski 2009). The markets include the cut-flower market for pre-petal syncarps, miniature fully formed fruit, and attractive cut foliage (F. Vidigal personal communication). The landscape or potted plant market will also take plants with ornamental fruit or foliage characteristics. For attractive blooms, *A. comosus* var. *bracteatus* is good for imparting a bright red coloration to the syncarp and *A. macrodontes* will impart a pink colour.

A. comosus var. *erectifolious* "Selvagem 6" is a good parent for obtaining smooth reddish leaves, including those in the crown. An example of this hybrid. For miniature fruit, *A. comosus* var. *ananassoides*is a good parent, as is *A. comosus* var. *erectifolious*. It is important that the small fruit has a strong attachment to a long (50 cm), thin stem and the crown is well formed with no side shoots. Of all the Ananas, *A. comosus* var. *ananassoides* displays considerable diversity in fruit and leaf colour and appearance. The collection of Ananas held by EMBRAPA (EmpresaBrasileira de PesquisaAgropecuária-Brazilian Agricultural

Research Corporation)holds accessions highly suited as parental stock (Souza *et al.,* 2006). Interspecific crosses also show ornamental interest and the potential for utilizing other genera might also exist. Successful intergeneric hybrids with Ananas are reported for Aechmea, Cryptanthus, Neoregelia (Anonymous, 2007) and Tillandsia (Valds*et al.,* 1998).

Novel Methods

During late 1990s, the commercial success of 'Golden Ripe', a new cultivar that stirred the world market of fresh pineapple and awakened the interest in cultivar diversification through hybridization.Introgression of resistance to diseases such as Phytophtora and Fusarium, the prevention of disorders such as internal browning (blackheart) and the control of specific traits such as early natural flowering, in elite cultivars. To avoid the uncertainty of segregation and recombination, genetic engineering appears to be a promising breeding strategy since it allows transferring a single gene, or a few genes, without substantially altering the initial genome. Efficient procedures for genetic transformation (Sripaoraya*et al.,*2001; Espinosa *et al.,* 2002) and *in vitro* regeneration and propagation (Escalona*et al.,* 1999; Firoozabady and Gutterson 2003; Sripaoraya*et al.,* 2003) have already been established. The first field and greenhouse trials of genetically transformed pineapple clones exhibiting reduced expression of polyphenol oxidase (PPO) and of 1-aminocyclopropane-1-carboxilate (ACC) synthase or expressing the bialaphos resistance (bar) gene have already been carried out (Rohrbach*et al.,* 2000; Sripaoraya*et al.,* 2001; Sripaoraya*et al.,* 2006; Botella and Fairbairn,2005;Trusov and Botella, 2006).

In this respect, the construction of dense genome maps of molecular markers is of paramount importance for the further isolation, *via* positional cloning, of genes of interest for pineapple improvement. This is of particular significance regarding those genes that are uniquely known and uniquely detected by their phenotypic expression in plants (e.g. resistance genes). A successful attempt to isolate protoplasts of the cultivar "Perolera" (Guedes*et al.,* 1996) was done but plant regeneration was not achieved.

Pineapple transformation, however, offers the possibility to make small targeted changes to the recipient plant's genome and is seen as an excellent strategy for genetic improvement. Methods involving the introduction of recombinant DNA to pineapple cells and tissues *via Agrobacterium tumefaciens*-mediated transformation and direct gene transfer through microprojectile bombardment are reported. Biolistics has been used to deliver genes conferring herbicide resistance (Sripaoraya*et al.,* 2001) and blackheart resistance (Ko*et al.,* 2006) into "Smooth Cayenne." Other groups focused on using Agrobacterium to introduce ACC synthase genes to control ripening (Firoozabady*et al.,* 2006; Trusov and Botella, 2006). Despite these advances, consumer resistance to transgenic fresh fruit is limiting wider use of this technology. Incorporation of only native genes from wild relatives and with expression only in plant parts not intended for consumption is the approach worth considering. In addition, before businesses and institutions will have freedom to operate with transgenic lines, intellectual property ownership must be ascertained, and strategies put in place to ensure plants are free from encumbrance, which would otherwise restrict the sale of product.

Genomics Resources Developed

The amount of genomic data in databases is still scanty, despite the economic importance of pineapple, but has been increasing in the last few years. A search for pineapple genomic data through the National Center for Biotechnology Information (NCBI) (http:// www.ncbi.nlm.nih.gov) found about 60 microsatelliteand other DNA marker loci from *A.*var. *bracteatus* and over 5,700 ESTs from *A.*var. *comosus.* About 140 SSR markers have also been published on EMBL database (http://srs.ebi.ac.uk), the main contributors being the Biotechnology Research Institute of Malaysian Sabah University for 76 SSRs (Kumar *et al.*, unpublished) and CIRAD in France with 50 SSRs (Blanc *et al.* unpublished). Also, recently an entire collection of ESTs was generated during an investigation into fruit ripening and nematode–plant interactions during root invasion (Moyle *et al.*, 2006) and has been made publicly available by an online pineapple bioinformatics resource named "PineappleDB" (http://www.pgel.com.au) (Geo Coppens d'Eeckenbrugge, *et al.*, 2011).

Genetic Diversity Analysis

Markers

Genetic diversity studies were every time in demand as it could reveal how much the species variation has achieved through out the breeding programmes.The genetic diversity among Ananas germplasm was initially investigated using **isozyme markers** (DeWald *et al.*, 1988;Garcia, 1988; Duval and Coppensd'Eeckenbrugge, 1992;Aradhya *et al.*, 1994). In the study of DeWald *et al.*, (1988), 15of 27 *A. comosus* cultivars were identified by five enzymaticsystems, two peroxidases and three phosphoglucomutases. Inthe study of Aradhya *et al.*, (1994), 161 pineapple accessions from the Hawaiian collection, including four different species of Ananas and one species of Pseudananas, were identifiedby six isozyme systems involving seven putative loci (Smith and Downs, 1979).

More recently, **DNA-based markers** have been used to studythe phylogenetic relationships between Ananas and relatedgenera. Restriction fragment length polymorphism (RFLP)markers were analyzed among 301 accessions of Ananas andrelated genera including 168 *A. comosus* accessions, suggesting that *A. comosus* has lower levels of polymorphism than wild Ananas species (Duval *et al.*, 2001). Similarly, based on amplified fragment length polymorphism (AFLP) markers pattern of Mexican germplasm collections, *A. comosus* accessions were reported to have a low level of diversity (Paz *et al.*,2005).

Chloroplast DNA (cpDNA) diversity of Ananas and relatedgenera were evaluated by PCR-RFLP (Duval *et al.*, 2003), suggesting that the genetic diversity of Ananas was relative to the geographical origin of the accessions but not the species. The seresults supported the pineapple classification by Coppensd'Eeckenbrugge and Leal (2003) and enable us to generate adendrogram for pineapple classification. The level of genetic diversity among commercial cultivars is still unclear.

Three commercial cultivar groups, 'Cayenne', 'Queen', and'Spanish', were investigated by **random amplified polymorphicDNA (RAPD).** 'Cayenne' and 'Queen' cultivars were groupedinto two separate clusters, whereas 'Spanish' failed to form amonophyletic group

(Sripaoraya *et al.*, 2001). However, major cultivar groups of the 148 *A. comosus* accessions of pineapple, such as 'Cayenne', 'Spanish', and 'Queen', could not be distinctively separated by AFLP (Kato *et al.*, 2005).

A study was conducted in the Phranakhon Rajabhat University, Thailand in 2012 studying the genetic diversity and genetic relationships among 15 accessions of pineapple (*Ananas comosus* (L.) Merr.) using **Inter simple sequence repeats** (ISSR) markers. Genomic DNA was extracted from fresh leaf samples. Nine ISSR primers were initially screened for analysis and four primers (ISSR1, ISSR3, ISSR 4 and ISSR 5) were chosen for further analysis. A total of 56 DNA fragments, varying from 100-2000 bp, were amplified, of which 27 (48.21%) were polymorphic. A dendrogram showing genetic similarities among pineapple was constructed which based on polymorphic bands using the SPSS program (version 18). ISSR analysis was found to be a rapid and suitable method for studying genetic diversity among indigenous 'Intrachit' and others.In another study, a selection of cultivated bromeliads were characterized *via* inter-simple sequence repeat (ISSR) markers with an emphasis on genetic diversity and population structure (Zhang, *et al.*, 2012).

Another study was conducted in China in 2013 using **microsatellite marker** based genetic diversity analysis. The two methods they used to develop pineapple microsatellite markers were 1) genomic library-based SSR development: using selectivelyamplified microsatellite assay, 86 sequences were generated from pineapple genomic library. 91 (96.8%) of the 94 Simple Sequence Repeat (SSR) loci were dinucleotide repeats (39 AC/GT repeats and 52 GA/TC repeats, accounting for 42.9% and 57.1%, resp.), andthe other three were mononucleotide repeats. Thirty-six pairs of SSR primers were designed; 24 of them generated clear bands ofexpected sizes, and 13 of them showed polymorphism. 2) EST-based SSR development: 5659 pineapple EST sequences obtained from NCBI were analyzed; among 1397 nonredundant EST sequences, 843 were found containing 1110 SSR loci (217 of them contained more than one SSR locus). Frequency of SSRs in pineapple EST sequences is 1SSR/3.73 kb, and 44 typeswere found (Feng*et al.*, 2013).

Leaf margin phenotype-specific restriction-site-associated DNA-derived marker analysis was employed to analyze three bulked DNAs of F1progeny from a cross between a 'piping-leaf-type' cultivar, 'Yugafu', and a 'spiny-tip-leaf-type' variety,'Yonekura' (cultivars of Japan). The parents were both *Ananas comosus* var. *comosus*. From the analysis, piping-leaf and spinytip-leaf gene-specific restriction-site-associated DNA sequencing tags were obtained and designated as PLSTsand STLSTs, respectively. SSR and CAPS markers are applicable to marker-assisted selection of leaf margin phenotypes in pineapple breeding (Urasaki*et al.*, 2015).

Evolution of CAM Photosynthesis

The pineapple possesses **Crassulacean acid metabolism (CAM)**, aphotosynthetic carbon assimilation pathway with high water-use efficiency.The sequencing of the genomes of pineapple varieties F153 and MD2 and a wild pineapple relative, *Ananas bracteatus* accession CB5 were done. The pineapple genome hasone fewer ancient whole-genome duplication event than sequenced grass genomes and a conserved karyotype with seven chromosomes from before the duplication event. The pineapple lineage has transitioned from C3 photosynthesis to CAM, with CAM-related genes exhibiting a diel expression pattern in photosynthetic tissues. CAM pathway genes were enriched with cis-regulatory elements associated with the regulation of circadian clock genes, providing the first cis-regulatory link between CAM and circadian clock regulation. Pineapple CAM photosynthesis evolved by the reconfiguration of pathways in C3 plants, through the regulatory neofunctionalization of pre-existing genes and not through the acquisition of neofunctionalized genes via whole-genome or tandem gene duplication(Ming, *et al.*, 2015).

Transgenic Pineapple in Delayed Flowering

In pineapples, unlike many other plant species, flowering can be induced by the gaseous plant hormone ethylene. Ithas been shown that prior to inflorescence emergence, theleaf basal-white tissue produces ethylene (Bartholomew, 1977; Min and Bartholomew, 1996). Use of ethylene and ethylene-releasing chemicals such as ethephon [(2-chloroethyl) phosphonic acid] has become a common practice for flowering induction among pineapple growers (Randhawa *et al.*, 1970; Reid and Wu, 1991; Manica*et al.*, 1994).

The key regulatory enzyme in the ethylene biosynthetic pathway is 1-amino-cyclopropane-1-carboxylate synthase(S-adenosyl-L-methionine methyl thioadenosine-lyase EC4.4.1.14) (ACC synthase) (Yu et al., 1979). Three genesfor ACC synthase have been cloned so far in pineapples and two of them have been characterized (Cazzonelli*et al.*, 1998; Botella *et al.*, 2000). ACACS1was shown to be expressed in fruits and in wounded leaves (Cazzonelli *et al.*, 1998), while ACACS2 expression is proposedto be associated with flowering (Botella *et al.*, 2000).It is shown that constitutive overexpression of an ACACS2 gene fragment causes methylation of the endogenous ACACS2 gene resulting in silencing. Continuous monitoring of the flowering dynamics of transgenic andcontrol plants showed that suppression of the ACACS2 gene resulted in significantly delayed flowering.

The ACC synthase gene ACACS2 could bea key element in the production of the ethylene burst that switches meristematic cells from vegetative to generative development in pineapple. It has been shown that silencingof ACACS2 in transgenic pineapple plants results in a significant flowering delay; however, it does not prevent it indefinitely.This proves that silencing of the ACACS2 gene using genetic engineering techniques can be successfully used tocontrol natural flowering in commercial situations, therefore addressing the major pineapple industry problem (Trusov and Botella, 2006).

Signal Transduction Studies

Plant receptor-like kinases (RLKs) can autophosphorylate serine and/or threonine residues and play animportant role in the perception and transmission of external signals (Shiu and Bleecker, 2001; Torii, 2004). Thede-phosphorylation of transmembrane receptor kinases catalyzed by phosphatases is an essential regulatory mechanisin receptor-mediated signaling (Shah *et al.*, 2002).Protein kinases play important roles in cellular signaling and metabolic regulation in plants (Shah *et al.*, 2001b). Inpineapple AcSERK1 plays an importantrole in the induction and development of Somatic embryos (Ma et al.,2012b) and that AcSERK2 is highly expressed only in embryogenic cells before the pro-embryonic stage (Ma *etal.,* 2012a). *AcSERK3* was expressed at high levels only in competentcells during somatic embryogenesis (SE) and there was no apparent difference in theexpression level between embryogenic and non-embryogenic callus. The highest expression was detected in roots.The His-tagged AcSERK3 fusion protein was expressed in *E. coli* and autophosphorylation was detected. Thus AcSERK3 plays an important role in callus proliferation and root development (Ma *et al.,* 2014).

References

Anonymous. 1981. Pineapples. Selected European Markets for Tropical and Off-season Fresh Fruits and Vegetables. UNCTAD/GATT, Geneva, 10-11.

Anonymous. 2010. 'MD-2' Pineapple named the American Society for Horticulture Science 2010 Outstanding Fruit Cultivar. Pine. News No. 17:2.15 Dec. 2011. (http://ishs-horticulture.org/ workinggroups/pineapple/PineNews17.pdf).

Antoni, M.G. and Leal, F.J. 1980. Species of the genus Ananas: origin and geographic distribution. Tropical Region. American Society for Horticultural Science. 24:103–106 (in Spanish).

Aradhya, M., Zee, F. and Manshardt, R.M. 1994. Isozyme variation in cultivated and wild pineapple. *Euphytica.* 79: 87–99.

Avise, J.C. 2000. Phylogeography: The History and Formation of Species; Harvard University Press: Cambridge, MA, USA,p. 447.

Bartholomew, D.P. 2009. 'MD-2' pineapple transforms the world's pineapple fresh fruit export industry. Pine. News No. 16:2–5. 15 Dec. 2011. (http://ishs-horticulture.org/ workinggroups/pineapple/PineNews16.pdf)

Bartholomew, D.P., Coppens de'Eeckenbrugge, G. and Chen, C.C. 2010. Fruit and nut register list No. 45. *Hort Science* 45:716–756.

Berg L. 2010.The Pineapple as Ornamental Motif in American Decorative Arts, *Historia.*

Bertoni, M.S. 1919. Contribution à l'étudebotanique des plantescultivées. I. Essaid'unemonographie du genre Ananas. *Anal. Cient. Parag.* (Serie II), 4: 250-322.

Bertoni, M.S. 1919. Contributions al'etudebotanique des plantescultivees. I. Essaid'une monographie du enre Ananas. *Anales Cientificos Paraguayos* (Ser.II): 4: 250-322.

Brewbaker, J.L., Gorrez, D.D. 1967. Genetics of self-incompatibility in the monocot genera, Ananas (pineapple) and Gasteria. *Am J Bot.* 54:611–616.

Broertjes, C. and Harten, V. A.M. Applied mutation breeding for vegetative propagated crops.

Cabral, J.R.S., A.P. de Matos and G.A.P. da Cunha. 1992. Selection of pineapple cultivars resistant to Fusariose in Brazil.In: Bartholomew, D.P. and Rohrbach, K.G. (Eds) First International Pineapple Symposium, Honolulu, Hawaii, USA, 2-6 November 1992.

Camargo, F. 1943. Vida e utilidade das Bromelia´ceas. Bol Tec InstitutoAgrono´mico do Norte, Belem, Para´, Brazil.

Carlier, J.D., Nacheva, D., Coppensd'Eeckenbrugge, G. and Leitão, J.M. 2006. Genetic mapping of DNA markers in pineapple. *ActaHort.*702: 79-86.

Chadha, K.L., Reddy, B.M.C. and Shikhamany, S.D. 1998. Pineapple, ICAR, New Delhi.

Chan, Y.K.1992. Recent advancements in Hybridization and Selection of Pineapple in Malaysia. In: Bartholomew, D.P. and Rohrbach, K.G. (eds) First International Pineapple Symposium, Honolulu, Hawaii, USA, 2-6 November 1992.

Clement, C.R. 1989. A center of crop genetic diversity in western Amazonia. *BioScience.* 39:624–631.

Clement, C.R. 1999. 1492 and the loss of Amazonian crop genetic resources. I: The relation between domestication and human population decline. *Econ Bot* 53: 188–202.

Clement, C.R., Cristo-Araújo M., Coppensd'Eeckenbrugge, G., Pereira, A.A. and Picanço-Rodrigues D. 2010. Origin and Domestication of Native Amazonian Crops. *Diversity.* 2: 72-106.

Collins, J.L. 1934. Introduction of pineapple plants into Hawaii and some brief accounts of pioneer pineapple growing. *Pineapple Quart.* 4: 119-120.

Collins, J.L. 1948. Pineapples in ancient America. *Science Monthly* (65), pp.372–377.

Collins, J.L. 1949. History, taxonomy and culture of the pineapple. *Economic Botany.* 3: 335–359.

Collins, J.L. 1951. Notes on the origin, history, and genetic nature of the Cayenne pineapple. *Pacific Science.* 5(1): 3–17.

Collins, J.L. 1960. The Pineapple: Botany, Cultivation and Utilisation. Interscience Publishers, New York.

Coppens de'Eeckenbrugge, G., Sanewski, G.M., Smith, M.K., Duval, M.F. and Leal, F. 2011. Ananas (In:) Kole,C. (Ed). Wild Crop Relatives: Genomic and Breeding Resources: Tropical and Subtropical Fruits. *Springer Science & Business Media, Science-* pp.21-41.

Coppensd'Eeckenbrugge, G., Duval, M.F. and Van Miegroet, F. 1993. Fertility and self-incompatibility in the genus Ananas. *Acta Horticulturae*: 334: 45-51.

Coppensd'Eeckenbrugge, G., Leal, F. and Duval, M.F. 1997. Germplasm resources of pineapple. *Hortic Rev.*21:133–175

Coppensd'Eeckenbrugge, G., Sanewski G. M., Smith, M. K., Duval, M F and Freddy, L. Ananas.(In): Chittaranjan, K. (Ed), 2011.Wild Crop Relatives: Genomic and Breeding Resources: Tropical and Subtropical Fruits. *Springer Science & Business Media,- Science-* pp.21-41

Duval, M.-F., Buso, G.C., Ferreira, F.R., Bianchetti, L. de B.; Coppensd'Eeckenbrugge, G.,Hamon, P. and Ferreira, M.E. 2003. Relationships in Ananas and other related genera using chloroplast DNA restriction site variation. *Genome*, 46: 990-1004.

Duval, M.F., Noyer, J.L., Perrier, X., Coppensd'Eeckenbrugge, G. and Hamon, P. 2001. Molecular diversity in pineapple assessed by RFLP markers. *Theor. Appl. Genet.*, 102: 83-90.

Emshwiller, E. 2006. Genetic data and plant domestication. (In) Zeder, M.A., Bradley, D.G., Emshwiller, E., Smith, B.D., (Eds.). Documenting Domestication: New Genetic and Archaeological Paradigms. University of California Press, Berkeley, CA, USA, pp. 99-122.

Feng, S., Tong, H., Chen, Y., Wang, J., Chen, Y., Sun, G., He, j. and Wu, Y., 2013. Hindawi Publishing Corporation, *BioMed Research International, Volume* 2013.11p.

García, M.L. 1988. Etude taxinomique du genre Ananas. Utilisation de la variabilitéenzymatique. Doctorate thesis, USTL, Montpellier.

Gibault, G. 1912. Histoire des légumes. Librairie Horticole. Paris.

Joy P.P., 2010. Benefits and Uses of Pineapple, Pineapple Research Station (Kerala Agricultural University), Vazhakulam, 686670.

Kerns, K.R. 1932. Concerning the growth of pollen tubes in pistils of Cayenne flowers. *Pineapple Quart* 1:133–137

Kerns, K.R. and Collins, J.L. 1937. The uses of acetylene to stimulate flower formation: A technique in pineapple breeding. Proc. *Haw. Acad. Sci., 11th Ann. Mtg.* 1935–1936. p. 25.

Kumar, N., 2006. Breeding of Horticultural Crops: Principles and Practices. New India Publishing Agency, India. pp.118-120.

Ladizinsky, G., 2012. Plant Evolution under Domestication, *Springer Science & Business Media*, 06-Dec- Science, p.7.

Lapade, A.G., Veluz, A.M.S. and Santos, I.S. 1995. Genetic improvement of the Queen variety of pineapple through induced mutation and in vitro culture techniques. Proc. Joint IAEA/FAO Internatl. Symp. "Induced Mutations and Molecular Techniques for Crop Improvement", IAEA, Vienna, 684-687.

Leal, F. 1990. On the validity of Ananas monstruosus. *Journal of the Bromeliad Society.* 40: 246–249.

Leal, F. and Coppensd'Eeckenbrugge, G. 1996. Pineapple. In: Janick, J. and Moore, J.N. (eds) Fruit Breeding. John Wiley & Sons, New York, pp. 515–606.

Leal, F. and Medina, E. 1995. Some wild pineapples in Venezuela. *J Bromel Soc* 45:152–158.

Leal, F., Coppensd'Eeckenbrugge, G. and Holst, B.K. 1998. Taxonomy of the genera Ananas and Pseudananas – a historical review. Selbyana: 19: 227-235.

Lin, B.Y., Ritschel, P.S. and Ferreira, F.R. 1987. Nu´merocromosso^mico de exemplares da famı´lia Bromeliaceae. *Rev Bras Fruit.*9: 49–55

Loeillet, D., Dawson, C. and Paqui, T.2011. Fresh pineapple market: From the banal to the vulgar. *Acta Hort.* 902:587–594.

Loison-Cabot, C. 1992. Origin, phylogeny and evolution of pineapple species. *Fruits.*47: 25–32.

Luther, H.E. 2002. An alphabetical list of bromeliad binomials, 8thed. The Bromeliad Soc Intl, Newberg, Oregon, 82 p (http://www.selby.org/clientuploads/ research/ Bromeliaceae/Binomial_BSI_2002.pdf)

Ma, J., He, Y., Hu, Z., Xu, W., Xia, J., Guo, C., Lin, S., Chen, C., Wu, C. and Zhang, J. 2014. Characterization of the third SERK gene in pineapple (*Ananas comosus*) and analysis of its expression and autophosphorylation activity *in vitro. Genetics and Molecular Biology,* 37(3): 530-539.

Majumder, S.K., Kerns, K.R., Brewbaler, J.L. and Johannessen, G.A. 1964. Assessing self-incompatibility by a pollen fluorescence technique. *Proc Am Hortic Sci.* 84:217–223.

Ming, R., VanBuren,R., Wai, C.M., Tang, H., Schatz, M.C., Bowers, J.E., Lyons, E., Wang, M.L., Chen,J., Biggers, E., Zhang, J., Huang, L., Zhang, L., Miao, W., Zhang, J., Ye, Z., Miao, C., Lin, Z., Wang, H., Zhou, H., Yim, W.C., Priest, H.D., Zheng, C., Woodhouse, M., Edger, P.P., Guyot, R., Guo, H.B., Guo, H., Zheng, G., Singh, R., Sharma, A., Min, X., Zheng, Y., Lee, H., Gurtowski, J., Sedlazeck, F.J., Harkess, A., McKain, M.R., Liao, Z., Fang, J., Liu, J., Zhang, X., Zhang, Q., Hu, W., Qin, Y., Wang, K., Chen, L. Y., Shirley, N., Lin, Y.R., Liu, L.Y., Hernandez, A.G., Wright, C.L., Bulone, V., Tuskan, G.A., Heath, K., Zee, F., Moore, P.M., Sunkar, R., Leebens-Mack, J.H., Mockler, T., Bennetzen, J.L., Freeling, M., Sankoff, D., Paterson, A.H., Zhu, X., Yang, X., Smith, A.C., Cushman, J.C., Paull, R.E. and Yu, Y.Q. 2015. The pineapple genome and the evolution of CAM photosynthesis. *Nature Genetics.* Advance online publication.

Morrison, S.E. 1963. Journals and Other Documents of the Life and Voyages of Christopher Columbus. Heritage Press, New York.

Muller, A. 1994. Contribution a` l'e´tude de la fertilite´ et de l'autofertilite´ dans le genre Ananas. Eng Thesis, ISTOM, Montpellier, France.

Noyer JL 1991. Etude pre´liminaire de la diversite´ ge´ne´tique du genre Ananas par les RFLPs. *Fruits* 46:372–375.

Noyer JL, Lanaud C, Duval MF, Coppens d'Eeckenbrugge G 1998. RFLP study on rDNA variability in Ananas genus. *Acta Hortic* 425:153–160.

Osei-Kofl F., Lokko, Y. and Amoatey, H.M. 1996. "Improvement of pineapple {Ananas comosus (L) Merr.) using biotechnology and mutation breeding techniques", Rep. Second FAO/IAEA Research Co-ordination Meeting on "In vitro Techniques for Selection of Radiation Induced Mutants Adapted to Adverse Environmental Conditions", IAEA, Vienna, 23-28.

Patin˜o, V.M. 2002. Historia y dispersio´n de los frutalesnativosdel Neotro´pico. CIAT, Cali, CO Duval, M.F., Coppensd'Eeckenbrugge, G., Ferreira, F.R., Cabral, J.R.S, Bianchetti, L.D.B. 1997. First results from joint EMBRAPA-CIRAD Ananas germplasm collecting in Brazil and French Guyana. *Acta Hortic.* 425:137–144.

Paull, R.E., Duarte, O., 2011. Tropical Fruits, CAB International, 2nd Ed., (1).pp. 327-365.

Paz,E.Y., Gil, K., Rebolledo, L., Rebolledo, A., Uriza, D., Martínez, O., Isidrón,M., Díaz, L., Lorenzo, J. C. and Simpson, J. 2012. Genetic diversity of Cuban pineapple germplasm assessed by AFLP Markers. Brazilian Society of Plant Breeding. *Crop Breeding and Applied Biotechnology.* 12: 104-110.

Perrotet, S. 1825. Catalogue raisonné des plantesintroduites dans les colonies francaises de Mascareigneet de Cayuenne, et de cellesrapports'eesvivantes des mersd'Asie et de la Guyane, au Jardin des Plantes de Paris. Memoires Society Linneas.3 (3): 89–151.

Pickersgill, B. 2007. Domestication of plants in the Americas: insights from Mendelian and molecular genetics. *Ann. Bot.*100: 925-940.

Purseglove, J.W. 1988. Bromeliaceae. Tropical Crops. *Monocotyledons*: 75-91.

Py, C., Lacoeuilhe, J.J. and Teisson, D.C. 1987. The Pineapple: Cultivation and Uses. G.P. Maisonneuve and Larose, Paris, 127-137.

Ray,P. K. 2002. Breeding Tropical and subtropical fruits, Springer - Verlag, Narosa Publishing House, pp.201-211.

Rodriquez, A.G. 1932. Influence of smoke and ethylene on the fruiting of pineapple (*Ananas sativus*). *Schult. J. Dept. Agr. Puerto Rico.* 16:5–18.

Rohrbach, K.G., Lealand, F. and Coppensd'Eeckenbrugge, G. 2003. History, Distribution and World Production. In: Bartholomew, D.P., Paull, R.E., Rohrbach, K.G. (Eds.) The Pineapple: Botany, Production and Uses. CAB International, New York, USA. pp. 1-32.

Schultes, R.E.1984. Amazonian cultigens and their northward and westward migration in pre-Columbian times. In: Stone, D. (ed) Pre-Columbian plant migration, Papers of the Peabody Museum of Archaeology and Ethnology. Vol 76. Harvard University Press, Cambridge, MA, USA, pp. 19–37.

Smith, L.B. and Downs, R.J. 1979. Bromelioideae (Bromeliaceae) Flora Neotropica. New York Botanical Garden, New York, USA.

Srivivas, R., Dore, S. and Chacko, E. K. 1981. Differentiation of plantlets in hybrid embryo callus of pineapple. *Scientia Horticult.* 15: 235-238.

Trusov, Y. and Botella, J.R., 2006. Silencing of the ACC synthase gene ACACS2 causes delayed flowering in pineapple [*Ananas comosus* (L.) Merr. *Journal of Experimental Botany,* 57:(14). pp. 3953–3960.

Urasaki, N., Goeku,S., Kaneshima,R., Takamine, T., Tarora,K., Takeuchi,M., Moromizato,C., Yonamine, K., Hosaka, F., Terakami,S., Matsumura,H., Yamamoto, T. and Shoda, M. 2015. Leaf margin phenotype-specific restriction-site-associated DNA-derived markers for pineapple (*Ananas comosus* L.). *Breeding Science*. 65: 276–284.

Vanijajiva, O. 2012. Assessment of genetic diversity and relationships in pineapple cultivars from Thailand using ISSR marker. *Journal of Agricultural Technology*. 8(5): 1829-1838.

Williams, D.D.F. and Fleisch, H. 1992. History of plant breeding in Hawaii. (In:) Bartholomew, D.P. and Rohrbach, K.G. (Eds) First International Pineapple Symposium, Honolulu, Hawaii, USA, 2-6 November 1992.

Zeder, M.A. 2006. Central questions in the domestication of plants and animals. *Evol. Anthropol.* 15: 105-117.

Zhang,F., Ge, Y., Wang, W., Yu, X., Shen, X., Liu, J., Liu, X., Tian, D., Shen, F. and Yu, Y. 2012. Molecular Characterization of Cultivated Bromeliad Accessions with Inter-Simple Sequence Repeat (ISSR) Markers. *Int. J. Mol. Sci.* 13: 6040-6052.

Online Sources

http://www.itis.gov/servlet/SingleRpt/SingleRpt?search_topic=TSN&search_value=42335-USDA PLANTS (2007-2010). 2010. *Ananas comosus*. Bromeliaceae of North America Update, database (version 2010)-24/11/2015, 8.00pm

https://www.daf.qld.gov.au/__data/assets/pdf_file/0007/66247/Ch1-The-Pineapple.pdf - 06/07/2015, 12.46 pm)

pineapplehttp://holidayandtraveleurope.blogspot.in/2013/03/italian-history-who-was-christopher.html-12/09/2015,4.04pm

http://library.ucf.edu/rosen/pineapple.php -10/07/2015, 2.00 pm.

http://www.kingoffruit.com.au/a-symbol-of-hospitality.html- 12/09/2015,11.45 am.

http://www.levins.com/pineapple.html SOCIAL HISTORY OF THE PINEAPPLE -10/07/15, 10.00 am.

http://foodnetworks.org/foods/fruit/kiwi-and-pineapple-image of world map-12/09/2015,12.03pm.

https://www.hort.purdue.edu/newcrop/morton/pineapple.html,12/09/2015,12.29 pm.

https://www.dolefruithawaii.com/Articles.asp?ID=142- 12/09/2015,12.35 pm

http://www.theepochtimes.com/n3/410939-reconsidering-history-as-we-enter-the-new-year/-12/09/2015,2.21 pm.

http://www.stlucieco.gov/media/history.htm- 12/09/2015, 3.00 pm

http://blogs.agu.org/georneys/2013/05/26/the-big-pineapple-in-bathurst-south-africa/-12/09/2015, 3.52 pm http://diannesutherland.blogspot.in/2014_02_01_archive.html-12/09/2015,4.12 pm

http://www.bigpineapple.com.au/attraction/-12/09/2015, 4.20 pm

http://www.prideofmaui.com/blog/activities/best-things-hawaiian-culture.html-12/09/2015, 4.23 pm

http://www.choosephilippines.com/do/festivals/1493/22nd-pinyasan-festival-in-daet/-12/09/2015,4.26 pm

http://americangardenhistory.blogspot.in/2014/04/maryland-indentured-garden-servant.html- pineapple history- 16 sep, 12.49 pm

http://bensladerealty.com/the-history-of-the-housewarming-party-and-traditional-gift-ideas/-16/09/2015,12.52 pm

http://www.beaufortonline.com/christopher-columbus-discoveredpineapples/16/09/2015, 2.19 pm

http://tolweb.org/treehouses/?treehouse_id=4383-30/11/2015, 11.51 am

17

Pomegranate

K Dinesh Babu, N V Singh, H B Shilpa, A. Maity,
N Gaikward, R K Pal and M Sankaran

Pomegranate (*Punica granatum* L.) is an important fruit crop of arid and semiarid regions of world. It is believed to be originated from Iran and belongs to the family Lythraceae (sub-family: Punicoideae). The genus Punica includes two species viz., *P. protopunica* (wild pomegranate) and *P. granatum* (cultivated pomegranate).The adaptability of the crop to extremes of temperature (-12 to +44°C), suitability to marginal lands with poor fertility, rocky lands with shallow depth etc., pave the way for its potential production in various ecosystems.

Exquisite fruit quality, enriched nutritional values, enormous medicinal values, huge demand in domestic and international market besides extended shelf life enables it to emerge as an eminent fruit crop of recent times. Use of improved varieties/ hybrids, quality planting material, high density planting, microirrigation system, fertigation, Integrated Disease and Insect Pest management etc have made feasible pomegranate cultivation, a commercial venture. India is one of the leading producers of pomegranate in world. In India, it is cultivated over 1.31 lakh ha with an annual production of 13.45 lakh tons and productivity of 10.27 tones /ha during 2013-14. The export of pomegranate fruits is around 35,000 tons/annum from India.

Introduction

Pomegranate (*Punica granatum* L.) is an economically important commercial fruit crop of arid and semiarid regions of the world (Babu *et al.*, 2012). It is a highly remunerative crop for replacing subsistence farming and alleviating poverty (Jadhav and Sharma, 2007). Pomegranate (*Punica grantum* L.) derives its scientific name from 'Pomum' (apple) and 'granatus' (grainy) meaning seeded apple. It is one of the oldest known edible fruits and is capable of growing in different agro climates, ranging from tropical to temperate regions of the world. It is believed to be originated from Iran (Pal and Babu, 2015).

This 'multi-seeded apple' is known by different common/vernacular names in various languages (Deshpande, 2008, Pal *et al.*, 2014).

Table1: Common names for pomegranate

S. No.	Language	Common Name
1.	English	Pomegranate
2.	Arabi	Rumman
3.	Spanish	Granada
4.	French	Grenade
5.	Persian	Gulnar
6.	Sanskrit	Dadima, Dantabiya, Dantabijak, Karak, LohitPushpam
7.	Hindi	Anar, Dalim, Dadim
8.	Marathi	Dalimb
9.	Bengali	Dalim, Bedana
10.	Gujarati	Dadam, Dadiyam
11.	Kannada	Dalimba, Daalimbae
12.	Malayalam	Mathalam
13.	Tamil	Madulai, Madhulai
14.	Telugu	Danimma, Daalimma, Danimmapandu
15.	Urdu	Anar

Pomegranate is a good source of protein, carbohydrate, minerals, antioxidants, vitamins- C, B and A. The fruit has been useful in controlling heart diseases, prostate cancer, diarrhoea, hyperacidity, tuberculosis, leprosy, abdominal pain, fever and many other medical complaints. There has been enormous increase in pomegranate area, production and export over the past decades owing to its immense medicinal/ therapeutic values and higher remuneration.

The fruit is consumed as table fruit as well as in processed forms. Pomegranate juice, Ready-to-serve (RTS) beverage, pomegranate wine, pomegranate seed oil, rind powder etc., are some of the value added products from pomegranate. Pomegranate is grown in different countries viz., Iran, Iraq, Afghanistan, Kazakhstan, Turkmenistan, Tajikistan, Armenia, Bangladesh, India, China, Myanmar, Vietnam, Thailand, USA, Spain, France, Portugal, Italy, Greece, Cyprus, Egypt, Turkey, Israel, Syria, *etc.*, It is also grown in East Asia for ornamental purpose. Among these countries, India, Iran, China, the USA and Turkey are the leading producers of pomegranate.The fruits of pomegranate are in great demand for fresh fruit and export in the national and international markets.

In India, it is cultivated over 1.31 lakh ha with an annual production of 13.45 lakh tons and productivity of 10.27 tons /ha during 2013-14 (NHB, 2014). The export of pomegranate fruits is around 35,000 tons/annum from India (Pal and Babu, 2015). India is the only country in the world, where pomegranates are available throughout the year from January to December. In India, Maharashtra and Tamil Nadu are the states producing pomegranates round the year (Pal and Babu, 2015).

Pomegranate has a deep association with the cultures of Mediterranean regions (Spain, Morocco, Tunisia, Greece, Turkey and Egypt) and Near East, where it is savored as a delicacy and is an important dietary component, revered in symbolism, and greatly appreciated for its medicinal properties.It is presumed that pomegranate was domesticated in the Middle East about 5000 years ago.

Appreciation and Symbolism

Pomegranate is known to be the symbol of fertility, abundance and prosperity. Both the Arabic name for pomegranate (rumman) and the Hebrew name (rimmon) are reported to originate as "fruit of paradise", which provides abundant demonstrations of its appreciation in these cultures. It was considered by the Greeks to the the "fruit of the dead" and provided sustenance to the residents of Hades (Lansky *et al.*, 2000). These tow considerations may demonstrate the amazing breadth of the pomegranate's potential consumer base. The fruit's unique flavors, with sweetness often counterbalanced by acidity, makes pomegranate easy to appreciate by most who try it. In addition to their use as a fresh fruit or fruit juice, the juice of the pomegranate also contributes distinctive character to many mid-eastern dishes, such as the Iranian fessenjan. As a practical contributor to the diet, these fruits were likely invaluable to early desert travelers as an easily carried, well-protected form of water (Morton, 1987).

In Zoroastrianism, the pomegranate symbolizes both fecundity and immortality and is an emblem of prosperity (Panthaky, 2006). Pomegranate has long been associated with love and was one of the symbols of the love goddess Aphrodite (Encyclopedia Britannica, 2006).

It is easy to imagine that the seediness of the pomegranate encouraged association with fertility. Perhaps this gave rise to the Greek myth in which Persephone must spend 6 months in the underworld after Hades forced her to eat six pomegranate seeds, but her return is celebrated with the coming of spring. A bit more mysterious is the rationale for Hebrew priests wearing vestments adorned with pomegranates (Exodus, 28:31) or the 480 BC attempt by King Xerxes to capture Greece with an army carrying spears adorned with pomegranates.

Linnaeus described the genus *Punica* for the first time in 1753. It is traditionally treated under the Punicaceae, a monogeneric family of two species *i.e.*, *Punica granatum* and *Punica protopunica*. The species *P. protopunica*, so called Socotra pomegranate, is endemic to the island of Socotra, Democratic Republic of Yemen (Guarino *et al.*, 1990). Punica nana, another form of *Punica granatum* is often treated as third species of Punica (Melgarejo and Martinez, 1992).

Two subspecies of *P. granatum* have been distinguished on the basis of ovary colour, a stable character which is retained when they are grown from seeds. Subspecies chlorocarpa is found mainly in the transcaucasus and subspecies porphyrocarpa mainly in central Asia (Anonymous, 1969).

P. granatum, commonly known as pomegranate has been used since the dawn of human civilization. Besides consumption as a raw fruit and for juice, pomegranate has tremendous medicinal potential and is used in traditional or herbal cures for many diseases like cancer, dirrhoea, diabetes, bloodpressure, leprosy, dysentery, tapeworm infection, hemorrhage, bronchitis, gum bleeding, dyspepsia and throat inflammation (Aviram and Dornfeld, 2001; Adams *et al.*, 2006; Lansky and Newman, 2007; Stover and Mercure, 2007).

Benthan and Hooker's system (1862-83)	**Engler's system (Melchior, 1964)**	**Takhtajan's system (1980)**	**Dahlgren's system (1980)**
Dicotyledons	Dicotyledoneae	Magnoliopsida	Dicotyledoneae
Polypetalae	Archichlamydeae	Rosidae	Myrtiflorae
Calciflorae	Myrtiflorae	Myrtanae	Myrtales
Myrtales	Myrtineae	Myrtales	Lythraceae
Lythraceae	Punicaceae	Myrtineae	Punica L.
Punica L.	Punica L.	Punicaceae	
		Punica L.	

Cronquist's system (1981)	**Young's system (Bedell and Reveal, 1982)**	**Thorne's system (1983)**	**APG-II system (2003)**
Magnoliopsida	Magnoliopsida	Dicotyledoneae	Magnoliopsida
Rosidae	Rosidae	Myrtiflorae	Rosidae
Myrtanae	Myrtanae	Myrtales	Myrtales
Myrtales	Myrtales	Lythrineae	Lythraceae
Lythraceae	Lythraceae	Lythraceae	Punica L.
Punica L.	Punica L.	Punica L.	

Origin

The pomegranate (*Punica granatum* L.) is believed to be originated from Iran (Primary centre of Origin). Besides, it is widely prevalent in Afghanistan, Pakistan and India, the Secondary Centres of Origin (De Candolle, 1967). From its origin in the area now occupied by Iran and Afghanistan, the pomegranate cultivation had spread to India, China and Mediterranean countries viz., Turkey, Egypt, Tunisia, Morocco, and Spain. Spanish missionaries brought the pomegranate to the America in the 1500s (Hodgson, 1917; La Rue, 1980).

Cytology

Pomegranate is basically diploid in nature with 2n=2x=16, 18 chromosomes (Smith, 1979; Darlington and Janakiammal, 1945). The number of chromosomes in the somatic complements of Dholka, Ganesh, Kandhari, Muscat White and Patiala were found to be 2n=16, while the variety Double Flower (ornamental type) had 2n=18 (Nath and Randhawa, 1959a). The chromosome number in Vellodu and Kashmiri varieties was found to be 2n=2x=18 (Raman, *et al.*, 1963).

History

The pomegranate is widely considered native in the region from Iran to northern India (Morton, 1987), with apparently wild plants in many forests of these areas. Others (Mars, 2000) suggest that it is native to the smaller area of Iran and vicinity, and spread by human movement to a much broader area in prehistory. In India, the fruits of the wild pomegranate have thicker rinds and extremely high acidity compared with cultivated types (Bist *et al.*, 1994). They are also reported to have much smaller arils (Kher, 1999). In Central Asia, the primary difference noted is the higher acidity in wild material (Kerimov, 1934). The pomegranate's origin in proximity to the ancient cultures of Mediterranean have provided a long, recorded history for pomegranate. Indeed, some have argued that the pomegranate is the apple of the biblical Garden of Eden, but this is disputed in a later review (McDonald, 2002).

The cultivation of pomegranate was started over 5,000 years ago. Pomegranate has been naturalized throughout the Mediterranean region (California Rare Fruit Growers, 1997). Edible pomegranates were cultivated in Persia (Iran) by 3000 BC (Anarinco, 2006) and were also present in Jericho in modern day Israel. By 2000 BC, Phoenicians had established Mediterranean Sea colonies in North Africa, bringing pomegranates to modern day Tunisia and Egypt. Around the same time, pomegranates become naturalized in western Turkey and Greece.

S. No.	Period of cultivation	Place of cultivation	Existence/ introduction	Remarks
1.	3000BC	Persia (Iran)	Existence; Cultivation of edible pomegranate	Anarinco, 2006
	3000BC	Jericho, Israel	Existence;	Anarinco, 2006
2.	2000BC	Mediterranean sea colonies of North Africa (Tunisia, Egypt)	Phoenicians introduced pomegranate	Anarinco, 2006
3.	2000BC	Turkey, Greece	Naturalization of pomegranate	Anarinco, 2006
4.	100BC	China	Dispersal of pomegranate	Anarinco, 2006
5.	800BC	Roman Empire, Spain	Spread of pomegranate	Morton, 1987
6.	100BC	Central and southern India		Morton, 1987
7.	1400 AD	Indonesia	Introduction	Morton, 1987
8.	1500 -1600AD	Central America, Mexico and South America	Introduced by Spanish	La Rue, 1980
9.	1700 AD	Florida, Georgia		Morton, 1987
10.	1770AD	West Coast, California		Seelig, 1970

Taxonomy

Botanically, the pomegranate is in the subclass Rosidae, order myrtales, which is home to a few other fruits such as guava (*Psidium guajava* L.) and Feijowa (*Feijowa* sp.). However, pomegranate is unusual in being one of the only two species in its genus, Punica, which is the sole genus in the family Punicaceae (ITIS, 2006). Recent molecular studies suggest a taxonomic reconsideration might place Punica within the Lythraceae (Graham *et al.*, 2005). The second species in *Punica*, *P. protopunica*, is found only on the island of Socotra, of the Arabian Peninsula, and is considered an ancestral species (Shilikina, 1973) or an independent evolutionary path (Kosenko, 1985).

The name Punica is the feminized roman name for Carthage, the ancient city in northeren Tunisia from which the best pomegranates came to Italy. It was initially known as *Malumpunicum*, the apple of carthage. But, Linnaeus selected the current name, with the specific epithet grantum, meaning seedy or grainy. Its common name in the United States, therefore, means 'seedy apple' (Encyclopedia Britannica, 2006). While considering naming,it is interesting to note that the fruit's name in French, grenade, provided the name for the weapon becaue of similarities in appearance (Encyclopedia Britannica, 2006).

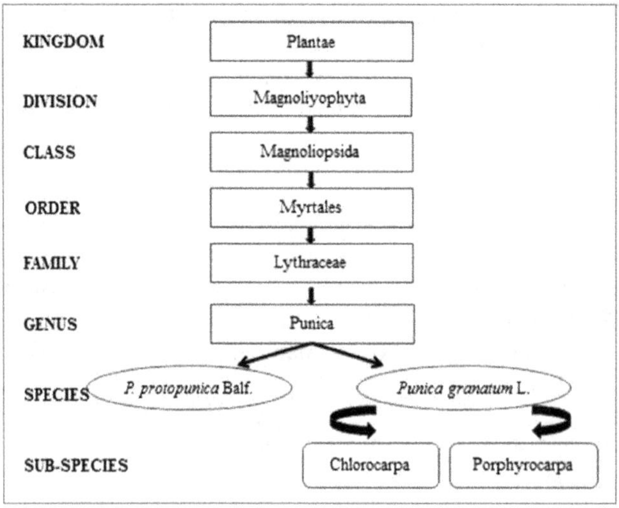

Fig 1: Botanical classification of pomegranate

Pomegranate belongs to the family Lythraceae (sub-family: punicoideae). The sub-family 'punicoideae' consists of single genus '*Punica*' with two species *Punica granatum* L. and *P. protopunica* Balf. The *P. granatum* is the cultivated pomegranate and has two subspecies *viz.*, chlorocarpa and porphyrocarpa. *P. protopunica* is wild and is endemic to Socotra islands.

Germplasm Collection

The genus 'Punica' has two species *viz.*, *Punica granatum* (cultivated pomegranate) and *Punica protopunica* (wild pomegranate). The cultivated pomegranate *P. granatum* is divided into two sub-species *viz.*, *P. granatum* subsp. chlorocarpa and *P. granatum* subsp. porphyrocarpa. The wild pomegranate (*P. protopunica)* is found in Socotra Island. Pomegranate grows as wild in Syria, Afghanistan, Central Asia and India (Saxena *et al.*, 1987). In India, wild pomegranate growing in Himachal Pradesh, Uttarakhant and Jammu and Kashmir is known as '*Daru*'. Chandra *et al.*, (2011) reported on the germplasm collection available in the field gene banks of India (Table 2).

Table 2: Germplasm collections of pomegranate in India

S. No.	Collection centres	State	Number of collections
1.	CCS Haryana Agricultural University, Bawal,	Haryana	09
2.	Rajasthan Agriculture University, Jobner	Rajasthan	09
3.	Punjab Agriculture University, Abohar	Punjab	19
4.	Tamil Nadu Agriculture University, Aruppukottai	Tamil Nadu	24
5.	Acharya N.G. Ranga Agriculture University, Anantpur	Andhra Pradesh	29
6.	ICAR-Central Arid Zone Research Institute (ICAR), Jodhpur	Rajasthan	34
7.	S.K.Nagar Agriculture University, SK Nagar	Gujarat	52
8.	Mahatma Phule Krishi Vidhyapeeth, Rahuri	Maharashtra	52
9.	ICAR-Indian Institute of Horticulture Research, Bengaluru	Karnataka	64
10.	ICAR-Central Institute of Arid Horticulture (ICAR), Bikaner	Rajasthan	152
11.	ICAR-National Research Centre on Pomegranate (ICAR), Solapur	Maharashtra	187

Germplasm Collection in the World

The important germplasm collection centres across countries are listed below.

S. No.	Country	Institute
1.	Afghanistan	Kabul University, Kabul
2.	Albania	Research Institute of Fruit Trees and Vineyard, Tirana
3.	Cyprus	Agricultural Research Institute, Nicosia
4.	France	CIRAD-FLHOR, Capesterre Belle-Eau
5.	Hungary	University of Horticulture and Food Industry, Budapest
6.	India	ICAR-National Research Centre on Pomegranate, Solapur ; ICAR-National Bureau of Plant Genetic Resources, Regional Station, Bhowali;
7.	Iran	University of Tehran, Karaj
8.	Israel	NeweYaar Research Center, Agricultural Research Organization, Bet-Dagan
9.	Italy	Institute of Agricultural Science, Verona
10.	Jordan	National Center for Agricultural Research and Extension, Amman
11.	Portugal	National Fruit Breeding Station, Alcobaca
12.	Russia	NI Vavilov Research Institute of Plant Industry, St. Petersburg

Contd...

S. No.	Country	Institute
13.	South Africa	Alternafruit SA (Pty) Ltd., Wellington
14.	Thailand	Chiang Mai, Bangkok
15.	Tunisia	U.R. Agrobiodiversite, Institute SuperieurAgronomique,Sousse
16.	Turkey	Alata Horticultural Research Institute, Erdemli-Mersin- Turkiye and Cukurova University, Turkey
17.	Turkmenistan	Turkmenian Experimental Station of Plant Genetic Resources, Garrygala
18.	USA	USDA-ARS National Clonal Germplasm Repository, Davis, California
19.	Uzbekistan	Schroeder Uzbeck Research Institute of Fruit Growing, Viticulture and Wine Production, Tashkent

Important varieties (variety/ parentage/ salient features)

Important varieties of pomegranate developed across countries:

Country	Variety	Characters	Remarks
Iran	760 genotypes &cvs in Yazd collection		Bahzadi Shahrbabaki, 1997
	Malas-e-Saveh Malas-e-Yazdi Rabab-e-Neyriz SisheKape-Ferdos Naderi-e-Budrood	Late ripening, medium to large size, thick red rind and red arils	Varasteh *et al.*, (2006)
	Alack	Early cv. That ripens in late Aug-early Sep, used for export	Iran Agro Food, 2007
	AlakShirin (sweet) AlakTorsh (sour)	Red, small sized, hard seeds	
	Maykhosh	Late export cv. (till end of Dec)	
China	87- Qing7	Early bearing, spur type mutant	Liu *et al.*, 1997
	Duanzhihong	Spur type cv. From Xingcheng, ripens in end of August, red skin, pinkish arils, 340g	(Liu, 2003)
Turkey	Hicaznar		
	Asinar	505g, large fruit, red arils, sweet-sour, soft seeds	Gozlekchi and Kaynak, 1997
	Eksilk	Sour(5%TA), red arils	Gozlekchi and Kaynak, 1997
	Emar	Dark red skin, red arils, sweet with low TA	Gozlekchi and Kaynak, 1997

Contd...

Country	Variety	Characters	Remarks
	Fellahyemez	Large pink arils, sweet with low TA, soft seeds	Gozlekchi and Kaynak, 1997
	Katirbasi	517g, large fruit, large red arils, sweet-sour	Gozlekchi and Kaynak, 1997
Spain	Valenciana	Small, early but not top quality	Costa and Melgarejo, 2000
	Mollar de Elche15	272g, deep red arils with soft seeds, sweet, low acid	Amoros *et al.*, 2000
	Mollar de Orihuela	414g, Red pink arils with soft seeds, sweet low acid	Amoros *et al.*, 2000
	PinonTierno de Ojas9	405g,Red pink arils with soft seeds, sweet, low acid	Amoros *et al.*, 2000
	Agridulce de Ojos4	524g, Red arils with hard seeds, bitter/ sweet, medium acid	Amoros *et al.*, 2000
USA	Golden Globe	Large fruit,Golden green fruit with pink blush, pink to red arils, small soft seeds, sweet	Karp, 2006
	Eversweet	Pink to red fruit with pink arils, soft seeds, sweet even when immature	Dave Wilson nursery, 2005; Karp, 2006
	Wonderful	Deep red arils, medium hard seeds, sweet-sour	Morton, 1987
	Early Wonderful	Deep red arils, medium hard seeds, sweet-sour, 2 weeks earlier than Wonderful	California Rare Fruit Growers, 1997
	Early Foothill	Deep red arils, medium hard seeds, sweet-sour, 2 weeks earlier than Wonderful	La Rue, 1980
	Granada	Deep red arils, medium hard seeds, sweet-sour, 1 month earlier than Wonderful	California Rare Fruit Growers, 1997
Georgia	ApsheronskiiKrasnyi, Burachnyi, Frantsis, KyrmyzKabukh, Lyaliya, Pirosmani, Rubin, Shirvani, Vedzisuri, ImeretisSauketeso	Splitting resistant cultivars	Vesadze and Trapaidze (2005)
Tunisia	Gabsi, Tounsi, Zehri, Chefli, Mezzi, Jebali, Garoussi, Kalaii, Zaghuoani, Andalousi, Bellahi	Gabsi- main cv., sweet; Tounsi- sweet, late ripening, Zehri- Sweet, ripens end of August-Sep, Chefli- Sweet, bold arils; Mezzi, Jebali, Baroussi – Sweet-sour, green skin; Kalaii- Sweet, bold arils; Zaghuoani, Andalousi- sweet	Mars and Marrakchi, (1999)

Contd...

Country	Variety	Characters	Remarks
Egypt	Arabi, Manfaloty, Nab ElGamal&Wardy	Manfaloty- sensitive to salt stress; Nab ElGamel- Saline tolerant	Abu-Taleb *et al.*, (1998) Saeed (2005)
Iraq	Ahmar (red), Aswad (black), and Halwa	Ahmar-red, Aswad-black	Morton, 1987
Saudi Arabia	Mangulati		Morton, 1987
Vietnam	Vietnamese	Evergreen cv. With orange flowers, bright red skincolour and small, juicy arils	Jene's Tropical Fruit, 2006)
Morocco	Gjeigi, Dwarf Evergreen, Grenade Jaune, Gordo de Javita, Djeibali, OnukHmam	17 clones and cultivars were reported	Oukabli *et al.*(2004)
Sicily, Italy	Dente di Cavallo, Neirana, Profeta, Racalmuto, Ragana, Selinunte	6 selections in Sicily	Barone *et al.* (2001)

Improvement

The pomegranate research in India was first started at the College of Agriculture, Pune in 1905 (Keskar *et al.*, 1993). Being cross pollinated crop, a lot of variability exists in seedling population, which can be utilized for further improvement programme. Genetic improvement in pomegranate has depended on the selection from seedling variability for centuries and their clonal propagation (Pareek and Sharma, 1993). Ganesh is a soft seeded selection from hard seeded Alandi. This cultivar has pinkish and sweet arils and soft seeds unlike the deep pink and sour aril and hard seeds in Alandi (Keskar, *et al.*, 1993). This variety was very popular till the last decade of the 20th century. After release of cultivar Bhagwa by MPKV, Rahuri, the major cultivated area of pomegranate was occupied by this cultivar and the area under Ganesh started declining.

Deciduous varieties (Sellimi, Roman, Chokab, Suffami, Wellingi, Ras-el-Baghi) from Baghdad, Palestine and other Mediterranean countries which were introduced to India failed to establish in warm climate of India. Similarly, introductions like Gulsha Rose Pink, ShirinAnar, Gulsha Red from erstwhile USSR sowed poor performance owing to their deciduous habit. Some introductions like Gulsha Rose Pink, Appuli, Guleshah Red, Lupania, BedanaSedana, Kandhari and Khog were used in breeding programmes to induce dark red aril pigment in the popular cultivars (Pareek and Sharma, 1993; Keskar, *et al.*, 1993). Later, a few commercial cultivars including soft seeded dark red aril types viz., Wonderful from USA, Males, Be Hastah, Alah, Agha Mohammad Ali, Post Sephid, Sirin from Iran and Rannyiz G-1-8-23, Rannyiz G-1-3-34, Cherenyj G-1-8-7 and few cultivars from Tunesia were introduced (Singh and Rana, 1993) and they showed variability in their morphological and qualitative traits. Hybridization has been exploited in edible pomegranate for improvement of some desirable traits (Babu *et al* 2014a) and improvement aspects of pomegranate has been described (Babu *et al.*, 2014c). The importance of some pomegranate varieties for ornamental purpose has also been reported (Babu *et al.*, 2014b).

Varietal improvement in pomegranate has been attempted both by selection of promising types from the indigenous ones and through controlled hybridization. Exotic introductions including Russian cultivars under the climatic conditions of Deccan plateau give sparse flowering and poor fruit set. Breeding objectives are to develop cultivars with high yield potential, better fruit quality with respect to fruit colour, soft seeds, red coloured arils and resistance to pests and diseases (Kumar, 2006) through different breeding methods *viz.*, selection, hybridization.

Some promising pomegranate varieties/ hybrids (Singh *et al.*, 2009) were developed through crop improvement programme in India (Table 6).

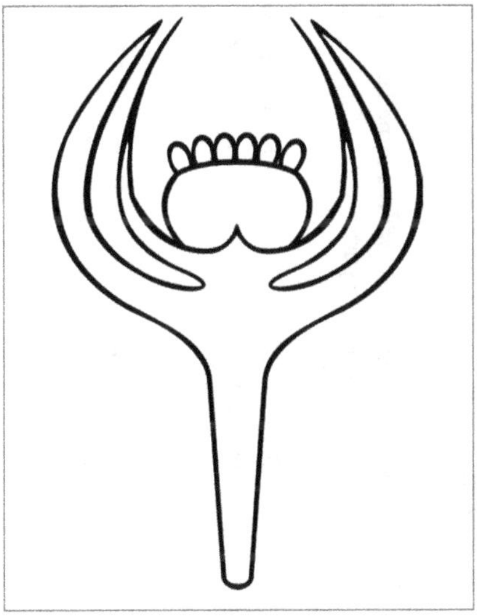

Male flower

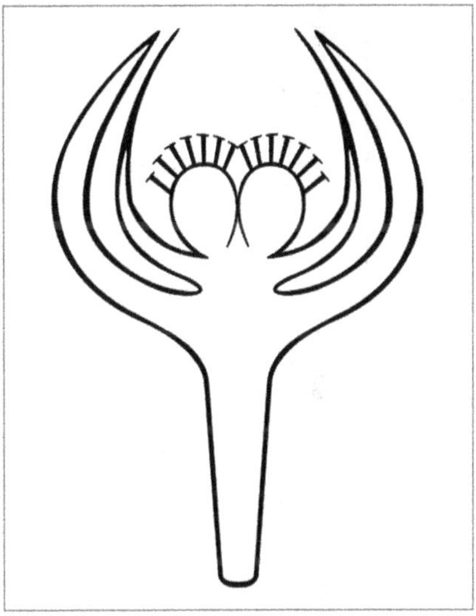

Male flower with pistillode

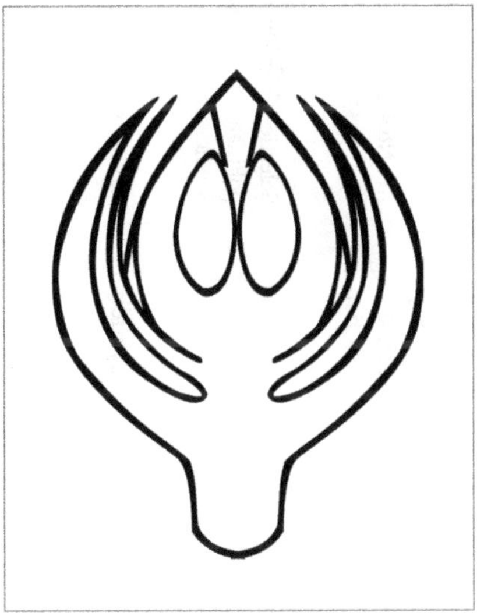

Female flower with staminodes

Bisexual flower

Table 6: Promising varieties/ hybrids developed in India

Variety	Year of release	Breeding methods	Parentage	Important traits	Institute
Ganesh	1936	Selection	Open pollinated seedlings of Alandi	Fruits are medium sized, with yellow, smooth surface and red tinge. The seeds are soft with pinkish aril. The juice tastes sweet. Soft seeded variety.	MPKV, Rahuri
CO-1	1983	Selection	-	It is a high yielding selection. Its fruits are medium-sized with attractive rind, soft seeds, higher pulp content and sweet taste.	TNAU, Coimbatore
G-137	1984	Clonal selection	Ganesh	Tree has spreading habit, fruit surface smooth, yellow with red tinge. Fruits are large sized with deep pink aril, sweet taste, soft seeds and prolific bearer; quality better than Ganesh.	MPKV, Rahuri
Yercaud-1	1985	Clonal selection	Acc No.455	Its fruits are medium sized, with easily peelable rind. The seeds are soft with attractive, deep purple aril.	TNAU, Coimbatore
Jyoti	1985	Selection	Selection from Open pollinated seedlings of Bassein Seedless and Dholka	The fruits are medium to large sized, having attractive, yellowish red, more fleshy and pink aril. Fruits are very sweet, soft seeded and taste good. It yields moderately.	UAS, Dharwad.
Mridula	1994	Hybridization	Ganesh x Guleshah Red	Plants are dwarf; Its fruits are medium-sized, rind smooth, dark red in colour. It has blood-red arils with very soft seeds, juicy and sweet in taste.	MPKV, Rahuri

Contd...

Variety	Year of release	Breeding methods	Parentage	Important traits	Institute
Ruby	1997	Hybridization	Multiple hybrid	It is a hybrid from a 3 way cross between Ganesh x Kabul x Yercaud and Gulesha Rose Pink which has soft and red arils with good flavour. The plants are dwarf, prolific bearer, providing uniformly red fruits.	ICAR-IIHR, Bengaluru
Amlidana	1999	Hybridization	Ganesh x Nana	Medium sized fruits, arils highly acidic (4.8%), short statured, suitable for anardana.	Ganesh x nana, Bengaluru
PhuleArakta	2003	Hybridization	Ganesh x Guleshah Red	Dark red rind; dark red arils; sweet in taste (Waskar et al., 2003)	MPKV, Rahuri
Bhagwa	2003	Hybridization	Ganesh x Guleshah Red	Attractive red rind and red arils, bold sized arils, sweet in taste; Predominant ruling variety (Waskar et al., 2003)	MPKV, Rahuri
PhuleBhagwa Super	2013	Selection	Selection among Bhagwa	Very attractive, dark saffron rind colour with glossiness on rind surface, good quality fruits, matures earlier than Bhagwa (Supe, et al., 2014)	MPKV, Rahuri

Conclusion

Pomegranate is an ancient fruit crop & has emerged as a predominant fruit crop in recent days. It is one of the first five fruit crops to be domesticated by mankind which has been richly embedded in human history. The development of soft seeded varieties has made a breakthrough in pomegranate breeding. There is an urgent need to introduce and conserve the wild species (*Punica protopunica*) in India. Besides, conservation of semi-wild forms, landraces, varieties and cultivars of pomegranate under Field Gene Banks deserve due attention for tapping the potential of biodiversity of pomegranate in the days to come. The screening of available biodiversity would pave the way for exploitation of resistance/ tolerance for important biotic (bacterial blight, wilt disease, *etc.*).

References

Anarinco. 2006. Pomegranate history. 1 Sept. 2006. www.anarinco.com/history.htm.

Babu, K.D., Sharma, J., Murthy, B.N.S., Chandra, R., Singh, N.V. and Pal, R.K. 2014a. Development of pomegranate varieties through breeding, In: AnarSampada - Lead papers and Abstracts of National Conference on Challenges and Opportunities for production and supply chain of pomegranate, UHS, Bagalkot,17-18 June, pp. 1

Babu, K.D., Sharma, J., and Pal, R.K. 2014b. Ornamental Pomegranates. In: Souvenir of National Conference on Challenges and Opportunities for production and supply chain of pomegranate, UHS, Bagalkot,17-18 June, pp. 1

Babu, K.D., Marathe, R.A., and Jadhav, V.T. 2012. Postharvest management of pomegranate, NRCP, Solapur p.116.

Bist, H.S., Srivastava R and Sharma G. 1994. Variation in some promising selection of wild pomegranate (*Punica granatum* L.). Hort. J. 7 :67-70.

California Rare Fruit Growers. 1997. Pomegranate. 1 Sept. 2006. www.crfg.org/pubs/ff/ pomegranate.html.

Chandra, R., Suroshe, S., Sharma, J., Marathe, R. and Meshram, D.T. 2011. Pomegranate growing manual, NRCP, Solapur, 56p.

Darlington, D.C. and Janaki Ammal, E.K. 1945. Chromosome atlas of cultivated plants.

De Candolle, A. 1967. *Origin of Cultivated Plants*, Hafner Publishers Co., New York, 468 pp

Deshpande, D.J. 2008. A handbook of herbal remedies. *Grobio (India)*, 496p.

Encyclopedia Britannica. 2006. Pomegranate. 1 Sept. 2006. http://search.eb.com/eb/ article-906074.

Hodgson, R.W. 1917. The pomegranate. *Calif. Agri. Extl. Sta. Bull.* 276. pp. 163-192

IT IS. 2006. Integrated Taxonomic Information System. 1 Sept. 2006.<www.itis.usda.gov/>.

Jadhav, V.T. and Sharma, J. 2007. Pomegranate cultivation is very promising. *Indian Horticulture.* pp.30.31

Keskar, B.G., Karale, A.R. and Kali, P.N. 1993. Improvement of pomegranate. (In): *Advances in Horticulture*, Vol. 1. (Eds.) Chadha, K.K. and Pareek, O.P., Malhotra Publishing House, New Delhi, pp. 399-05.

Kher, R. 1999. A note on the physico-chemical characters of the wild pomegranate (*Punica protopunica* L.). *Ann. Biol. Ludhiana* 15 :231-232.

Kerimov, A. 1934. Biochemical study of the subtropical fruit trees of Azerbaijan. *Bul. Appl. Bot.* 5:325-348.

Kosenko, V.N. 1985. Palynomorphology of representatives of the family Punicaceae. *Bot. Z.* 70 : 39-41.

Kumar, N. 2006. Breeding of horticultural crops – Principles & Practices, New India Publishing Agency, New Delhi. pp.131-132

La Rue, J.H. 1980. Growing pomegranate in California. DANR publication leaflet 2459. 1 Sept, 2006. <http://fruitsandnuts.ucdavis.edu/crops/pomegranate_factheet.shtml#b>.

Mars, M. 2000. Pomegranate plant material : Genetic resources and breeding, a review. Options Mediterraneennes Ser. A 42:55-62.

McDonald, J.A. 2002. Botanical determination of the Middle Eastern tree of life. *Econ. Bot.* 56 : 113-129.

Morton, J. 1987. Fruits of warm climates. Miami, FL.

Nath, N. and Randhawa, G.S. 1959a. Studies on cytology of pomegranate. *Indian J. Hort.* 16 : 210-15.

Pal, R.K. and Babu, K.D. 2015. Pomegranate. In: Managing postharvest quality and losses in horticultural crops. Vol-2. Fruit crops (Chadha, K.L. and Pal, R.K. Eds.), Daya Publishing House, pp. 517-533.

Raman, V.S., Kesavan, P.C, Manimekalai, G., Alikhan, W.M. and Rangaswami, S.R. 1963. Cytological studies in some tropical fruit plants – Banana, annona, *guava*, guava and pomegranate, *South Indian Horticulture*, 11 : 27-33.

Seelig, R.A. 1970. Fruit and vegetable facts and pointers : Pomegranates. United Fresh Fruit Association, Washington, DC.

Shilikina, I.A. 1973. On the xylem anatomy of the genus *Punica* L. Bot. Z. 58 : 1628-1630.

Singh, H.P., Parthasarathy, V.A. and Prasath, D. 2009. Studium Press (India) Pvt. Ltd., pp. 48-51.

Smith, P.M. 1979. Minor crops. (In) : Evolution of crop plants (Ed.) Simmonds, N.W., Longman, p.320.

[illegible] important element of [illegible] (Ed.) [illegible] Pal, J.K. and [illegible] O.S. [illegible]

[illegible] 1996. [illegible]

[illegible] A [illegible] chemical characteristics [illegible] of pomegranate [illegible]

[illegible] Azerbaijan [illegible]

[illegible] of the [illegible]

[illegible] Breeding of horticultural crops [illegible] Principles and Practices. New India Publishing Agency, New Delhi, pp. 131-152.

[illegible] pomegranate [illegible] DAMR [illegible] Shape [illegible] microsatellite [illegible] pomegranate [illegible]

[illegible] genetic resources and breeding [illegible] Review [illegible]

[illegible] De C. [illegible] of the Middle Eastern [illegible]

[illegible] Mitra, [illegible]

[illegible] Studies on cytology of pomegranate. Indian J. [illegible] 10: 27 [illegible]

[illegible] post [illegible] quality and losses [illegible] (Chadha, K.L. et al. [illegible] eds.) [illegible]

[illegible] Mahmoud [illegible] W.M. and Kangasabapathi, S.P. [illegible] studies in some [illegible] plants [illegible] anther [illegible] South [illegible] 11 [illegible]

[illegible] Fruit [illegible] facts and pointers [illegible] pomegranate [illegible] United Fresh [illegible] Delhi [illegible]

[illegible] genus Punica. J. Bot. [illegible]

Smith, P. [illegible] 2009. [illegible] Press (India) Pvt. Ltd., pp [illegible]

[illegible] Evolution of [illegible] Simmonds, N.W. [illegible]

18

Radish

Shirin Akhtar and Abhishek Naik

Radish (*Raphanus sativus* L.) is one of the mostly grown root vegetables. It is extensively cultivated in Europe, USA, China, Japan and Korea. In India it is grown everywhere especially in Uttar Pradesh, Bihar, West Bengal, Assam, Punjab, Haryana, Himachal Pradesh and Gujarat.

The common name 'radish' is derived from the Latin 'radix' meaning root. The common name 'radish' has been used over many centuries. In India it is commonly known as Muli, Mula or Mullangi. The name *Raphanus* has been derived Greek and means "quickly appearing" referring to the rapid germination of these plants. The old name used for the genus was *Raphanistrum*, derived from the same Greek root (Jan and Badar, 2012). The bigger sized Oriental radishes are also called 'diakens'.

This root vegetable can be enjoyed either raw or cooked, as well as in juice form. It makes an excellent addition to salad dishes. Radishes offer many health and nutritional benefits. Not only are the roots of these cruciferous vegetables nutritious, but so are their leaves. The leaves of the radish plant actually contain more Vitamin C, protein and calcium than their roots. They have been used to treat kidney and skin disorders, fight cancer and even soothe insect bites. Radish is rich in folic acid, Vitamin C and anthocyanins. These nutrients make it a very effective cancer-fighting food. It is said that radish is effective in fighting oral cancer, colon cancer and intestinal cancer as well as kidney and stomach cancers. Radishes contain Vitamin C, zinc, B-complex vitamins and phosphorus. All of these are very effective in treating skin disorders such as rashes and dry skin. Mashed raw radish can be used as a soothing and refreshing face pack. Dieters can benefit greatly from radishes, as they are low in calories, cholesterol and fat. They also contain a lot of roughage and contain a lot of water. Because of its high roughage content, it is also very useful in treating both piles and constipation. Radish juice helps to soothe the digestive system and detoxify the body. Radish helps to relieve congestion within the respiratory system, making it an excellent reliever for asthmatics and those who suffer from bronchial infections and sinus problems. It is beneficial for both the gall bladder and liver functions, as it acts as a cleanser. It contains sulphur based chemicals, which regulate the production and flow of bilirubin and bile, enzymes and acids and also help remove excess bilirubin from the blood. This makes it an excellent detoxifying agent for the body. It protects and soothes the gallbladder and liver, while protecting them from infections. It is also highly effective in treating jaundice, as it is able to halt the destruction of red blood cells and to increase the supply of oxygen to the blood. For this treatment however, the black radish is preferred. Radish is a natural diuretic. This makes them effective in preventing and fighting urinary tract infections. Radish juice helps to cure the burning feeling during urinary tract or bladder infections, as it is an excellent kidney cleanser.

Cytotaxonomic Background

Radish and its wild relatives are members of the mustard family Brassicaceae and all have 2n = 18 chromosomes (Lewis-Jones *et al.* 1982). Cultivated radish (*Raphanus sativus*) is an ancient crop having multiple, independent origins in Eurasia and Eastern Asia from several wild species including *Raphanus raphanistrum*, *R. maritimus*, *R. landra* (Crisp 1995, Yamagishi and Terachi, 2003). Radish has a number of inter-compatible wild and cultivated relatives either within the same genus or closely related genera, including wild radish (*Raphanus raphanistrum*), *Brassica rapa*, *B. oleracea*, *B. nigra*, *Sinapis arvensis* and *S. alba*. Variable levels of successful hybridization among these species have been observed by various researchers (Liu *et al.* 2003, Scheffler and Dale, 1994) and cultivated radish has been found to be most successfully and spontaneously crossed with *R. raphanistrum*. Inter-specific hybridization most easily produces *R. raphanistrum* x *R. sativus* F_1 hybrid offspring (Snow *et al.* 2001) relative to other potential inter-specific hybridization events (Lelivelt *et al.* 1993). When *R. sativus* is the pollen donor, the relative amounts of crop and conspecific pollen affects the chances of occurrence of the *R. sativus* x *R. raphanistrum* hybridization. Both distance between crop and wild radish and frequency of wild and cultivated radish affect the rate of hybridization (Klinger and Ellstrand, 1994; Lee and Snow, 1998).

Early History

Radish is an anciently cultivated annual or biennial vegetable. It is an ancient crop that appears to have multiple origins from several wild species (Crisp, 1995; Yamagishi and Terachi, 2003). It most likely originated in the area between the Mediterranean and the Caspian Sea (Crisp, 1995). Kaneko *et al.* (2007) suggested that radish originated from coastal regions along the Mediterranean and Black Seas. Its roots may be traced from the wild radish in southwest China (Cheo *et al.* 1987). It is possible that radishes were domesticated in both Asia and Europe. It was domesticated in Europe in pre-Roman times (Lewis *et al.*, 1982). According to Herodotus (c. 484-424 BC), radish was one of the important crops in ancient Egypt nearly 5000 years ago, as depicted by its pictures on the walls of the Pyramids about 4000 years ago. The benefits of radishes have been known since the time of pyramid construction in ancient Egypt. The Egyptians made sure that the labourers were fed a diet rich in radish, garlic and onion, which modern researchers have found to be extremely rich in raphanin, allicin and allistatin. These powerful natural antibiotics would certainly help to prevent outbreaks of disease in the over-crowded conditions of the work camps (Shuttleworth, 2010). Cultivated radish and its uses were reported in China nearly 2000 years ago (Li, 1989) and in Japan radishes were known some 1000 years ago (Crisp, 1995) before the establishment of the "Silk Road" that permitted extensive trade with central Asia. According to Snow and Campbell (2005) radish was independently domesticated in ancient Egypt over 5000 years ago and in China over 2000 years ago.

Most scholars believed that cultivated radish (*R. sativus* L.) originated from wild radish (*R. raphanistrum* L.) while others thought that *R. sativus* was derived by hybridization between *R. maritimus* and *R. landra* (Kitamura, 1958, Kaneko *et al.*, 2007). Wide variations gradually evolved in different places all over the world during the process of spreading, such as fresh edible cherry radish (*R. sativus* L. var. *radicola* Pers), oil radish (*R. sativus* L. var. *oleifera*), feed radish (*R. sativus* L. var. *caudatus*), black radish (*R. sativus* L. var. *niger*) and large root radish (Daikon) (*R. sativus* L. var. *longipinnatus* Bailey) (Shen *et al.*, 2013) Considerable ecological and morphological differences exist among the cultivated radishes in the different regions of the world; hence they are presumed to have originated from more than one source. The important wild species of radish, namely, *R. raphanistrum* L., *R. maritimius* Smith, *R. landra* Morett. and *R. rostratus* DC, all found in the Mediterranean region are involved in origin of the European types. The Japanese types are likely to have been derived from *R. sativus* var. *raphanistroides* Makino or *R. raphanistroides* Sinsk. that occur in the coastal regions of Japan (Sadhu, 1999). Banga (1976) suggested that the small European types of radish originated much later than the larger types. At the end of the 16th century, a long white type probably originated from the large type (Helm, 1957). The globular forms first developed in the 18th century, first the white ones and then the red one (Sadhu, 1999).

The early domestication of radishes, evolutionary processes and human selection of preferred types have led to significant variations in size, colour and taste of this vegetable crop. Among these, the small-rooted radishes are grown in temperate regions of the world and harvested throughout the year (Crisp, 1995). Larger-rooted cultivars such as Chinese radish are predominant in East and South-east Asia (Schippers, 2004). Today, radishes are grown throughout the world.

Recent History

The maximum diversity of culinary and morphological types of radishes is found in Asia, particularly in China and Japan at present. Deciphering the origins of cultivated radish is complicated due to several forms of the crop and uses in different parts of the world, and by its long history of dispersal around the globe. Crisp (1995) recognized several categories of radishes. European radishes have small swollen roots (partly hypocotyl and partly root) and are grown primarily in short-season, temperate regions to be eaten as fresh vegetables. Large-rooted daikon radishes are cultivated mostly in both temperate and tropical regions of Asia. Daikon radishes are eaten raw, as a cooked vegetable, canned, or pickled. There are another two less common types of radishes, one bred for their leaves (fodder radish) or seed pods, and the latter selected for use as either oil-seed crops or vegetables. "Rat-tail" radish has been selected for both its leaves and its edible immature seed pods, which are up to 80 cm long. In modern-day Pakistan, some traditional landraces of radish are grown for both their immature seed pods and their swollen roots (Rabbani *et al.*, 1998). Thus, there is a huge variety of ways in which the fruit, seeds, sprouts, leaves, and roots are grown for traditional dishes around the world.

Pistrick (1987) and later on Specht (2001), classified cultivated radishes (*Raphanus sativus* L.) into three groups:

1. convar. *oleifera* (*Raphanus sativus* var. *oleiformis* Pers.), also called *R. sativus* Leaf Radish Group (Wiersema and León 1999). These are the oilseed and fodder radishes, which are grown in Southeast Asia and in Europe for the purpose of leaf fodder as well as green manure.
2. convar. *caudatus* (*Raphanus sativus* var. *caudatus* (L.) L. H. Bailey), also known as *R. sativus* Rat-Tailed Radish Group (Wiersema and León, 1999). It is the rat-tail radish (also known as mougri, radis serpent) grown for its edible immature green or purple seed pods and leaves. This type is grown in Southeast Asia.
3. convar. *sativus (Raphanus sativus* var. *sativus*), also known as *R. sativus* Small Radish Group (Wiersema and León 1999). In this group all forms are with edible roots, leaves and germinated radish sprouts. Many varieties belong to this group but generally are of the small type (commonly called radish, small radish, turnip radish, petit rave).

The last group, convar. *sativus*, which is composed of swollen root is the primary commercial crop, widespread and economically important as root crop. The seeds have economic value for seed supply sales and the consumption of germinated radish sprouts.

Raphanus sativus L. var. *niger* J. Kern, also known as *R. sativus* Chinese Radish Group with the common names Chinese radish, Japanese radish, and Oriental radish has been recognized by Wiersema and León (1999) as fourth cultivated group.

Radishes have been classified in different ways according to Zhu *et al.* (2008):

Based on root size

a. Small-rooted (sometimes referred to as var. *radicula*)
b. Large-rooted types or Daikon (including names such as var. *nigra, niger, sinensis, acanthiformis* or *longipinnatus*)

Based on use

a. Fresh edible cherry radish (*R. sativus* L. var. *radicola* Pers)
b. Oil radish (*R. sativus* L. var. *oleifera*)
c. Feed radish (*R. sativus* L. var. *caudatus*);

Based on geography:

a. European
b. Chinese
c. Indian
d. Japanese

Based on the adaptation to growing seasons and regions:

a. Spring or summer radish (var. *longipinnatus* Bailey)
b. Winter radish (var. *longipinnatus* Bailey)
c. All-season radish (var. *radiculus* Pers.)

Based on temperature requirement for flowering radishes has classified into:

a. Temperate / European type: They require chilling temperature for bolting. They do not set seeds under tropical conditions. These are quick growing radish grown in short duration, about 25-30days. The roots produced are less pungent and smaller in size. It is generally low yielder and roots become pithy immediately after attaining marketable stage.
b. Tropical / Asiatic type: They do not require chilling temperature for bolting and set seeds freely under tropical conditions. These are more pungent types and remain edible for a longer period in field even after attaining marketable stage. These are slow growing types, produce large roots and generally are high yielders.

An elaborate study by Yamagishi and Terachi (2003) revealed that the most likely ancestors of cultivated radish are *Raphanus raphanistrum*, *R. maritimus*, *R. landra*, or their earlier progenitors by comparision of configurations of two mitochondrial gene regions, *cox1* and *orfB*, among three wild species and cultivated radish from Europe, Asia, and Japan. Mitochondrial genes are useful for tracing phylogenetic histories because they are maternally inherited and they seldom recombine. Five mitochondrial haplotypes were found among accessions of *R. sativus*. Cultivated varieties that share haplotypes with wild relatives are likely to be descended from one or more of these wild species, or their progenitors.

Reconstructing the phylogenetic history of a crop is complicated because wild accessions may include wild or feral relatives that have hybridized with the crop during the past millennia. Also, conclusions from this approach can be influenced by the diversity of wild and crop accessions under study. Bearing these caveats in mind, it appears that cultivated radishes have originated multiple times from wild taxa that were very similar to *R. raphanistrum*, *R. maritimus*, and *R. landra*, which is consistent with the earlier evidence for independent domestication in Eurasia and eastern Asia. *Raphanus raphanistrum*, *R. maritimus*, and *R. landra* are able to hybridize with each other and with *R. sativus*, all having chromosome number, 2n = 18 (Lewis-Jones *et al.*, 1982), and in recent times they have been classified as subspecies of *R. raphanistrum* (Chater, 1993; Jalas, 1996). Since *R. sativus* sometimes hybridizes with *R. raphanistrum* in the field, several authors have suggested that these taxa also should be consolidated into a single species (Bett and Lydiate, 2003; Crisp, 1995; Snow *et al.*, 2001). Perfect collinearity between genomes in the positions of 144 informative RFLP markers on radish's nine chromosomes has been observed in a study involving crop-wild hybrids (Bett and Lydiate, 2003). The close taxonomic relationships among the putative wild ancestors of radish, suggest that the progenitors of cultivated radish were wild forms of radish that shared many attributes with weedy *R. raphanistrum* (*R. raphanistrum* subsp. *raphanistrum*).

Domestication of radish from its wild relatives involved selection for a larger roots with better flavour, often with a red or purple skin, along with high seed production for propagation of the crop. Selection for a larger root probably resulted in delayed flowering and a tendency to be biennial. As with many other crops, selection for easier seed harvest might also been a criterion for its domestication. In weedy *R. raphanistrum*, tough, hard-to-crack fruits protect the seeds from any kind of damage, bird predation, *etc.* Fruits of *R. raphanistrum* are shed gradually as they mature on the mother plant, and there is no splitting of the fruit to release the seeds. Instead, the fruit breaks into distinct sections, each of which encapsulates a single seed. In this way, the fruit wall of *R. raphanistrum* acts as a protective seed coat for the seeds, which lack the strong, impervious seed coat found in many other annual species. However, in domesticated *R. sativus*, seeds are contained within indehiscent fruits, but the fruits remain firmly attached to the plant even after senescence and they are not divided into sections. In many modern varieties of radish, the spongy seed pods can be easily crushed by hand, thereby allowing the seeds to be extracted from the seed pod more efficiently.

Based on recent studies using chloroplast single sequence repeats (cpSSRs), Yamane *et al.* (2009) postulated three independent domestication events which include black Spanish radish and two distinct cpSSR haplotype groups. One of the haplotype groups is geographically restricted to Asia, presenting higher cpSSR diversity than cultivated radish from the Mediterranean region or wild radish types. This implies that Asian cultivated radish cannot be traced back to European cultivated forms which spread to Asia, but might have originated from a still unknown wild species that is different from the wild ancestor of European cultivated radish (Yamane *et al.* 2009). Early domestication, evolutionary processes and human selection of preferred types have led to significant variations in size, colour and taste of this vegetable crop. Among the different types, small-rooted radishes are grown in temperate regions of the world and harvested throughout the year (Crisp, 1995), while larger-rooted cultivars such as Chinese radish are predominant in East and Southeast Asia (Schippers, 2004).

Radish being an open-pollinated crop that is self-incompatible, breeding goals have been improvement of root colour, taste, texture, size and shape, as well as agronomic performance, day-length requirements to prevent premature bolting, seed yield, and other traits (Banga, 1976; Crisp, 1995).

Varieties grown in temperate climates germinate in early spring and are tolerant to cool temperatures. The spatial scale at which radishes are grown ranges from small kitchen gardens to large, industrial-scale farms. Like other members of the Brassicaceae family, the plants produce glucosinolates that attributes to its peppery flavour and may help in herbivore defence and allelopathy (Agrawal, 1999; Irwin *et al.*, 2003). Asiatic radishes have been selected for resistance to various pathogens, such as *Fusarium*, *Albugo candida*, *Peronospora parasitica*, and viruses (Crisp, 1995) and Chinese and Japanese radishes sometimes have been used as sources of resistance genes (Williams and Pound, 1963). Disease resistance was not a priority for the small-rooted European varieties that were generally grown in the spring and the roots were harvested before diseases become prevalent. In the recent times, European varieties have been selected for multiple plantings per season, involving both annual and biennial lifespans. Besides, resistance to fungal disease pathogens (e.g., *Fusarium*, *Rhizoctonia* spp.) have become breeding goal in development of improved European radish.

Development of F_1 hybrids and exploitation of heterosis has improved the yield and overall performance of radishes in recent times. Cytoplasmic male sterility (CMS) found in many Japanese and Chinese cultivars (Ogura, 1968), has been for production of hybrid seed in large-rooted cultivars (Crisp, 1995). Lately, this *Ogura* type of CMS has been used to produce F_1 hybrids of small-rooted European varieties. Open-pollinated seed production without use of CMS is still very common in radishes. A great deal of genetic diversity is found in open-pollinated radish varieties, as well as other related *Raphanus* populations (Ellstrand and Marshall, 1985). Genetic diversity is especially high in regions where landraces are grown (e.g, Japan, Pakistan), because plants that are grown for seed often cross-pollinate with other cultivated and feral varieties (Rabbani *et al.*, 1998; Yamaguchi and Okamoto, 1997).

Conventional breeding in radish is expected to become more sophisticated with the use of genomics-guided strategies. The first genetic map of the *Raphanus sativus* genome was published in 2003 (Bett and Lydiate, 2003), that paved the way for the use of marker-assisted breeding and comparative mapping with *Arabidopsis* and *Brassica* species.

Transformation in radish has been done using a floral dip method (Curtis and Nam, 2001). However, transgenic radish has a relatively small market limiting its acceptance by seed companies till date. Radish is rarely an important staple crop, and hence future breeding may be focused on the aesthetic value of its colour patterns and shapes, for greater acceptability in both novel and traditional food markets (Crisp, 1995). Breeders also have focused on transferring a late-flowering trait to commercially important varieties, but conventional breeding techniques have been unsuccessful, producing low quality hybrid (Lee, 1987). Curtis *et al.* (2002) obtained late-flowering radishes by expressing an antisense GIGANTEA gene fragment from *Arabidopsis*. This trait is a possible candidate for the development of transgenic radishes.

Bae *et al.* (2012) developed an efficient protocol for the transformation of radish hairy root cultures using cotyledon explants infected by *Agrobacterium rhizogenes* R1000, a strain with the binary vector pBI121. The transgenic root cultures were further used investigation of the molecular and metabolic regulation of anthocyanin biosynthesis and evaluation of the genetic engineering potential of this species.

Recently, a genomics and genetics database of radish called RadishBase (http://bioinfo.bti.cornell.edu/radish) has been developed in order to query, analyze and integrate these radish resources efficiently (Shen *et al.*, 2013). The database contains radish mitochondrial genome sequences, expressed sequence tag (EST) and unigene sequences and annotations, biochemical pathways, EST-derived single nucleotide polymorphism (SNP) and simple sequence repeat (SSR) markers, and genetic maps. The tools and interfaces in RadishBase allow efficient mining of recently released and continually expanding large-scale radish genomics and genetics data sets, including the radish genome sequences and RNA-sequence data sets.

De-domestication: Crop Radish Turning into Invasive Weed?

Over the last century, the crop radish *Raphanus sativus* and the wild radish *R. raphanistrum* have coalesced into a hybrid lineage through bidirectional hybridization (Ridley *et al.*, 2008) that has displaced all natural populations of both parents in California (Ridley and Ellstrand, 2009). This lineage is commonly referred to as "California wild radish", and is now found throughout a major portion of naturally disturbed coastal areas, human-infested inland sites in California (DiTomaso and Healy 2006; Panetsos and Baker 1967), as well as south into Baja California, Mexico and north into Oregon (Hegde *et al.* 2006; Ridley 2008). Interspecific *Raphanus* hybrids also appear in Europe, but they are not invasive (Stace 1975).

Evolution plays an important role in the invasion process (Barrett *et al.*, 2008; Cox, 2004; Keller and Taylor, 2008; Lambrinos, 2004; Lee, 2002). Rapid evolutionary change can bring substantial change in introduced populations, and even lead to the creation of novel, highly invasive genotypes (Dlugosch and Parker, 2008; Ellstrand and Schierenbeck, 2006). Hybridization, particularly, intraspecific hybridization has been attributed to enhance invasiveness in several species, including both plants and animals (Facon *et al.* 2005; Lavergne and Molofsky 2007) by creation of genotypes that are more reproductively successful than the pure progenitor genotypes, probabaly due to fixed heterosis, the purging of deleterious alleles or the transfer of adaptations (Ellstrand and Schierenbeck 2006; Rieseberg *et al.* 1999). Apart from that increased genetic diversity relative to progenitor populations might also affect the success of hybrid-derived populations by enhancing their ability to respond to selection and to adaptively evolve (Ellstrand and Schierenbeck 2006; Lee 2002; Parker *et al.* 2003; Sakai *et al.* 2001). Interspecific hybridization of such type may lead to extinction of one or both hybridizing taxa via genetic assimilation (Chapman and Burke, 2006).

Conclusion

Raphanus is an important model system, much used by plant ecologists, evolutionists and geneticists for studying pollination biology, life history variation, ecological genetics, floral evolution and plant-herbivore interactions (Ridley and Ellstrand, 2009). It is an important root crop with great diversity and distribution, and there is much scope of its improvement. Breeders are constantly involved in this process through conventional breeding techniques and recently through the involvement of molecular tools. However, the rapidity with which extinction via hybridization can occur, it may be said that this crop is under extinction threat. Genetic markers should now be utilized, particularly based on morphological grounds, in cases where persistent gene flow appears to be eroding species boundaries. It is a challenge before the breeders and the biotechnologists to prevent the extinction of one or two species of *Raphanus.*

References

Agrawal, A.A. 1999. Induced responses to herbivory in wild radish: Effects on several herbivores and plant fitness. *Ecology.* 80: 1713-1723.

Bae, H., Kim, Y.B., Park, N.I., Kim, H.H., Kim, Y.S., Lee, M.Y. and Park, S.U. 2012. Agrobacterium rhizogenes-mediated genetic transformation of radish (*Raphanus sativus* L. cv. *Valentine*) for accumulation of anthocyanin. *Plant Omics Journal.* 5(4): 381-385.

Banga O. 1976. Radish. (In): Evolution of Crop Plants. (Ed.) Simmonds, N.W. Longman, New York. pp. 60-62.

Barrett, S.C.H., Colautti, R.I. and Eckert, C.G. 2008. Plant reproductive systems and evolution during biological invasion. *Mol Ecol* 17:373–383. doi:10.1111/j.1365-294X.2007.03503.x.

Bett, K.E. and Lydiate, D.J. 2003. Genetic analysis and genome mapping in *Raphanus. Genome.* 46: 423-30.

Chapman, M.A. and Burke, J.M. 2006. Radishes gone wild. *Heredity.* doi:10.1038/sj.hdy.6800899.

Chater, A.O. 1993. *Raphanus* L. (In): Tutin, T.G. *et al.*, Flora Europea, Ed. 2, Cambridge University Press, Cambridge. p. 417.

Cheo, T.Y., Guo, R.L., Lan, Y.Z., Lou, L.L., Kuan, K.C. and An, Z.X. 1987. Angiospermae, Dicotyledoneae, Cruciferae. *In*: Flora Reipublicae Popularis Sinicae. ed.Cheo, T.Y. Vol 33. Science Press, Beijing (China). pp. 1-483.

Cox, G.W. 2004. Alien species and evolution: the evolutionary ecology of exotic plants, animals, microbes and interacting native species. Island Press, Washington, p. 400.

Crisp P. 1995. Radish, *Raphanus sativus* (Cruciferae). (In): Evolution of Crop Plants, 2nd Edition, (Eds.) Smartt, J. and Simmonds, N.W. Harlow, Longman Scientific and Technical. pp. 86-89.

Curtis, I.S. and Nam, H.G. 2001. Transgenic radish (*Raphanus sativus* L. *longipinnatus* Bailey) by floral-dip method - plant development and surfactant are important in optimizing transformation efficiency. *Transgenic Research.* **10:** 363-71.

Curtis, I.S., Nam, H.G., Yun, J.Y. and Seo, K.H. 2002. Expression of an antisense GIGANTEA (GI) gene fragment in transgenic radish causes delayed bolting and flowering. *Transgenic Research.* 11: 249-56.

DiTomaso, J.M. and Healy, E.A. 2006. Weeds of California and other western states. University of California Division of Agriculture and Natural Resources, Oakland.

Ellstrand, N.C. and D.L. Marshall. 1986. Patterns of multiple paternity in populations of *Raphanus sativus. Evolution.* 40: 837-842.

Ellstrand, N.C. and Schierenbeck, K.A. 2006. Hybridization as a stimulus for the evolution of invasiveness in plants? *Euphytica.* 148:35–46. doi:10.1007/s10681-006-5939-3.

Facon, B., Jarne, P., Pointier, J.P. and David, P. 2005. Hybridization and invasiveness in the freshwater snail Melanoides tuberculata: hybrid vigour is more important than increase in genetic variance. *Journal of Evolutionary Biology.* 18:524–535. doi:10.1111/j.1420-9101.2005.00887.x.

Hegde, S.G., Nason, J.D., Clegg, J.M. and Ellstrand, N.C. 2006. The evolution of California's wild radish has resulted in the extinction of its progenitors. *Evolution : International Journal of Organic Evolution.* 60:1187– 1197.

Helm, J. 1957. Über den typus der art *Raphanus sativus* L., deren Gliederung und Synonymine. *Kulturplanzen* 5.

Irwin, R.E., Strauss, S.Y., Storz, S., Emerson, A. and Guibert, G. 2003. The role of herbivores in the maintenance of a flower color polymorphism in wild radish. *Ecology.* 84: 1733-43.

Jalas, J., Suominen, J, and Lampinen, R. 1996. Raphanus. Atlas Florae Europaeae -Distribution of Vascular Plants in Europe. Vol. 11. Helsinki University Printing House, Helsinki. pp. 290-293.

Jan, M. and Badar, A. 2012. Effect of crude extract of *Raphanus sativus* roots on isolated trachea of albino rat. *Pakistan Journal of Physiology.* 8(1): 23-26.

Kaneko, Y., Kimizuka-Takagi, C., Bang, S.W. and Matsuzawa, Y. 2007. Radish. *In:* Genome Mapping and Molecular Breeding in Plants. Ed. Kole, C. Springer, New York. pp. 141–160.

Keller, S.R. and Taylor, D.R. 2008. History, chance and adaptation during biological invasion: eparating stochastic phenotypic evolution from response to selection. *Ecology Letters.* 11: 852–866. doi:10.1111/j.1461-0248.2008.01188.x.

Kitamura, S. 1958. Cultivars of radish and their change. (In:) Japanese Radish. (Ed.) Nishiyama, I. Japan Society for the Promotion of Science, Tokyo. pp. 1–19.

Klinger, T. and Ellstrand, N. C. 1994. Engineered genes in wild populations: weed-crop hybrids of *Raphanus sativus. Ecological Applications.* 4: 117–120.

Lambrinos, J.G. 2004. How interactions between ecology and evolution influence contemporary invasion dynamics. *Ecology.* 85:2061–2070. doi:10.1890/03-8013.

Lavergne, S. and Molofsky, J. 2007. Increased genetic variation and evolutionary potential drive the success of an invasive grass. *Proceedings of Natioal Academy of Sciences, USA.* 104: 3883–3888. doi: 10.1073/pnas.0607324104.

Lee, C.E. 2002. Evolutionary genetics of invasive species. *Trends in Ecology and Evolution.* 17: 386–391. doi:10.1016/S0169-5347(02)02554-5.

Lee, S.S. 1987. Bolting in radish. (In:) Improved Vegetable Production in Asia, No. 36, eds. Asian and Pacific Council, Food and Fertilizer Technology Center, Taipei. pp. 60-70.

Lee, T. N., and Snow.A.A. 1998. Pollinator preferences and the persistence of crop genes in wild radish populations (*Raphanus raphanistrum*, Brassicaceae). *American Journal of Botany*. 85: 333–349.

Lelivelt, C. L. C., Lange, W. and Dolstra, O. 1993. Intergeneric crosses for the transfer of resistance to the beet cyst nematode from *Raphanus sativus* to *Brassica napus*. *Euphytica*. 68: 111–120.

Lewis-Jones, L.J., Thorpe, J.P. and Wallis, G.P. 1982. Genetic divergence in four species of the genus *Raphanus*: Implications for the ancestry of the domestic radish *R. sativus*. *Biological Journal of the Linnaean Society*. 18: 35-48.

Liu, F., Zhao, H., Yao, L., Zhang, J. J. and Cao, M. Q. 2003. Gene flow from transgenic chinese cabbage (*Brassica campestris* ssp *chinensis* var. *utilis*) to related Cruciferae species through outcrossing. *Acta Botanica Sinica*. 45: 681–687.

Ogura, H. 1968. Studies on the new male sterility in Japanese radish with special reference to utilization of this sterility toward the practical raising of hybrid seeds. *Memoirs of Faculty of Agriculture Kagoshima University*. 6: 39-78.

Panetsos, C.A. and Baker, H.G. 1967. The origin of variation in "wild" *Raphanus sativus* (Cruciferae) in California. *Genetica*. 38: 243–274. doi:10.1007/BF01507462.

Parker, I.M., Rodriguez, J. and Loik, M.E. 2003. An evolutionary approach to understanding the biology of invasions: local adaptation and general-purpose genotypes in the weed Verbascum thapsus. *Conservation Biology*. 17:59–72. doi:10.1046/j.1523-1739.2003.02019.x.

Pistrick, K. 1987. Untersuchungen zur Systematik der Gattung Raphanus L. *Kulturpflanze*. **35:** 225-321.

Rabbani, M.A., Murakami, Y., Kuginuki, Y. and Takayanagi, K. 1998. Genetic variation in radish (*Raphanus sativus* L.) germplasm from Pakistan using morphological traits and RAPDs. *Genetic Resources and Crop Evolution*. 45: 307-16.

Ridley, C.E. 2008. Hybridization and the evolution of invasiveness in the California wild radish (*Raphanus sativus*). In: Botany and Plant Sciences. University of California Riverside, Riverside, p. 138.

Ridley, C.E. and Ellstrand, N.C. 2009. Evolution of enhanced reproduction in the hybrid-derived invasive, California wild radish (*Raphanus sativus*). *Biological Invasions*. 11: 2251–2264.

Ridley, C.E., Kim, S.C. and Ellstrand, N.C. 2008. Bi-directional history of hybridization in California wild radish *Raphanus sativus* (Brassicaceae) as revealed by chloroplast DNA. *American Journal of Botany*. 95:1437–1442.

Rieseberg, L.H., Archer, M.A. and Wayne, R.K. 1999. Transgressive segregation, adaptation and speciation. *Heredity*. 83:363–372. doi:10.1038/sj.hdy.6886170.

Sadhu, M.K. 1999. (**In**): Vegetable crops. (Ed.) Bose, T.K., Som, M.G. and Kabir, J. pp.470-491.

Sakai, A.K., Allendorf, F.W., Holt, J.S., Lodge, D.M., Molofsky, J., With, K.A., Baughman, S., Cabin, R.J., Cohen, J.E., Ellstrand, N.C., McCauley, D.E., O'Neil, P., Parker, I.M., Thompson, J.N. and Weller, S.G. 2001. The population biology of invasive species. *Annual Review of Ecology, Evolution, and Systematics*. 32:305–332. doi:10.1146/annurev.ecolsys.32.081501.114037.

Scheffler, J. A., and P. J. Dale. 1994. Opportunities for gene transfer from transgenic oilseed rape (*Brassica napus*) to related species. *Transgenic research*. 3: 263–278.

Shen, D., Sun, H., Huang, M., Zheng, Y., Li, X. and Fei, Z. 2013. RadishBase: A Database for Genomics and Genetics of Radish. *Plant and Cell Physiology*. 54(2): e3(1–6). doi:10.1093/pcp/pcs176.

Snow, A.A., Uthus, K.L. and Culley, T.M. 2001. Fitness of hybrids between weedy and cultivated radish: implications for weed evolution. *Ecological Applications*. 11: 934-43

Specht, C.E. 2001. Raphanus. (In): Encyclopedia of Agricultural and Horticultural Crops. Hanelt, P., (Ed.). Mansfelds 6 vols. Springer-Verlag, Heidelberg Vol. 3. pp. 1476-1481.

Stace, C.A. 1975. Hybridization and the flora of the British Isles. Academic Press, London, p. 626.

Wiersema, J.H. and León, B. 1999. World Economic Plants – A Standard Reference. CRC Press, USA.

Williams, P.H. and Pound, G.S. 1963. Nature and inheritance of resistance to Albugo candida in radish. *Phytopathology*. 53: 1150-1154.

Yamagishi, H. and Terachi, T. 2003. Multiple origins of cultivated radishes as evidenced by a comparison of the structural variations in mitochondrial DNA of *Raphanus*. *Genome*. 46: 89-94.

Yamaguchi, H. and Okamoto, M. 1997. Traditional seed production in landraces of daikon (*Raphanus sativus*) in Kyushu, Japan. *Euphytica*. 95:141-147.

Yamane, K., Lü, N. and Ohnishi, O. 2009. Multiple origins and high genetic diversity of cultivated radish inferred from polymorphism in chloroplast simple sequence repeats. *Breeding Science*. 59: 55–65.

Zhu, D.W., Wang, D.B. and Li, X.X. 2008. Chinese crops and wild relatives, Vegetable crops Volume 1. Chinese Agricultural Press, Beijing.

19

Saffron

Ali Izanloo and Mohammad Ali Behdani

The cultivated saffron (*Crocus sativus* L.) is a perennial geophyte and genetically sterile plant that is only vegetatively propagated via its corms, which undergo a period of dormancy. It is widely cultivated in various parts of the world from Spain to China. Iran is the leading producer and exporter of saffron, which it alone accounts for about 90 % of the total harvest areas and 80% of the total export worldwide (Ghorbani, 2007).

Saffron is well known for its dried stigmata which are used as a spice (saffron), as a dye and traditional medicine. Saffron contains more than 150 volatile, non-volatile and aroma-yielding compounds (Abdullaev, 2002). Based on biochemical analyses of dry stigma of saffron extracts, carotenoids, namely crocin and crocetin and the monoterpene aldehydes picrocrocin and safranal are the most important active carotenoid secondary metabolites of saffron. Crocin and crocetin are responsible for its colouring power, picrocrocin for the bitter taste and safranal for the aroma (Lozano *et al.*, 1999; Carmona *et al.*, 2006). The demand for saffron is increasing worldwide not only because of its production and use, but also for its various medicinal properties of the metabolic components, which have anticarcinogenic and antioxidant effects (Abdullaev and Espinosa-Aguirre, 2004; Magesh *et al.*, 2006; Lim, 2014; Samarghandian and Borji, 2014; Bhandari, 2015). In addition, saffron is experiencing an increasing research interest because of the numerous mysteries surrounding its evolution and origins (Fernández, 2006). Much information is available on the use of saffron as a dye, aroma and for medicinal purposes, but there is a lack of information on its evolution and origins. There are different opinions about origin of saffron. Vavilov (1951) proposed Middle East (Minor Asia, Turkestan and Iran) as centre of origin of saffron, whereas recent contributions identified Greece (Crete) as place where saffron domestication has been started and wild *C. cartwrightianus* Herb. was introduced as the closest wild relative of saffron (Negbi, 1999; Frello and Heslop-Harrison, 2000; Grilli Caiola *et al.,* 2004). Although, Greece has been mentioned as probable origin of saffron, its site of origin is still a matter of dispute, and it is not clear whether its domestication occurred at more sites simultaneously or at different times (Fernández, 2004; 2006; Grilli Caiola and Canini, 2010). However, the botanical origin of saffron is not yet clear.

The increase in grown saffron is related to the possibility of obtaining higher yields per hectare and the high quality of commercial saffron. The solution to these two problems needs genetic improvement of the plant, the amelioration of cultivation practices, as well as controlled conditions during the preparation, storage and marketing of the spice. The genetic improvement of saffron by crossing requires knowledge of the original species or wild relatives as well as of the ways and site of origin of the cultivated plant. Many studies have been dedicated to solving this puzzle although with disputed results, the exact parents of saffron remaining an open question (Grilli Caiola and Canini, 2010). Therefore, the objective of this manuscript is to review the literatures about the phylogenic and evolutionary studies on *Crocus sativus* evolution and origin.

History

The earliest apparent reference to saffron cultivation goes back to about 2300 BC (Srivastava *et al.*, 2010). A definite identification of saffron crocuses dates from about 1700-1600 BC, in the form of a fresco painting, in which crocus-gatherers can be observed, in the Palace of Minos at Knossos in Crete. Other important records are found in the palace of Akrotiri in Thera (now Santorini, Greece, 1700-1450 BC) where frescoes represent young women collecting crocuses and offering them to a divinity (Figure 1). These crocuses, growing in rows of a regular pattern, are believed to be cultivated; unfortunately, it is not possible to determine with certainty which *Crocus* species (*C. sativus* or *C. cartwrightianus*) had inspired these paintings. From the same period of time, saffron cultures existed at the Nile delta of Egypt (Kandeler and Ullrich, 2009). Historically, saffron was first harvested from the wild *Crocus cartwrightianus,* a mutant of which—*Crocus sativus,* distinguished by its elongated stigmas—was observed, selected and domesticated on Crete during the Late Bronze Age (Negbi, 1999). Therefore, it is believed that wild precursor of domesticated saffron was *Crocus cartwrightianus* (Srivastava *et al.*, 2010).

The term used in ancient Greek for Crocus is 'koricos', whereas the Romans used the term 'crocum'. By contrast, 'saffron' probably originates from the Arabic word 'zafaran' or 'zaafar' meaning yellow. The Arabic 'safran' is quite similar in various other languages.

The Genus *Crocus*

Crocus genus is a member of *Iridaceae* family. The classification of genus *Crocus* based on morphology was made in 1982 by Brian Mathew in his book *The Crocus*. The genus

Fig. 1: Wall painting of crocus pickers, eastern wall of upper floor, Xeste 3, Akrotiri, Thera (Doumas, 1992).

Crocus is divided into two subgenera; subgenus *Crocus* with extrose anthers and subgenus *Crociris* with introse anthers. *Crociris* contains only *Crocus banaticus* and the subgenus *Crocus* comprises all the remaining species. The subgenus *Crocus* is further divided into two sections: section *Crocus* and section *Nudiscapus*, and each is again divided into Series a–f and g–o, respectively.

Mathew (1982) described 81 species of *Crocus*. Since then, more new species have been described (Petersen *et al.*, 2008; Kerndorff *et al.*, 2012; Kerndorff *et al.*, 2013). Recently, the genus *Crocus* consists of about 100 recognized species (Ruksans, 2010; Harpke *et al.*, 2013) distributed from southwestern Europe, through central Europe to Turkey and southwestern parts of Asia, as far east as western China with the centre of species diversity in Asia

Minor and on the Balkan Peninsula (Harpke *et al.*, 2013). The genus *Crocus* is especially well represented in arid countries of south-eastern Europe and Western and Central Asia. Many crocuses are known as popular ornamentals. The taxonomy of *Crocus* is extremely complicated due to the lack of clear distinctive characters, the wide range of habitats and the heterogeneity of the morphological traits and cytological data (Norbak *et al.*, 2002).

Crocus is a genus of perennial geophytes, which its morphology is modified to fit for a short period of active growth after a long resting period. Phytogeographically, the majority of species occur within the Mediterranean floristic region, extending eastwards into the Irano-Turanian region. Both of these areas are characterized by cool to cold winters with autumn-winter-spring precipitation and warm summers with very little rainfall. The latter region experiences much colder winters and generally less rainfall. The genus *Crocus* is well adapted to such conditions, the plants being in active growth from autumn to late spring and surviving the summer drought below ground by means of a compact corm (Mathew, 1999). *Crocus* grows in mountainous areas from sea level up to 2500 m, mostly on limestone ground in winter rain areas around the Mediterranean Sea. The largest diversity of species is found in Turkey, but species are found all the way from Portugal and Spain in west to Western China in east. Characteristically, most species have a relatively limited geographical distribution adapted to specific habitats (Larsen, 2011). Generally, the complexity of the evolutionary history of the genus *Crocus* suggests an intensive species hybridization and explosive speciation in *Crocus* evolution that could be on the basis of the origin of Saffron.

Cytogenetics of *Crocus*

Crocus is a highly complex genus with a wide range of chromosome numbers (Brighton et al., 1973). *Crocus* has a relatively large genome size (Frizzi *et al.*, 2007) as well as huge cytological variation (Sik *et al.*, 2008) and some species do even show interspecific variation on chromosome numbers (Candan *et al.*, 2009). The very wide range of chromosome numbers in *Crocus*, 2n=6, 8, 10, 12, 14, 16, 18, 19, 20, 22, 23, 24, 26, 28, 30, 32, 34, 44, 48 and 64 have been reported (Brighton *et al.*, 1973). Additionally, a vast amount of variation in morphology has been observed. The large morphological and cytological diversity of *Crocus* represents a history of massive hybridization and speciation. Inter- and intraspecific variation suggests that the genus during evolution has been exposed to massive adaption demands and selection pressure.

The karyotype of *C. sativus* has been studied by a number of authors, and, while contemporary accessions from different countries usually reveal a common karyotype without major differences, other karyotypes have been described in the former literature (Brighton *et al.*, 1973; Ghaffari, 1986).

Series *Crocus*

Series *Crocus* consists of 10 species (Table 1) characterized by a red-orange style deeply divided into three long stigmatic branches and yellow anthers. All species are autumn flowering from October to November (Jacobsen and Orgaard, 2004). Species are all diploids, except the triploid *C. sativus*. The species are distributed from southern Italy (*C. thomasii*) to western Iran (*C. pallasii* subsp. *haussknechtii*) (Mathew, 1982, 1999).

Table 1: All species belonging to the Series f, Crocus

Number	Species of Series *Crocus*	Chromosome numbers
1	*C. pallasii* Goldb.	2n = 12, 14, 16
2	*C. matheweii* H. Kerndorff & E. Pasche	2n = 16
3	*C. thomasii* Ten.	2n = 16
4	*C. cartwrightianus* Herbert	2n = 16
5	*C. sativus* L.	2n = 3x = 24
6	*C. moabiticus* Bornm. & Dinsm. ex Bornm.	2n = 14
7	*C. oreocreticus* B. L. Burtt	2n = 16
8	*C. asumaniae* B. Mathew & T. Baytop	2n = 26
9	*C. hadriaticus* Herbert	2n = 16
10	*C. naqabensis* Al-Eisawi & Kiswani	2n = 14

Although concern in the *Crocus* genus is mainly related to *C. sativus*, there is also growing interest in other ornamental and wild related species. Many *Crocus* species could be used as a source of food colorants and pharmaceutical, and are also rich in high added value compounds possessing biological activity (fungicidal, antioxidant or insecticidal) that can be extracted from corms, tepals and leaves. Furthermore, the recorded tolerance to summer drought and winter cold, together with their showy flowers, makes wild species interesting as ornamental in areas with severe climatic conditions. Wild species can also be used as sources of useful genes in improvement programmes of the cultivated species (de Castilla-La, 2015).

Crocus sativus

Crocus sativus is grown for production of the most expensive spice (saffron) in the world. The production of saffron is an important household income in several regions of Asia and Mediterranean areas (Negbi, 1999). Today, almost all saffron is grown in Mediterranean in the West and Near East countries encompassing Iran, Turkey and Kashmir and Kishtwar in Jammu, India, in the East. Elsewhere in other continents, except Antarctica, comparatively insignificant amounts are being produced (Lim, 2014). In Iran, *C. sativus* is a major crop which is of great sociological and economical importance (Mollafilabi, 2004; Kafi and Showket, 2007). In Iran, 80000 hectares are under saffron cultivation, 97% of which is situated in the Khorasan province, having a dry climate, hot summers and relatively cold winters (Mollafilabi, 2004). Iran is the largest producer and exporter of saffron, which is responsible for more than 90 % of the world saffron production (Agayev *et al.*, 2007; Ghorbani, 2007). The world production of saffron has been estimated be to approximately 300 tons per year. Saffron has the potential to grow on low-productive soils, with warm and dry summers and relatively mild and dry winters (Kumar *et al.*, 2008). Saffron is therefore a potentially valuable crop in areas where the climate makes plant production problematic. Due to its unique biological, physiological and agronomic traits, saffron is able to exploit marginal land and to be included in low-input cropping systems, representing an alternative viable crop for sustainable agriculture (Gresta *et al.*, 2008). For a long time, saffron had been neglected by researchers and farmers and it was considered as an orphan crop. However in the last few years it is gaining a more interesting role in low-input agricultural systems and as an alternative crop. In addition, more attentions have been taken to cultivation of saffron in areas where the risk of water scarcity and the frequency of drought and high temperature has increased due to climate change. The idea of growing saffron in some deprived regions has been proposed as way to substitute crops for the socio-economic development of rural area (Behdani, 2011). Moreover, saffron is a very attractive crop for organic and low input agriculture considering that no irrigation, chemical fertilization or chemical weed treatments are applied in some environments in which it is cultivated (Ghorbani and Koocheki, 2007; Gresta *et al.*, 2008).

Clonal Origin of *C. sativus*

C. sativus is a male-sterile triploid (2n = 3x = 24) (Karasawa, 1933), initially assumed to be autotriploid, although a growing amount of evidence supports alloploidy as the most probable mechanism of origin. This cultivated species is not known to occur in the wild (Mathew, 1999) and the origin of it has consequently been a subject of discussion (Grilli Caiola and Canini, 2010). It is debated whether *C. sativus* originated from only one species or whether it is of hybrid origin (Zubor *et al.*, 2004; Castillo *et al.*, 2005) and a final prove for the parental origin of *C. sativus* is still missing (Larsen, 2011).

It is assumed that an ancient hybridization plus domestication event had occurred, which resulted in the unique clone, known as *C. sativus*. Therefore, it can be hypothesised that all saffron crocus clones are similar to each other and all corms of *C. sativus* are of clonal origin. This hybridization event has given rise to corms of all the world's saffron crocuses, solely by vegetative propagation (Jacobsen and Orgaard, 2004).

A number of molecular studies have been carried out in order to prove the clonal origin of *C. sativus*. Chloroplast and microsatellite markers did not reveal any differences between 28 *C. sativus* isolates coming from 10 different geographic regions of the world (Fluch *et al.*, 2010). Caiola *et al.* (2004) made RADP analysis on *C. sativus* accessions from five different countries and found all accessions to be identical. Likewise, RAPD, ISSR and microsatellite markers were tested on 44 accessions of *C. sativus* from 11 different countries (Rubio-Moraga *et al.*, 2009) but again no polymorphism was found, which supports the hypothesis that all plants of *C. sativus* grown worldwide are of clonal origin. Nonetheless, morphological deviates of *C. sativus* have been described, *e.g. C. sativus* 'Cashmerianus' (Yau *et al.*, 2006) which is grown in Kashmir. However, it has not been proved that this cultivar is genotypically different from other *C. sativus*. Thus, the phenotypical variations that now and then are described within *C. sativus* are most likely due to differences in climate and cultivation practices.

Origin Site of *C. sativus*

The localization of the hybridisation event of saffron origin has not been ascertained so far. There are two different views on the geographical origin of *Crocus sativus.* The Mediterranean region (Greece) is one of the most probable sites of origin of saffron; another possible site is located in the the Western–Central Asia (Turkey–Iran–India area), where saffron cultivation is reported to be thousands of years old (Grilli Caiola and Canini, 2010). Botanical research by Mathew (1999) appeared to suggest Crete, Eastern Greece, as the centre of origin, disproving the view of a Western or Central Asia origin. However, more recent studies suggested that saffron may have originated in Mesopotamia (Alavi-Kia *et al.* 2008). According to some authors (Alberini, 1990; Winterhalter and Straubinger, 2000) saffron originated first in Iran and Kashmir, from where the Phoenicians introduced it to the Greek and Romans. Later on, it was brought by the Arabs to Spain.

Genetic Diversities in *C. sativus*

Genetic diversity in crop species is of essential importance for crop improvement, which relies on new genes, new regulation of genes and new gene combinations. Ancestral species are a major source of genetic diversity, and traits of interest may be introduced as chromosomal segments through direct crossing or through genetic manipulation techniques in crop improvement programmes (Alsayied *et al.*, 2015). Therefore, in order to establish biodiversity conservation strategies and crop improvement programmes, it is important to ascertain the levels of genetic diversity within the species *C. sativus*. Since saffron is a triploid species and it is vegetatively propagated by means of corm multiplication, the occurrence of genetic variation is very limited, with the exception of a few somatic mutations that are not easily detectable (Fernández, 2004).

For most crops, domestication is seen as a bottleneck reducing genetic variation; further artificial selection can also causes reduction in genetic diversity. Given the high levels of polymorphism between the species and even individual accessions, minimal if any variation is evident in *C. sativus*, despite accessions from a broad geographical range being included. Moreover, the decrease of land surface dedicated to saffron cultivation in many areas has possibly resulted in genetic erosion that adds up to the limited genetic variation for *C. sativus* due to its sterile habit (De-Los-Mozos-Pascual *et al.*, 2010). Probably, as today *C. sativus* exists only as a cultivated species due to its high male-sterility (Zanier and Grilli Caiola, 2001), selection of the best corms for saffron production may have influenced the amount of variation in *C. sativus* (Beiki *et al.*, 2010).

However, the presence and extent of genetic variation in this species is a matter of controversy, as testified by several contradictory articles providing contrasting results about the genetic variation of the species. Some authors have concluded that there is little or no genetic variation in saffron (Grilli Caiola *et al.*, 2004; Moraga *et al.*, 2009; Fluch *et al.*, 2010). Interestingly, Rubio-Moraga *et al.* (2009) stated that saffron crocus is a monomorphic species, deriving their conclusion from the fact that no clear polymorphic signals were found when analysing accessions from various countries with different molecular markers (RAPD, ISSR, SSR). in contrast, other recent studies with a range of DNA markers have indicated limited genetic diversity within the species (Sik *et al.*, 2008; Beiki *et al.*, 2010; Nemati *et al.*, 2012; Babaei *et al.*, 2014; Nemati *et al.*, 2014; Mir *et al.*, 2015). Low genetic differences by using EST-derived SSR (Fluch *et al.*, 2010; Nemati *et al.*, 2014) have been identified. Single Nucleotide Polymorphisms (SNPs) in different accessions, which are often heterozygous, have been detected by direct sequencing (Fernández, unpublished data). Siracusa *et al.* (2013) detected genetic differentiation between samples from different geographical areas (Europe and Asia) using AFLP markers, although no evidence of any significant phenotypic variation of samples of different geographical provenience was noted. Interestingly, the same authors highlighted the presence of intra-accession variability by analysing four plants for each accession. These works clearly demonstrate that the situation is still unclear, having evidenced alternatively no-variability, low variability, and variability in *C. sativus*.

Remarkably, phenotypic variations have been frequently observed in the field by researchers and saffron producers. morphological differences including flower size, tepal shape and colour intensity with lobed tepal and more intense colour of tepals in plants in some accessions have been already reported (Grilli Caiola *et al.,* 2001). Variants of saffron with an increased number of stigmas have been reported with a frequency of $1.2x10^{-6}$ of the rare type flowers (Estilai, 1978). Interestingly, such phenotypic variations are occasionally unstable and can change from one growing season to another (Fernández, 2004). However, phenotypic differences in saffron have been attributed to the environmental influences such as the climate and cultivation practices. The quality of saffron is, therefore, more likely to be defined by growing conditions such as soil, water, temperature and altitude, collection and processing techniques, and is less dependent on the origin of the corm (Agayev *et al.*, 2009; Maggi *et al.*, 2011; Torelli *et al.*, 2014). Nonetheless, Agayev *et al.* (2009) showed a rapid and stable response of *C. sativus* to clonal selection. The best performing plants (more productive corms of saffron) were identified and selected from a large number (many thousands) of plants from various Iranian populations, being able to create true "cultivars". The efficacy of clonal selection suggested genetic diversity in the species. Despite the existence of different saffron ecotypes or commercial varieties, the actual genetic variability present in *C. sativus* at worldwide scale still remains unknown.

Parental Species of *C. sativus*

The genetic origin of *C. sativus* is not clear and information on saffron ancestors is not univocal. There is several data support the alloploidy of *C. sativus*. Since *C. sativus* has 2n = 3x = 24 chromosomes and morphologically belongs to Series *Crocus*, it is believed that the ancestors of the species should be searched between one or two diploid species of series *Crocus* with 2n = 16. Therefore, most ancestral studies carried out so far, have focused on the relationship between species within Series *Crocus*. Different species of *Crocus* series *Crocus* have been suggested as potential ancestors of *C. sativus*. In the case of allotriploid saffron, *C. cartwrightianus, C. hadriaticus, C. oreocreticus* (Jacobsen and Ørgaard, 2004; Agayev *et al.*, 2010) or *C. thomasii* and *C. pallasii* or *C. cartwrightianus* and *C. pallasii* (Tammaro, 1990) have been proposed as candidate ancestral species, with each contributing the basic set of x = 8 chromosomes (Alsayied *et al.*, 2015). *C. cartwrightianus* and *C. hadriaticus* are present currently in Greece but not in overlapping areas (Frello and Heslop-Harrison, 2000). Other possible parents e.g., *C. thomasi* from Italy and Croatia, *C. mathewii* from Turkey, and *C. pallasii* ssp. *haussknechtii* from Iran, cannot be excluded (Frello and Heslop-Harrison, 2000; Grilli Caiola *et al.*, 2001; Grilli Caiola *et al.*, 2004). Brighton (1977) in a kariological study suggested that possible ancestors of *C. sativus* are *C. cartwrightianus* or *C. thomasii.*

The diploid species *C. cartwrightianus* is more similar to *C. sativus* morphologically and cytologically, although, the largest difference is that the flowers of *C. sativus* are double-sized compared to those of *C. cartwrightianus* (Mathew, 1982). Quantitative and qualitative DNA analysis studies (Brandizzi and Grilli Caiola, 1998; Grilli Caiola *et al.*, 2004) indicated that the DNA composition of *C. sativus* was more similar to that of *C. cartwrightianus* rather than to that of *C. thomasii*. Recent AFLP[1] analysis confirmed that the quantitative and qualitative traits of *C. cartwrightianus* and *C. thomasii* DNA are compatible with *C. sativus* (Zubor *et al.*, 2004).

Most cytological, cytogenetical, biochemical, and molecular data have led to the hypothesis that *C. cartwrightianus* is the most probable progenitor of *C. sativus* (Mathew 1999; Grilli Caiola 2004; Zubor *et al.* 2004; Frizzi *et al.* 2007; Larsen *et al.*, 2015). In addition, Petersen et al. (2008) have recently found a sample of *C. sativus* to be sister of a sample of *C. cartwrightianus*. Petersen *et al.* (2008) analysed five plastid regions; their analysis included 86 recognized species of the genus and their study also found C. cartwrightianus to be closely related to C. sativus. Overall, all studies so far have suggested *C. sativus* and *C. cartwrightianus* to be the closest related species. *C. cartwrightianus* is an autumn flowering species with a blue–violet flower having red three branched stigma, which grows on the Cyclades as well as on Crete (Figure 2) (Kandeler and Ullrich, 2009). Therefore, *C. cartwrightianus* seems to be a probable parental, providing two out of the three genomes, but the other parental species remains unclear (Fernández, 2004).

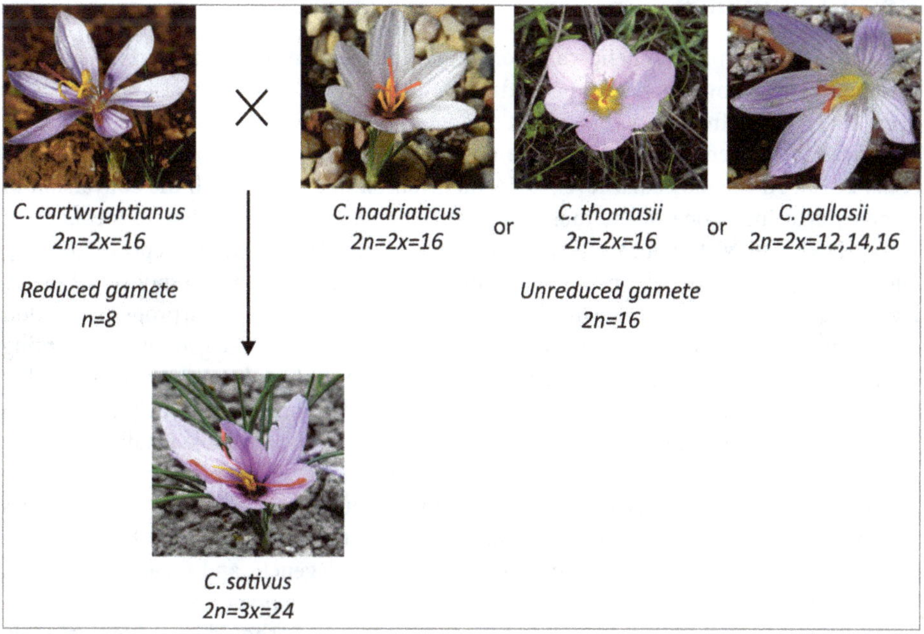

Fig. 2: A possible mechanism of hybridization and probable parents of *C. sativus*

Further, other diploid Crocus species, such as *C. thomasii*, *C. hadriaticus*, *C. oreocreticuus*, and *C. pallasii* have also been considered as possible progenitors of saffron. Based on IRAP[2] markers analysis, Iranian *C. almehensis* and *C. michelosnii* were also reported as the closest relatives of saffron and probably the possible wild ancestors of this cultivated species (Alavi-Kia *et al.*, 2008). Sanei *et al.* (2007) reported that *C. pallasii* subsp.

1. Amplified Fragment Length Polymorphisms
2. Interretrotransposon Amplified Polymorphism

haussknechtii is one of the closest relatives of *C. sativus* based on karyotype data, Erol *et al.* (2014) also found the maximum similarity between an accession of *C. pallasii* subsp. *pallasii* and *C. sativus*. Alsayied *et al.* (2015) with the IRAP analysis indicated that the most likely ancestors of saffron were *C. cartwrightianus* and *C. pallasii* subsp. *Pallasii*. Hence, *C. pallasii* subsp. *Pallasii* was suggested to be a candidate ancestor for *C. sativus*. Plastid, ribosomal and nuclear single-copy gene sequences, focused on those used for phylogenetic analysis (Harpke *et al.*, 2013) have also suggested *C. cartwrightianus* and *C. pallasii* as ancestral species of *C. sativus*.

It is assumed that a possible fertilization event resulted in an allotriploid plant, which was selected and later and identified as *Crocus sativus*. It is likely that *C. cartwrightianus* and *C. hadriaticus* or *C. pallasii* or *C. thomasii* have hybridized at some point during evolution, where the two species may have co-existed (Figure 2). It probably descends through fertilization of a normal reduced egg cell (haploid, n) with an unreduced male (diploid, 2n) gamete (Caiola and Canini, 2010).

Recent Evidence

Although a limited genetic variation in *C. sativus* have been reported, phenotypic variations from flower size, shape and color intensity to quality carachteristics have been observed in frequency. Epigenetic changes are meiotically or mitotically heritable modifications in gene function that are not due to changes in DNA sequence (Bonasio *et al.*, 2010). At the molecular level, DNA methylation of cytosine with the conversion to 5-methylcytosine is one of the most widespread epigenetic modifications. As epigenetic marks, DNA methylation may alter strongly the structure of chromatine, deeply influencing eukaryotic gene expression and the development and environmental responses of plants likewise. Recent works have shown that epigenetics plays a key role in the proper regulation of self-incompatibility, sex determination, shoot regeneration and genomic imprinting (Saze *et al.*, 2012). In recent years, an increasing number of studies have evidenced that some epigenetic marks can persist through DNA replication and can be stably transmitted to the following generations (Zhang and Hsieh, 2013). Stable, heritable epialleles (whose expression depends on their epigenetic status) causing heritable phenotypic variations are known in plants. So, scientists have begun to argue about the possible role for epigenetics in crop breeding, transgene regulation and epialleles creation (Zhang and Hsieh, 2013). Considering that gene expression can be influenced by both genetic and epigenetic changes, epigenetics could be a plausible cause of the alternative phenotypes observed in saffron crocus (Busconi *et al.*, 2015). However, a true epigenetic mark should be stable and stably transmitted to the progeny through cellular divisions for several cycles (Richards, 2011). This phenomenon is known as Transgenerational Epigenetic Inheritance (TEI). Busconi et al. (2015) performed a molecular marker analysis using MS-AFLP[3] or MSAP AFLP markers (Siracusa *et al.*, 2013) in the World Saffron and Crocus Collection (WSCC). They found a little genetic variability and high epigenetic variability inside the saffron crocus germplasm. They hypothesised that genetic variability in *C. sativus* is most likely a consequence of spontaneous mutations and all of the accessions shared a very similar genetic constitution. This offers new opportunities for basic research, and saffron may become an election crop

3. Methyl Sensitive Amplified Fragment Length Polymorphism

for epigenetic studies, in order to investigate how a same genotype can adapt to different environmental conditions. The successful clonal selection results obtained by Agayev *et al.* (2009), which highlighted the presence of phenotype variability between and within accessions for different morphologic and agronomic traits can also be explained based on stable epigenetic marks transmitted to the offspring through TEI.

Prospects

Since *Crocus sativus* exhibit little or no genetic variation, there would be a problem for breeders and horticulturists to breed for qualitative and quantitative traits of interests. All grown *C. sativus* are male-sterile and share an identical gene-pool which is problematic in relation to breeding strategies and future crop improvement. In order to incorporate genetic variation in saffron crops it is therefore necessary to find the right answer to the question about the ancestral origin of *C. sativus*. A deeper insight in the origin of *C. sativus* would furthermore provide important knowledge about general hybridization mechanisms, which might help to understand complicated speciation patterns of *Crocus* and other genera with alike complex patterns and of triploids in general. It has been stated more than once that *C. cartwrightianus* is the closest relative to *C. sativus*; however, *C. hadriaticus* and other species of Series *Crocus* have been suggested. Wide genetic diversity is of importance for the development of improved varieties. In order to broaden the genetic variation within-species, it is important to study the hybridization compatibility between species of Series *Crocus*, and it will certainly be valuable to resynthesize *C. sativus* as a triploid from crossing *C. cartwrightianus* with other species of Series *Crocus* for improvement of saffron cultivars, exploitation of genetic diversity and conservation of the Crocus germplasm. Therefore, breeding strategies involving hybridization with related species are important tools in order to improve yield and quality of saffron (Agayev *et al.*, 2007). Horticultural traits of different *Crocus*-species can without problems using traditional breeding approaches in combination with the *in vitro* techniques (controlled hybridization). This potential cross-compatibility, together with *in vitro* methods which raise successful seed set, may open the door to breeding programmes for the genetic improvement of the saffron crocus. Further studies on the origin of *C. sativus* should use improved molecular approaches, combined with both morphological and cytological studies. These approaches should be applied on species as well as hybrids.

Other approaches for widening the genetic variability would be mutation induction. Induced mutations play an important role in creating new variations in clonally propagated plants, such as saffron, because mutations in such plants may easily get stabilized and can be manipulated. So mutation breeding offers a scope for induction of variability in saffron for its subsequent utilization. In addition, in vitro mutagenesis by combining both tissue culture techniques and induced mutation strategy can accelerate crop improvement and germplasm innovation.

References

Abdullaev FI 2002. Cancer Chemopreventive and Tumoricidal Properties of Saffron (*Crocus sativus* L.). Experimental Biology and Medicine 227: 20-25.

Abdullaev FI and Espinosa-Aguirre JJ 2004. Biomedical properties of saffron and its potential use in cancer therapy and chemoprevention trials. *Cancer Detection and Prevention* 28: 426-432.

Agayev Y, Fernandez J-A and Zarifi E 2009. Clonal selection of saffron (*Crocus sativus* L.): the first optimistic experimental results. *Euphytica* 169: 81-99.

Agayev Y, Shakib A, Soheilivand S and Fathi M 2007. Breeding of saffron (*Crocus sativus*): possibilities and problems. *Acta Horticulturae.*

Alberini M 1990. Saffron: sap ore colors Lo zafferani. *In* Proceeds of the International Conference on Saffron (*Crocus sativus* L.), L`quila, Italy, pp. 39-45.

Alsayied NF, Fernández JA, Schwarzacher T and Heslop-Harrison JS 2015. Diversity and relationships of *Crocus sativus* and its relatives analysed by inter-retroelement amplified polymorphism (IRAP). *Annals of Botany* 116: 359-368.

Babaei S, Talebi M, Bahar M and Zeinali H 2014. Analysis of genetic diversity among saffron (*Crocus sativus*) accessions from different regions of Iran as revealed by SRAP markers. *Scientia Horticulturae* 171: 27-31.

Behdani MA 2011. Saffron (*Crocus sativus*). (**In**) KV Peter, (Ed), Future crops, Vol 1. DAYA Pub., New Delhi.

Beiki AH, Keifi F and Mozafari J 2010. Genetic Differentiation of Crucus Species by Random Amplified Polymorphic DNA. *Genetic Engineering and Biotechnology Journal* 18: 1-10.

Bhandari PR 2015. *Crocus sativus* L. (saffron) for cancer chemoprevention: A mini review. *Journal of Traditional and Complementary Medicine* 5: 81-87.

Bonasio R, Tu S and Reinberg D 2010. Molecular signals of epigenetic states. *Science* 330: 612-616.

Brandizzi F and Grilli Caiola M 1998. Flow cytometric analysis of nuclear DNA in *Crocus sativus* and allies (Iridaceae). *Plant Systematics and Evolution* 211: 149-154.

Brighton CA, Mathew B and Marchant CJ 1973. Chromosome Counts in the Genus Crocus (Iridaceae). *Kew Bulletin* 28: 451-464.

Busconi M, Colli L, Sánchez RA, Santaella M, De-Los-Mozos Pascual M, Santana O, Roldán M and Fernández J-A 2015. AFLP and MS-AFLP Analysis of the Variation within Saffron Crocus (*Crocus sativus* L.) Germplasm. PLoS ONE 10: e0123434.

Candan F, Sik L and Kesercioglu T 2009. Cytotaxonomical studies on some *Crocus* L. taxa in Turkey. *African Journal of Biotechnology* 8.

Carmona M, Zalacain A, Sánchez AM, Novella JL and Alonso GL 2006. Crocetin Esters, Picrocrocin and Its Related Compounds Present in *Crocus sativus* Stigmas and *Gardenia jasminoides* Fruits. Tentative Identification of Seven New Compounds by LC-ESI-MS. *Journal of Agricultural and Food Chemistry* 54: 973-979.

Castillo R, Fernández J-A and Gómez-Gómez L 2005. Implications of carotenoid biosynthetic genes in apocarotenoid formation during the stigma development of *Crocus sativus* and its closer relatives. *Plant Physiology* 139: 674-689.

De-Los-Mozos-Pascual M, Fernández JA and Roldán M 2010. Preserving Biodiversity in Saffron: The Crocusbank Project and the World Saffron and Crocus Collection. *In*, Ed 850. International Society for Horticultural Science (ISHS), Leuven, Belgium, pp. 23-28.

de Castilla-La JdC 2015. Descriptors for Crocus (*Crocus* spp.).

Doumas C 1992. The wall paintings of Thera. The Thera Foundation. *In*. Athens: Petros M. Nomikos.

Erol O, KAYA HB, ŞIK L, Tuna M, Can L and Tanyolac MB 2014. The genus Crocus, series Crocus (Iridaceae) in Turkey and 2 East Aegean islands: a genetic approach. *Turkish Journal of Biology* 38: 48-62.

Estilai A 1978. Variability in saffron (*Crocus sativus* L.). *Experientia* 34: 725.

Fernández J-A 2004. Biology, biotechnology and biomedicine of saffron. *Recent Research Developments in Plant Science* 2: 127-159.

Fernández J-A 2006. Genetic resources of saffron and allies (*Crocus* spp.). *In* II International Symposium on Saffron Biology and Technology 739, pp. 167-185.

Fernández J-A, Santana O, Guardiola J-L, Molina R-V, Heslop-Harrison P, Borbely G, Branca F, Argento S, Maloupa E and Talou T 2011. The World Saffron and Crocus collection: strategies for establishment, management, characterisation and utilisation. *Genetic Resources and Crop Evolution* 58: 125-137.

Fluch S, Hohl K, Stierschneider M, Kopecky D and Kaar B 2010. *Crocus sativus* L.-molecular evidence on its clonal origin. *Acta Horticulturae*: 41-46.

Frello S and Heslop-Harrison J 2000. Repetitive DNA sequences in *Crocus vernus* Hill (Iridaceae): the genomic organization and distribution of dispersed elements in the genus Crocus and its allies. *Genome* 43: 902-909.

Frizzi G, Miranda M, Pantani C and Tammaro F 2007. Allozyme differentiation in four species of the Crocus cartwrightianus group and in cultivated saffron (*Crocus sativus*). *Biochemical Systematics and Ecology* 35: 859-868.

Ghaffari SM 1986. Cytogenetic studies of cultivated *Crocus sativus* (*Iridaceae*). *Plant Systematics and Evolution* 153: 199-204.

Ghorbani M 2007. The Economics of Saffron in Iran. *Acta Hort.* (ISHS) 739: 321-331.

Ghorbani R and Koocheki A 2007. Organic saffron in Iran: prospects and challenges. *In*, Ed 739. *International Society for Horticultural Science (ISHS), Leuven*, Belgium, pp. 369-374.

Gresta F, Lombardo G, Siracusa L and Ruberto G 2008. Saffron, an alternative crop for sustainable agricultural systems. A review. Agronomy For Sustainable Development 28: 95-112.

Grilli Caiola M and Canini A 2010. Looking for Saffron's (*Crocus sativus* L.) Parents. *Functional Plant Science and Biotechnology* 4: 1-14.

Grilli Caiola M, Caputo P and Zanier R 2004. RAPD Analysis in *Crocus sativus* L. Accessions and Related *Crocus* Species. *Biologia Plantarum* 48: 375-380.

Grilli Caiola M, Di Somma D and Lauretti P 2001. Comparative study of pollen and pistil in *Crocus sativus* (Iridaceae) and allied species. *Annals of Botany* 1: 93-103.

Harpke D, Meng S, Rutten T, Kerndorff H and Blattner FR 2013. Phylogeny of Crocus (Iridaceae) based on one chloroplast and two nuclear loci: Ancient hybridization and chromosome number evolution. *Molecular Phylogenetics and Evolution* 66: 617-627.

Jacobsen N and Orgaard M 2004. Crocus cartwrightianus on the Attica Peninsula. *Acta Horticulturae*: 65-70.

Kafi M and Showket T 2007. A comparative study of saffron agronomy and production systems of Khorasan (Iran) and Kashmir (India). *Acta Horticulturae* 739: 123.

Kandeler R and Ullrich WR 2009. Symbolism of plants: examples of European-Mediterranean culture presented with biology and history of art: JANUARY: Crocus. *Journal of Experimental Botany* 60: 6-8.

Karasawa K 1933. On the triploidy of *Crocus sativus*, L. and its high sterility. *The Japanese Journal of Genetics* 9: 6-8.

Kerndorff H, Pasche E, Blattner F and Harpke D 2013. A new species of *Crocus* (Liliiflorae, Iridaceae) from Turkey. Stapfia 99: 141-144.

Kerndorff H, Pasche E, Harpke D and Blattner F 2012. Seven new species of *Crocus* (Liliiflorae, Iridaceae) from Turkey. na.

Kumar R, Singh V, Devi K, Sharma M, Singh M and Ahuja PS 2008. State of art of saffron (*Crocus sativus* L.) agronomy: a comprehensive review. *Food Reviews International* 25: 44-85.

Larsen B 2011. Origin of Crocus sativus (Iridaceae): Inter- and intraspecific variation, population structure and hybridization compatibility within Series Crocus. University of Copenhagen, Copenhagen.

Lim TK 2014. Crocus sativus. (In) Edible Medicinal and Non Medicinal Plants. Springer Netherlands, pp. 77-136.

Lozano P, Castellar MR, Simancas MJ and Iborra JL 1999. A quantitative high-performance liquid chromatographic method to analyse commercial saffron (*Crocus sativus* L.) products. *Journal of Chromatography A* 830: 477-483.

Magesh V, Singh J, Selvendiran K, Ekambaram G and Sakthisekaran D 2006. Antitumour activity of crocetin in accordance to tumor incidence, antioxidant status, drug metabolizing enzymes and histopathological studies. Molecular and Cellular Biochemistry 287: 127-135.

Maggi L, Carmona M, Kelly SD, Marigheto N and Alonso GL 2011. Geographical origin differentiation of saffron spice (*Crocus sativus* L. stigmas) - Preliminary investigation using chemical and multi-element (H, C, N) stable isotope analysis. *Food Chemistry* 128: 543-548.

Mathew B 1982. The Crocus. Timber Press, Portland, Oregon.

Mathew B 1999. Botany, Taxonomy and Cytology of C. Sativus L. and Its Allies. *In* Saffron. CRC Press, pp. 19-30.

Mir JI, Ahmed N, Khan MH, Mokhdomi TA, Wani SH, Bukhari S, Amin A and Qadri RA 2015. Molecular Characterization of Saffron-Potential Candidates for Crop Improvement. *Not. Sci. Biol.* 7: 81-89.

Mollafilabi A 2004. Experimental findings of production and echo physiological aspects of saffron (*Crocus sativus* L.). *Acta Horticulturae*: 195-200.

Moraga AR, Castillo-López R, Gómez-Gómez L and Ahrazem O 2009. Saffron is a monomorphic species as revealed by RAPD, ISSR and microsatellite analyses. BMC Research Notes 2: 1-5.

Negbi M 1999. Saffron Cultivation: Past, Present and Future Prospects. (In) M Negbi, Ed, SAFFRON *Crocus sativus* L., Vol 8. CRC, p. 152.

Nemati Z, Mardi M, Majidian P, Zeinalabedini M, Pirseyedi SM and Bahadori M 2014. Saffron (*Crocus sativus* L.), a monomorphic or polymorphic species? 2014 12: 10.

Nemati Z, Zeinalabedini M, Mardi M, Pirseyediand S, Marashi S and Nekoui S 2012. Isolation and characterization of a first set of polymorphic microsatellite markers in saffron, *Crocus sativus* (Iridaceae). *American Journal of Botany* 99: e340–e343.

Norbak R, Brandt K, Nielsen JK, Orgaard M and Jacobsen N 2002. Flower pigment composition of Crocus species and cultivars used for a chemotaxonomic investigation. *Biochemical Systematics and Ecology* 30: 763-791.

Petersen G, Seberg O, Thorsøe S, Jørgensen T and Mathew B 2008. A Phylogeny of the Genus Crocus (Iridaceae) Based on Sequence Data from Five Plastid Regions. *Taxon* 57: 487-499.

Richards EJ 2011. Natural epigenetic variation in plant species: a view from the field. *Current Opinion in Plant Biology* 14: 204-209.

Rubio-Moraga A, Castillo-López R, Gómez-Gómez L and Ahrazem O 2009. Saffron is a monomorphic species as revealed by RAPD, ISSR and microsatellite analyses. *BMC Research Notes* 2: 1-5.

Ruksans J 2010. Crocuses: a complete guide to the genus. Timber Press.

Samarghandian S and Borji A 2014. Anticarcinogenic effect of saffron (*Crocus sativus* L.) and its ingredients. *Pharmacognosy Research* 6: 99-107.

Sanei M, Rahimyan H, Agayev Y and Soheilivand S 2007. New Cytotype of *Crocus pallasii* subsp. haussknechtii from West of Iran. *Acta Horticulturae* 739: 107.

Saze H, Tsugane K, Kanno T and Nishimura T 2012. DNA methylation in plants: relationship to small RNAs and histone modifications, and functions in transposon inactivation. *Plant and Cell Physiology* 53: 766-784.

Sik L, Candan F, Soya S, Karamenderes C, Kesercioglu T and Tanyyolc B 2008. Genetic variation among *Crocus sativus* L. species from western Turkey as revealed by RAPD and ISSR marker. *Journal of Applied Biological Science* 2: 73-78.

Siracusa L, Gresta F, Avola G, Albertini E, Raggi L, Marconi G, Lombardo GM and Ruberto G 2013. Agronomic, chemical and genetic variability of saffron (Crocus sativus L.) of different origin by LC-UV–vis-DAD and AFLP analyses. *Genetic Resources and Crop Evolution* 60: 711-721.

Srivastava R, Ahmed H, Dixit RK, Dharamveer and Saraf SA 2010. *Crocus sativus* L.: A comprehensive review. *Pharmacognosy Reviews* 4: 200-208.

Tammaro F 1990. *Crocus sativus* L. cv di Navelli (L'Aquila saffron): environment cultivation, morphometric characteristic, active principles, uses. (In) F Tammaro, L Marra, (Eds), Lo zafferano: Proceedings of the International Conference on Safforn (*Crocus sativus* L.), L'Aquila, Italy 1989, pp. 47-96.

Torelli A, Marieschi M and Bruni R 2014. Authentication of saffron (*Crocus sativus*, L.) in different processed, retail products by means of SCAR markers. *Food Control* 36: 126-131.

Vavilov NI 1951. The origin, variation, immunity and breeding of cultivated plants. Translated from Russian by K.S. Chester. *Chron. Bot.* 13: 1-366.

Winterhalter P and Straubinger M 2000. Saffron-Renewed interest in an ancient spice. *Food Reviews International* 16: 39-59.

Yau S, Nimah M and Toufeili I 2006. Yield and quality of red stigmas from different saffron strains at contrasting Mediterranean sites. *Experimental Agriculture* 42: 399-409.

Zanier R and Grilli Caiola M 2001. Self incompatibility mechanisms in the Crocus sativus aggregate (Iridaceae): a preliminary investigation. *Ann. Bot. ns (Italy)* 1: 83-90.

Zhang C and Hsieh T-F 2013. Heritable Epigenetic Variation and its Potential Applications for Crop Improvement. *Plant Breeding and Biotechnology* 1: 307-319.

Zubor A, Suranyi G, Gyori Z, Borbély G and Prokisch J 2004. Molecular biological approach of the systematics of *Crocus sativus* L. and its allies. *Acta Horticulturae.*: 85-94.

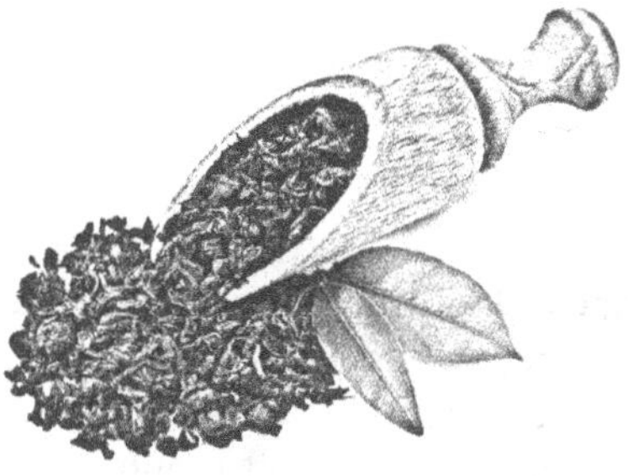

20

Tea

Pradip Baruah and N. Muraleedharan

Tea is the most popular beverage in the world and its consumption is next only to water. The scientific findings on its beneficial health properties and as a drink of general well being are making it even more popular. Tea is cultivated in more than 50 countries of the world from Georgia at 43° N latitude to Nelson (South Island) in New Zealand at 42° S latitude. Tea drinking is spread across the world over the years, and it has become a part of social custom, healthy life style and often, a habit. Tea was considered a health drink, a divine remedy for various ailments since the initiation of tea drinking, and the Taoists called it 'elixir of immortality' and considered 'teaism is Taoism'.

Tea drinking originated in China about 5,000 years ago and the plant was first cultivated in South-East China. The word 'tea' is derived from 't'e' of the Chinese Fukian dialect and in Cantonese, tea is known as 'Cha'. The origin of tea is obscured by a maze of legends. The first authentic reference to tea is found in an ancient Chinese dictionary which was revised about the year 350 A.D. by P'O, a Chinese scholar. The first book exclusively on tea was published in 780 A.D. by Lu Yu (733-804), respected as the Sage of Tea. The book 'Ch'a Ching' or 'The Classic of Tea' in three volumes is divided into ten chapters, each describing various kinds of tea, cultivation, manufacturing methods, etc. and gives information on the tea growing districts of China (Ukers, 1935). Use of tea as a beverage commenced towards the end of the sixth century in China. The habit of tea drinking later spread to Japan approximately in 593 A.D. where it gained tremendous popularity and became an integral part of Japanese culture. It is believed that the 'Tea Ceremony' was started in Japan at about 1159 A.D. Japanese *Chado* or *Sado* ("way of tea") of *Cha-no-Yu* (hot water tea) is a time honoured institution in Japan, rooted in the principle of Zen Buddhism.

Tea drinking spread to other parts of the world only in the middle of seventeenth century. In 1497, opening of a sea route to the East by the Portuguese facilitated large scale trading between Europe and the Orient. The Dutch bought tea from Japan and the first consignment of tea in Europe was from the Island of Hirado in 1610 A.D. The Dutch dominated the tea trade to Europe for more than a century and then the British emerged as the largest trader. Till the middle of the nineteenth century, China continued to be the main supplier of tea (Baruah, 2008). Tea had been popular in erstwhile Russia also and they were aware of it way back in 1567. Tea became fashionable and popular as a beverage in Holland and England in mid of 1660s, and it emerged as the most popular beverage in England. Green tea was first used in England in 1715, but black tea had been the popular one in the Europe in those days (Deka, Taparia, 1999).

The discovery of a plant similar to the tea plant growing wild in the jungles of Assam in North East India, its identification as tea plant in 1834 and beginning of tea cultivation in India since then are part of the history of tea growing. Tea drinkers worldwide liked the new Indian and Ceylon teas, which resulted in sharp decline of China's tea exports by the year 1900, and by the same year one million acres of tropical jungle were felled in India and Ceylon and planted with tea (Weatherstone, 1992).

It was only in the nineteenth century that tea could be grown and manufactured successfully anywhere in the world except in China and Japan. Tea was first introduced in Java (Indonesia) in 1684 with tea seeds imported from Japan. But tea growing on a large scale started only in the beginning of the 19th century with Assam type seed plants. By the end of 18th century, tea cultivation was tried in different parts of the world, particularly in South Asia. In former Indo-China (now Laos, Cambodia and Vietnam), tea probably grew wild as in China, and the local people were used to it since long. However, only in about 1900 A. D. the French revived its organised cultivation. In Taiwan, tea grew wild but commercial cultivation started only by about the middle of 19th century (Bezbaruah, 1999).

Ceylon (Sri Lanka) is a major tea growing area in the world where tea seeds were first tried in 1839. The trial plantings did not get much attention as coffee cultivation had been very successful there till the leaf rust disease destroyed the coffee plantations. The first commercial tea plantation was undertaken in Ceylon in 1867 and in course of time more and more areas were planted with tea. By the turn of the century in 1900 over 120,000 hectares were planted with tea (Weatherstone, 1992).

Place of Origin of Tea, Kinds of Tea, their Spread and Taxonomy

The place of origin of tea is a matter of speculation and still not fully settled. It is widely believed that the place of origin is China. While wild tea plants of Assam and Cambod races were recorded from Assam, Manipur, Mizoram, Burma, Thailand and the entire Annamite chain from the extreme north of the gulf of Tonkin to South Vietnam and Laos since the early part of 19th century, it could not be decisively ascertained if the plants were really wild or relics of migratory tribes inhabiting the region (Barua, 1989).

While focusing the origin of tea, three races of tea are to be understood. The botanical name of tea plant is *Camellia sinensis* (L) O. Kuntze . It has three races, *viz.* Assam tea plant, *Camellia assamica* (Masters), China tea plant, *Camellia sinensis* L and the Cambod or Southern form, *Camellia assamica* sub sp. *Lasiocalyx* (Planch.MS) (Barua, 1989). Based on morphological characterstics of size and shape of the leaves, these races can be differentiated. India, China and Vietnam are the three countries of the world where the three main cultivated varieties of tea are believed to have originated- Assam in India (Indo-Burma region), China in South China and South Vietnam for Cambod or Southern form. It is believed that the three races of tea dispersed to three different areas from its place of origin. The tea plant probably originated in the region around the point of intersection of latitude 29^0 North and longitude 98^0 East, near the source of the river Irrawaddy, which is the meeting ground of Assam, North Burma, South-West China and Tibet. The great rivers of South-East Asia, Yangtze Kiang, Mekong, Irrawaddy and Luhit flowing through this region dispersed tea to different areas from the place of origin (Barua, 1989).

The first scientific description of the tea plant was by Kaempfer in 1712, though it was Linnaeus who in 1753 gave its first taxonomic description, under the name *Thea sinensis* while describing two ornamental species, *Camellia sasanqua* and *C. japonica.* Later he distinguished two kinds of tea and named them *Thea viridis* and *Thea bohea*, the former for producing green tea and the latter for making black tea. The name *T. sinensis* was dropped. Soon it was realized that green and black tea can be produced from both the plants and the name *T. viridis* was dropped. These nomenclatural changes had created considerable confusion among the botanists working on tea and different names were used to describe the wide range of cultivated tea plants. Barua (1989) recorded that the Tea Research Institutes of India, Ceylon and Java approached the International Botanical Congress for a decision on the correct nomenclature of tea and the Botanical Congress in its session held in Amsterdam in 1935 decided to unite the two genera *Thea* and *Camellia* into a single genus *Camellia.* A committee appointed by the Congress decided *Camellia sinensis* (L.) to be the correct name of the tea plant. Technically *Camellia sinensis* (L.) O. Kuntze is the full name of the commercially cultivated tea plant. However, the position of the two taxa represented by the China and Assam plants was not decided. The large leaf Assam plants were considered as a variety *Assamica.* Sealy (1958) distinguished two small leaf forms, *macrophylla* and *parvifolia.* Wight (1962) did not agree with the classifications and suggested a species rank for the Assam plant and proposed the names *C. sinensis* L. for China plants and *C. assamica* (Masters) for the Assam plants. An account of a third race of tea, called the 'Cambod' or southern form was described by Watt (1908). However, White (1962) did not give a specific rank for this form and treated it as a sub species of *C. assamica* and assigned the name *C. assamica* sub species *lasiocalyx* for the southern or Cambod form.

It was in 1823, the tea plant was discovered in Assam when Major Robert Bruce came to know about the existence of tea growing wild in Assam (Griffiths, 1967). According to Baildon (1877) and Hannangan (1987), Robert Bruce was informed about the tea plants growing wild in Assam by a local noble man, Maniram Dutta Barua known subsequently as Maniram Dewan as he saw the indigenous plants and he also introduced Bruce to the Singpho tribe Chief Bessa Gam (Barua, 1992). However, it was Robert's brother Charles Alexander Bruce who collected the plants due to death of Robert Bruce and his efforts ultimately led to recognition of the tea plants of Assam as true indigenous tea in 1934 after years of dispute on its genuineness (Baruah, 2014). C. A. Bruce extensively explored the forests of Assam particularly in the country of the Singphos, on the south side of the Brahmaputra, along and down the river Buri Dihing and and found tea plants growing wild there and at other places like Phakial, Tingri, *etc.*

India is a major tea growing country in the world producing about 25 per cent of the total production of tea in the world which was second to the largest tea producer China in the world. In the total production of tea in the world, the major producers are China, India, Kenya and Sri Lanka. Assam is the largest tea producing state accounting for 53.04 per cent of total production from an area of 55.62 per cent of the total area under tea in India. The commercial production of tea started in Assam in India in 1839 after its discovery in the forests of Assam in 1823 and official recognition of the wild variety of tea plants of Assam as real tea in 1834.In India tea is produced in 16 states across the country and the major tea producing states besides Assam are West Bengal, Tamil Nadu, Tripura, Kerala, Himachal Pradesh, Arunachal Pradesh, etc. The teas of Assam, Darjeeling of West Bengal, Nilgiris of Tamil Nadu, Kangra of Himachal Pradesh have unique characteristics. While Assam tea is famous all over the world for its strong liquoring character, Darjeeling tea is famous for its unique aroma.

In South India, cultivation of tea is concentrated along the Western Ghats, in Kerala & Tamil Nadu Tea planting on a commercial scale started in the Nilgiris by 1853 and in Kerala (Travancore) by 1859. Tea cultivation picked up by the end of the last century in South India. The China type bushes and their hybrids were initially used in the South Indian plantations but a gradual shift to Assam hybrids took place over the years(Barua, 1989).

West Bengal is an important tea growing state of India and the three tea growing areas are Dooars, Terai and Darjeeling. The tea industry developed on an extensive scale in Darjeeling in 1856 (Baruah, 1998). Tea cultivation spread to Terai from Darjeeling.

In North India, tea plantation was initially tried in Dehra Doon and Kangra at the foothills of the western Himalayas with China seeds since 1830s and cultivation continued till 1857. The total area under tea at Dehra Doon and Kangra had declined considerably over the years along with production of green and black tea.

India produces various types of tea, viz. black CTC, black orthodox, green tea, oolong tea, white tea, handmade tea, traditional *phalap*, instant tea, tea bag, value added tea with medicinal plants etc. and also some amount of diversified products like tea cola, confectionaries, etc.

Propagation, Breeding and Productivity

In the early years of tea cultivation seeds from promising plants were selected, on the basis of visual criteria. Later, the same criteria were used for the selection of good cultivars by vegetative propagation. The objectives of tea breeding is the development of cultivars with high yield, quality and tolerance to drought. Very little attention had been paid for resistance to pests and pathogens, through initial screening would have ruled out the high susceptibility to any major pest or pathogen. The productivity of bushes is directly related to the production of dry matter and its partioning. Plants with large leaf favours production of heavier shoots, but medium sized semi erect leaves are more productive in terms of dry matter production. It is also important that the characters of high productivity remain stable under different agro-climatic conditions, soil types and altitudes. Genotype-environment interaction assumes greater relevance, in view of climate change. Highly significant correlations between total leaf area and yield have been reported and therefore, this aspect should also be considered in selection programme. Quality of tea is the other most important character breeders and growers look forward to. Leaf pubescence is closely related to quality especially for the production of orthodox type of tea. Similarly, frequency of occurrence of Calcium oxalate crystals in the leaf petiole is also related to quality. The content of theaflavins and the presence of sweet aroma compounds imparting a high flavor index are features that contribute to quality.

Tea plants are self sterile and cross fertile. Therefore, seed grown plants have higher genetic diversity since they are produced by uncontrolled hybridization. Crosses between genetically diverse clones have resulted in the development of progenies with improved hybrid vigour. Today, biclonal and polyclonal seeds are getting wider acceptance as planting material. These clonal seed stocks with their tap root system and genetic variability may help them to withstand climate changes better than the clonal plants, obtained through vegetative propagation.

Productivity of tea is determined by an array of factors such as weather, soil type, cultivars and agronomic practices such as manuring, pruning, harvesting and plant protection. Among the producing countries Kenya, enjoying equatorial climate has the highest tea productivity, followed by India and Sri Lanka.

Manufacturing and Types of Tea

Tea can be classified in many ways. According to the most commonly used classification (based on manufacturing process), three types of tea are manufactured; black tea, green tea, and oolong tea. The black and green teas are further divided into many grades, each representing a distinct size, style and density. Tthere is no uniform or exact system of grading and the grades of tea vary considerably from one country of origin to another.

Black tea is of three types – orthodox and CTC. While CTC (crush, tear, curl) was invented in Assam in 1930, orthodox tea is the traditional method of manufacturing where rollers are used to roll and twist the plucked leaves after withering. In CTC manufacturing, the plucked leaves are put in to a CTC machine after withering where they are turned to small crushed bits by the sharp teeth of roller machine. Green tea mainly is of orthodox type and numerous varieties of green tea are available. Oolong tea is an intermediate between black and green tea with regard to fermentation.

Tea contains numerous chemical constituents having diverse chemical properties and many of these have pharmaceutical and therapeutic values, and some of them with nutraceutical value. These chemical components are being exploited to produce different categories of products to target various groups of consumers. The various diversified products developed are instant tea, tea tablet, tea cola, tea toffee and confectionary products of tea like tea biscuit, tea cake, etc. (Baruah, *et al, 2005*).

Speciality teas are those teas with distinctively different characteristics from those of regular tea. The speciality tea product varieties may be classified into four broad categories, viz. specific origin, specific occasion, specific added flavour and historical association. 'Earl Grey', originally a recipe from China flavoured with the essence of bergamot, is most popular of all speciality teas (Gill, 1992).

Prospects of Tea

Tea is endowed with an array of chemical compounds, such as carbohydrates, proteins, polyphenols, caffeine, theanine, enzymes, vitamins and minerals. From the pharmacological and therapeutic points of view, polyphenols and caffeine are the most important, though vitamins and minerals are also considered medically significant. Scientific data have indicated that the beneficial effects of tea are due to its polyphenolic contents that may affect carcinogen metabolism, free radical scavenging activity and formation of DNA adducts. Researchers have proved that tea drinking can reduce the risks of heart diseases, cancers of lung, breast, stomach, liver, colon and prostrate, hypertension and stroke. They have proved that the antioxidant properties of tea can inactivate influenza virus and bacteria associated with gastric ulcer and also decrease cholesterol and low density lipoprotein. The pattern of research activity on the health effects of tea, is primarily centered on the cardiovascular system, cancer therapy and gastro intestinal tract (Muraleedharan, 2009).

Tea cultivation is a major livelihood of millions of people in the continents of Asia and Africa. In most of the tea growing countries tea productivity and production are adversely affected by both biotic and abiotic stresses resulting from climatic abrasions. It is necessary to develop cultivars which will be adaptable to varying climatic conditions, including tolerance to drought, pests and diseases which is vital to the sustainability of tea plantations. Other management practices will include the development of integrated nutrient management involving the use of organic manures, including vermicompost, soil and soil moisture conservation and irrigation, wherever possible. In countries such as India and Sri Lanka shade trees are an integral part of the tea plantations and these shade trees reduce the ambient temperature by 2-3°C and also contribute to the organic matter through their leaf drop and loppings.

Pests and diseases are responsible for considerable crop loss in most of the tea growing countries. The change in climate is also bringing about changes in the pest and disease scenario. New pests, especially geometrid caterpillars like *Hyposidra talaca* are shifting to tea from forests whereas the incidence and intensity of infestation of sucking pests like thrips, *Helopeltis* and scale insects and mealy bugs are on the increase. A large number of insect parasitoids and predators are active in the ecosystem while certain microbes like *Beauveria bassiana*, *Metarhizium anisopilae* and *Paecilomyces fumosoroseae* are being employed for the control of pests. Non chemical pest control strategies will have to play a major role in the management of tea pests. Climate change may not only affect the productivity of the gardens but also the quality of tea. It is known that high temperature and high CO_2 concentration may lead to a decrease in the content of protein and mineral nutrients in the plants. In tea the content of secondary metabolites like caffeine, catechins and volatile flavour compounds are affected by the changes in climate. Research on tea plant improvement programme with assistance from biotechnologists will help to develop new cultivars which can adapt to the changes in climate change, leading to sustainable cultivation, and provide livelihood to the millions of people engaged in the growing and processing of tea.

References

Baildon, S. 1877. Tea in Assam. Quoted in *Science and Practice in Tea Culture*, Dr D. N. Barua, 1989.

Barua, D. N. 1989. *Science and Practice in Tea Culture*. First publish, Tea Research Association, Calcutta-Jorhat, 7-34.

Barua, D. N. 1992. *The Tea Industry of Assam: Its Historical Background*. Paper presented at Seminar on Tea Industry of Assam on Jorhat College Silver Jubilee, Jorhat, 22.02.1992.

Baruah, Pradip 1998. *Darjeeling Tea* (In Bengali). Dainik Sangbad 12.03.1998, Agartala,.

Baruah, Pradip 2008. *The Tea Industry of Assam: Origin and Development*. First publish, EBH Publishers (India), Guwahati, pp. 23-59.

Baruah, Pradip 2014. *Types of Tea, Value Addition and Product Diversification: Indian Experience*. Key note speaker presentation at International Conference on Tea Science and Development, theme: A Sustainable Tea Industry for Social, Economical and Technological Development, Karatina University, Kenya, 24th-27th September, 2014.

Baruah, Sabitri, P Tamuly, P Bordoloi, A K Bordoloi, S Sabhapandit, R Gogoi, J N Kalita, L P Bhuyan, M N Gogoi andM Hazarika 2005. Technology for Product Diversification and Value Added Items of Tea, *Proceedings of 34th Tocklai Conference, 2005,* Tea Research Association, Tocklai Experimental Station, Jorhat, Assam, pp. 63-69.

Bezbaruah, H. P. 1999. *Origin and History of Development of Tea*. Global Advances in Tea Science, first edition 1999, Aravalli Books International (P) Ltd., New Delhi, pp. 383-391.

Deka, Aniruddha and Taparia Madan 1999. *Dictionary of Tea*. First publish, Computech India, Jorhat, Assam, pp. 258-272.

Encyclopedia Britannica, 1994-2002.

Gill, M. 1992. *Speciality and Herbal Teas*. Tea: Cultivation to Consumption, first edition, Chapman & Hall, London, pp. 513-534.

Griffiths, Sir Percival 1967. *The History of the Indian Tea Industry*. Wiedenfeld and Nicolson, London: 33-58.

Hannangan 1987. Darjeeling Plantations (From Old Files). *The Assam Review and Tea News*, 76 (2): 34.

Muraleedharan, N. 2009. Tea : Quality, standards and health benefits. *Planters' Chronicle*, 105 (2):5-16.

Sealy, J. R. 1958. A revision of the genus *Camellia. The Royal Horticultural Society, London.*

Ukers, W. H. 1935. *All About Tea*, Vol. 1, Tea and Coffee Trade Journal Co., New York.

Watt, G. 1908. *The commercial products of India*, John Murray, London.

Weatherstone, J. 1992. *Historical Introduction.* Tea: Cultivation to Consumption, first edition, Chapman and Hall, London, pp. 1-23.

White, G. 1962. Tea classification revised, *Current Science* 31:298-299.

21

Watermelon

B R Choudhary

Watermelon [*Citrullus lanatus* (Thunb.) Matsam. & Nakai] is an important crop belonging to Cucurbitaceae family with ($2n=2x=22$). It is one of the most widely cultivated crops in the world (Huh *et al.*, 2008). Its global consumption is greater than that of any other cucurbit. Watermelon is grown for its fleshy, juicy and sweet fruit. Mostly eaten fresh, they provide a delicious and refreshing dessert in hot weather. Now it is no longer just a summer fruit and is becoming an everyday fruit like apples, bananas and oranges. The major nutritional components of the fruit are carbohydrates (6.4g/100g), vitamin A (590 IU) and lycopene (4100µg/100g), an anti-carcinogenic compound found in red fleshed watermelon. Lycopene has been classified as useful in the human diet for prevention of heart attacks and certain types of cancer. Watermelon seeds are rich in fat (about 45% edible oil) and protein (30-40%). The seeds are powdered and backed like bred in India. The seed kernels are also used in various sweets and other delicacies. The unripe fruits are also cooked as a vegetable in some parts of India.

Botanical Description

Watermelon differs from other economically important cucurbits by having pinnatified leaves. The hairy stem is thin, angular and grooved bearing branched tendrils. Watermelon is an annual vine or creeper. It has an extensive root system so that it can be well adapted for river bed cultivation to utilize subterranean moisture. Stems are branched and prostrate. It is monoecious (both male and female flowers occur on the same vine) in sex expression with very high male to female (7:1) sex ratio, but the ratio ranges from 4:1 to 15:1. However, andromonoeciuos (staminate and perfect flowers on the same plant) sex form is also found. The plant is classified as naturally cross pollinated crop. The pistillate flowers normally occur in every 7th leaf axil while the intervening axils are occupied by the staminate flowers. Flowers are unisexual, cross pollinated and light yellow in colour. Corolla is showy yellow in colour, united and 5 in number. Stamens are three and attached to the calyx tube. Ovary is inferior with green or yellow colour. The edible part of the fruit is the endocarp (planceta). Fruit is pepo (berry with hollow cavity, thick rind and seeds with parietal placentation). Botanically it varies in shape, size, colour and taste (Choudhary *et al.*, 2012). Fruits are large, round to oblong or cylindrical fruits measure as long as 60 cm. Flesh is red or pink or yellow in colour on full ripening. The rind may be dark green to light green in colour and it may be with stripes or without stripes. Seeds are many, dark brown to black in colour. The seed coat is warty.

The whole period of bud developmental stage is completed in 12-16 days by the male and in 11-13 days by female bud. The rate of increase in size is associated with bud development and this increase in size is more rapid in earlier stages *i.e.* up to 8-12 days, thereafter it decreases and during the post developmental period remains more or less constant. The anthesis starts in the early hours of morning at 5.30 am and continues up to 7.30 am with peak between 6.30-7.00 am. The dehiscence of anthers starts $1^{1/4}$ hours before anthesis *viz.,* 4.45 am and continues up to 6.30 am. Pollen is visually evident in sticky masses adhering to anther. The stigma becomes receptive 2 hours before anthesis and continues up to 3 hours after anthesis. From the date of pollination and fertilization it takes 30-40 days to full maturity and ripening of fruits.

Species, Origin and Distribution

The genus *Citrullus* consists of eight species and sub-species. Watermelon, the only cultivated species of the genus (Mallick and Masui, 1986). Watermelon is native to Tropical Africa chiefly the Kalahari Desert (the current nations of Namibia and Botswana) where wild forms are still found (De Candolle, 1882). Whitaker and Davis (1962) also describe the existence of a secondary diversification centre in India. Watermelon is thought to have been domesticated in Africa at least 4000 years ago and now grown worldwide, particularly in regions with long, hot summers (Robertson, 2004). Watermelon is also mentioned in the Bible as a food eaten by the ancient Israelites while they were in bondage in Egypt (Freedman and Myers, 2000). In the 7th century, watermelons were being cultivated in India and by the 10th century had reached China. Now it spread throughout the tropics and the Mediterranean (Tindall, 1983). In the New World, cultivation began in Massachusetts as early as 1629 (Mohr, 1986). Watermelon was brought to America by Spanish and quickly became very popular crop (Robinson and Decker-Walters, 1997). A good amount of variability is also found in South East and North Africa, Southern and South Central Asia, Iran, Pakistan and India. China is the largest watermelon producer in the world and the other major watermelon producing countries are Turkey, Iran, Egypt, the United States, Mexico and Korea.

Whitaker (1933), Shimotsuma (1960) and Anghel (1969) considered wild watermelon (*Citrullus colocynthis*) as probable ancestor of watermelon, based on cytogenetical investigations, intercrossing compatibility and dissemination in Africa and Asia. However, it has been suggested on the basis of chloroplast DNA investigations, that the cultivated and wild watermelon diverged independently from a common ancestor, possibly *C. ecirrhosus* from Namibia (Dane and Liu, 2007). Morphologically the *colocynth*, has similar characters with *vulgaris*, but fruits are bitter and seeds are small. However, the bitter forms of *C. lanatus* were considered the probable ancestor of watermelon by others. *C. lanatus* is one of four known diploid (n=11) species that belong to the xerophytic genus *Citrullus* Schard. ex Eckl. & Zeyh., found in temperate regions of Africa, Central Asia and the Mediteranean (Whitaker and Davis, 1962; Jeffrey, 1975 and Whitaker and Bemis, 1976).

Shimotsuma (1965) classified watermelon as *Citrullus lanatus* (Thumb.) Mansf., *C. ecirrohsus*, *C. colocynthis* (L.) Schard. The possible ancestor of watermelon is *C. colocynthis* (L.). The genus *Citrullus* Schard. ex Eckl. & Zeyh. contains the following four diploid species:

S.No.	Species	Distribution	Growth habit	Reference
1.	*Citrullus lanatus* (Thunb.) Matsam. & Nakai includes the cultivated watermelon (*Citrullus lanatus* var. *lanatus*) also called '*egusi*' melon, and the preserving melon (*C. lanatus* var. *citroides*) also called '*tsamma*' melon.	Tropical and subtropical climates worldwide including West and South Africa.	Annual	Whitaker and Bemis, 1976
2.	*C. colocynthis* (L.) Schard – 'bitter apple'	Sandy areas throughout northern Africa, southwestern Asia and the Mediterranean.	Perennial	Zamir *et al.* 1984; Burkill, 1985 and Jarret *et al.*, 1997
3.	*C. ecirrhosus* Cong.	Desert regions of Namibia.	Perennial	Meeuse, 1962
4.	*C. rehmii* De Winter	Desert regions of Namibia.	Annual	Meeuse, 1962 and De Winter, 1990

According to Fursa (1981), the cultivated species *C. lanatus* includes three subspecies: (i) *lanatus*, (ii) *vulgaris* which has two varieties, var. *vulgaris* and var. *cordophanus*, and (iii) *mucocospermus*.

Bitter apple (*colocynth*) is a perennial cucurbit and known as *Indrayan* in Hindi. It is native of arid regions and occupies the vast area extending from the west coast of northern Africa (Senegambia, Morocco and the Cape Verde Islands), eastward the Sahara, Egypt, Arabia, Persia and Baluchistan and throughout India. At the Red Sea, near Kossier, it occurs in immense quantity. In India, it is found most abundantly in north-western plains, especially in Barmer, Bikaner, Jaisalmer and Jodhpur districts of Rajasthan and Gujarat in wild form.

Bitter apple is considered a very close relative to watermelon. It is perennial with prostrate or climbing angular stems and bifid tendrils; leaves ovate or triangular, deeply 3-lobed; flowers monoecious, yellow, solitary, axillary; fruit pepo, 4-10cm in diameter, smooth, globose, green mottled with yellow blotches, pulp bitter, spongy; seeds numerous, white or light brown. The roots are large, fleshy leading to a high survival rate due to the long tap root. It possesses extreme abiotic resistant attributes which may be exploited in improvement programme of watermelon.

Fruit contains 15% pulp, 62% seed and 23% rind. The mesocarp contains glucose (1.3% on flesh basis). The processed mesocarp may be good sources of pectin. The juice of the fruit contains citrullin, citrullene and citrullnic acid. The fruits of bitter apple also contain cucurbitacin B and its glycoside, cucurbitacin I. The peel free flesh of ripe fruits contains yellow bitter oil. The seeds are used for edible purposes as well as to extract oil. Seeds contain 16.7% yellow coloured semi-dry oil rich in linoleic acid. The bitter taste and the powerful medicinal properties of the pulp are due to the presence of amorphous glucoside colocynthin. Walz (1891) obtained from an alcoholic extract of *colocynth* an ether soluble crystalline and tasteless substance insoluble in water, which is called colocynthin. *Colocynth* possesses several medicinal properties and is used in treating bilious derangements of chronic constipation and dropsy, fevers and cases requiring purgatives. The pulp is also used for varicose veins and for piles. It is diuretic, expectorant, and is also used in remedies against cancer and wounds.

References

Anghel, I. 1969. Studies on the cytology and evolution of the *Citrullus vulgaris. Comun. Bot.*, 11:49-55.

Burkill, H.M. 1985. The Useful Plants of West Tropical Africa. Royal Botanic Gardens, Kew, 2nd edn.

Choudhary, B.R., Pandey, S. and Singh, P.K. 2012. Morphological diversity analysis among watermelon (*Citrullus lanatus* (Thunb) Mansf.) genotypes. *Prog. Hort.*, 44(2):321-326.

Dane, F. and Liu, J. 2007. Diversity and origin of cultivated and citron type watermelon (*Citrullus lanatus*). *Genet Resour. Crop Evol.*, 54:1255–1265.

De Candolle, A. 1882. Origin of Cultivated Plants. New York. pp. 262ff, s.v. "Watermelon".

De Winter, B. 1990. A new species of *Citrullus* (Benincaseae) from the Namib. *Bothalia*, 20: 209-211.

Freedman, D.N. and Myers, A.C. 2000. Eerdmans Dictionary of the Bible. Amsterdam University Press. p. 1063.

Fursa, T.B. 1981. Intraspecific classification of water-melon under cultivation. *Kulturpflanze*, 29:297-300.

Huh, Y.C., Solmaz, I. and Sari, N. 2008. Morphological characterization of Korean and Turkish watermelon germplasm. 1 Cucurbitaceae 2008, *Proceedings of the IXth EUCARPIA meeting on genetics and breeding of Cucurbitaceae* [Pitrat, M. (Ed.)], INRA, Avignon (France), May 21st-24th.

Jarret, R.L., Merrick, L.C., Holms, T., Evans, J. and Aradhya, M.K. 1997. Simple sequence repeats in watermelon [*Citrullus lanatus* (Thunb.) Matsum. & Nakai]. *Genome*, 40:433-441.

Jeffrey, C. 1975. Further notes on Cucurbitaceae: III. Some African taxas. *Kew Bull.*, 30: 475-493.

Mallick, M.F.R. and Masui, M. 1986. Origin, distribution and taxonomy of melons. *Scientia Horticulturae*, 28: 252- 261.

Meeuse, A.D. 1962. The Cucurbitaceae of Southern Africa. *Bothalia*, 8: 1-111.

Mohr, H.C. 1986. Watermelon breeding. In: Bassett, M.J. (Ed.). Breeding vegetable crops. Westport, Connecticut: AVI Publishing Co. Westport, Conn. pp.37-66.

Robertson, H. 2004. *Citrullus lanatus* (Watermelon, Tsamma). Museums Online South Africa. Iziko Museums of Cape Town Online Publication: http:/museums.org.za/bio/index.htm.

Robinson, R.W. and Decker-Walters, D.S. 1997. Cucurbits. New York Cab International. 226p. (Crop Production Science in Horticulture n°.6).

Shimotsuma, M. 1960. Cytogenetical studies in the genus *Citrullus*. IV. Intra and inter-specific hybrids between *C. colocynthis* Schr. and *C. vulgaris* Schrad. *Jpn. J. Genet.*, 35:303-312.

Shimotsuma, M. 1965. Cultivated plants and their relatives. Kyoto University, Science, Kyoto.

Tindall, H.D. 1983. Vegetables in the tropics. The Macmillan Press Limited, London, pp. 150-152.

Walz, H. 1891. Fluckiger Pharma Cognosie des Pflanzenreichs, 3rd edn.

Whitaker, T.W. 1933. Cytogenetical and phylogenetic studies in the Cucurbitaceae. *Bot. Gaz.*, 93:780-790.

Whitaker, T.W. and Bemis, W.B. 1976. Cucurbits. In: Simmonds N.W. (ed.), Evolution of crop plants. Longman, London. pp. 64-69.

Whitaker, T.W. and Davis, G.N. 1962. Cucurbits-Botany, cultivation and utilization. Interscience Publishers, Inc. New York.

Zamir, D., Navot, N. and Rudich, J. 1984. Enzyme polymorphism in *Citrullus lanatus* and *C. colocynthis* in Israel and Sinai. *Plant Syst. Evol.*, 146: 133-137.

Variation in ovary of watermelon

Plant and fruits of *Citrullus colocynthis*

Pentalobate leaf **Non-lobed (entire) leaf**

Plant and fruits of *Citrullus colocynthis*

Monoecious **Andromonoecious**

Sex forms in watermelon

Variation in seed traits of watermelon

Variation in fruit shape in watermelon

Variation in fruit shape and drediness in watermelon

Variation in fruit traits of watermelon

www.ingramcontent.com/pod-product-compliance
Lightning Source LLC
Chambersburg PA
CBHW060246230326
41458CB00094B/1466
9781787150010